SANTA RITA DEL COBRE

MINING THE AMERICAN WEST

SERIES EDITORS

DUANE A. SMITH
ROBERT A. TRENNERT
LIPING ZHU

SANTA RITA DEL COBRE

A COPPER MINING COMMUNITY IN NEW MEXICO

Christopher J. Huggard AND Terrence M. Humble

University Press of Colorado

Published by the University Press of Colorado
5589 Arapahoe Avenue, Suite 206C
Boulder, Colorado 80303

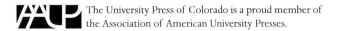 The University Press of Colorado is a proud member of
the Association of American University Presses.

The University Press of Colorado is a cooperative publishing enterprise supported, in part, by
Adams State College, Colorado State University, Fort Lewis College, Metropolitan State College
of Denver, Regis University, University of Colorado, University of Northern Colorado, and
Western State College of Colorado.

∞ This paper meets the requirements of the ANSI / NISO Z39.48-1992 (Permanence of Paper).

Library of Congress Cataloging-in-Publication Data

Huggard, Christopher J., 1962–
  Santa Rita del Cobre : a copper mining community in New Mexico / Christopher J. Huggard and
Terrence M. Humble.
    p. cm. — (Mining the American West)
  Includes bibliographical references and index.
  ISBN 978-1-60732-152-1 (hardcover : alk. paper) — ISBN 978-1-60732-153-8 (ebook) —
ISBN 978-1-60732-249-8 (pbk. : alk. paper) 1. Santa Rita (N.M.)—History. 2. Copper mines and
mining—New Mexico—Santa Rita—History. I. Humble, Terrence M., 1941– II. Title.
  F804.S33H84 2011
  978.9'692—dc23
                    2011038338

In memory of our fathers,

Pat Humble and O. Orland Maxfield,

and

to the People of Santa Rita

Contents

Illustrations

Preface

This collaborative project began many years ago. In some ways, it started the day Terry Humble was born in 1941 in Santa Rita. His passion for his hometown was cultivated while growing up in the copper camp. He attended the local schools and the Community Church. He played in the open pit, piloted his bicycle in the concrete water flume, and frequented other places at the mine works. He joined Boy Scout Troop 105 and as a rambunctious teenager

camped in the beautiful Gila wilderness and with his buddies climbed the iconic Kneeling Nun. The fiery red-head was intent on experiencing the natural and cultural features of southwestern New Mexico. This homegrown Santa Ritan, however, learned early in his life that his beloved town would not always be there for him and his family and friends. The steady growth of the open pit ensured the death of the copper town. After a stint in the US Navy in the late 1950s and early 1960s in Tennessee, Alaska, and California, this son of a miner became alarmed about the demise of Santa Rita.

On his return to New Mexico in 1966 to work as a hard-rock miner in Magdalena, Terry brought a new family with him. He had betrothed Artemisa "Micha" Bernal, a Mexican woman with three children from a previous marriage, and they began their bilingual family, eventually having three daughters of their own. He also brought back with him a certificate as a journeyman diesel mechanic he had earned at the Los Angeles Trade Technical College. With diploma in hand, he decided to move back to Grant County, and after stints mining underground at various lode mines, he landed a position with the Kennecott Copper Corporation as a heavy equipment diesel mechanic at the Chino Mine in Santa Rita. He worked there for eight years as a mechanic and truck shop foreman before going to Toquepala for a four-year hiatus with the Southern Peru Copper Corporation. He returned to Chino in 1978, working for Kennecott and later Phelps Dodge until his retirement in 2001.

Terry realized soon after his return to New Mexico that he needed to do something to preserve the memory of Santa Rita. "I became interested in local history," he recently wrote to me, "especially concerning Santa Rita, NM, about 1967 when the town was being moved out due to pit expansion; I realized I was witnessing the end of an era. I began interviewing all of the old timers (both in English and Spanish), writing down their recollections. I also began making copies of all Santa Rita pictures I was able to find and collected any references to Santa Rita/Chino Copper Company history." A trained mechanic, Terry began his new vocation as a self-taught historian who had a passion for his hometown that was disappearing before his eyes. It was his dedication to Santa Rita's memory, in fact, that brought us together.

In 1990, I completed my coursework for a doctorate in history at the University of New Mexico in Albuquerque. I knew I was going to write a dissertation on the history of mining and the environmental consequences under the direction of the late Gerald D. Nash. I suggested to Professor Nash in one of our early conferences that I complete a historical study of the Kelly-Magdalena mining area just west of Socorro, New Mexico. Principally underground silver and lead mines, the operations there had been relatively minor in comparison to the state's greatest mining district in Grant County. Dr. Nash understood this distinction and gently redirected my attention to the most prolific mineral field in the state. Soon after I completed my written exams, I made my first visit to the area, following leads from Nash as well as Jolane Culhane, a fellow graduate student. Jolane, who had lived in Silver City for many years and was soon to become a history professor at Western New Mexico University, introduced me to Pat Humble, who had been a longtime underground miner and knew the local history scene, holding "court" at his antique shop on Little Walnut Road. I soon learned, however, that Pat knew of someone who had been far more involved in preserving the history of mining, especially at Santa Rita. It was his son, Terry.

Soon after Mr. Humble's suggestion that I contact his son, I made a visit to Terry and Micha's home on Guinevan Street in Bayard, the town that "replaced" Santa Rita. They were incredibly gracious and invited me into their living room for ice tea and a discussion of my project, a history of the economic and environ-

mental impact of mining in southwestern New Mexico. After showing me some of his records and photographs, Terry made the fateful decision to introduce me to a plethora of copies and notes of company records he had obtained before Phelps Dodge destroyed the original records. After Phelps Dodge purchased the Chino Mine in 1987, officials of the company ordered workers to dump them into the tailings water at Hurley. I learned about their final destination on that first visit to the Humble home when Terry took me to a shed in his backyard. The shelves and tables were spilling over with box after box of the records he had transposed concerning the day-to-day operations in Santa Rita. For an aspiring professional historian, I felt like a prospector who had discovered the mother lode; Terry was in essence offering me a grubstake so I could develop my "mine." Although Terry was a bit covetous of his treasure, he willingly entrusted those precious pieces of paper to me. I collected as many as he allowed me to take at a time and made hundreds of pages of copies. I was mucking documentary "ore" from a new friend's "ore" shed. His trust touched my heart and initiated a friendship that has evolved into one of two brothers-in-history arms.

I completed my dissertation in 1994 after several additional visits to the Humble home and shed, as well as to several other repositories, such as Special Collections at Western New Mexico University in Grant County and elsewhere in the state. Since I was interested in the broader history of mining in southwestern New Mexico, I gave very limited coverage to Santa Rita with hopes of returning someday to complete a much more in-depth study. In the meantime, I reworked several of the chapters for publication as journal articles and book chapters, always keeping in mind that Santa Rita's story should someday be told in full. Various other projects also intervened in my career.

All the while, Terry continued his mission to preserve the memory of Santa Rita by continuing to collect and examine historical records and photographs. Throughout the 1990s and 2000s, he became increasingly active in memorializing his hometown and Grant County, generally. He continued to interview locals, both Chicanos and Anglos, and then published two articles in the *Mining History Journal* and *Outlaws & Lawmen* to preserve particular aspects of Santa Rita's story. He also became an active member of the Silver City Corral of Westerners International, the Fort Bayard Historical

Society, and the Mining History Association, earning a well-deserved reputation as the principal historian of Santa Rita. I call him "Mr. Santa Rita."

In 2007, my college, NorthWest Arkansas Community College, awarded me a semester-long sabbatical in my position as a professor of history. I proposed completing another history project and then got the idea that I could use some of my research time to revisit the story of Santa Rita. That is when I visited Terry again early that summer with a proposal. I suggested to my friend, whom I had seen only a couple of times since 1994, that we do a photographic history of Santa Rita for the mining history series of the University Press of Colorado. He then informed me that he had been collecting more records and photographs since I had last visited. I was astonished to discover he had purchased hundreds more historical pictures and had compiled a massive tome he titled "Copper Town." The project took an immediate turn with this realization. "Copper Town," it ends up, was a year-by-year compilation of information on Santa Rita from the early nineteenth century through the 1950s. He had produced more information on Santa Rita than had ever been gleaned. He had examined historic newspapers, various companies' reports, hospital and school records, and other materials. He had conducted personal interviews and had collected transcribed interviews by others to use with many additional sources for his year-by-year account. After a couple of hours of discussion, we both decided that between "Copper Town" and my own recent research we had plenty of information to produce a full-length narrative to go along with a collection of selected photographs.

In the meantime, I had made additional visits to historical repositories in Santa Fe and Boulder, Colorado. I found new information and illustrations at the Palace of the Governors Photograph Archives at the Museum of New Mexico and then carefully examined the records of Local 890 of the International Union of Mine, Mill, and Smelter Workers housed in the history archives at Norlin Library at the University of Colorado at Boulder. We found unpublished photographs housed in the Rio Grande Historical Collection at New Mexico State University, the New Mexico Bureau of Mines and Mineral Resources Photograph and Slide Collection, and the Silver City Museum Photograph Collection. We also had access to several new studies, in particular, Ellen Baker's insightful book on Mine Mill and the Salt of the

Earth strike and film; Melissa Burrage's outstanding thesis on A. C. Burrage, who founded the Chino Copper Company; Bob Spude's compelling history of late nineteenth-century Santa Rita; and Sheila Steinberg's sociological study of Santa Ritans' memories of their lost town. Helen Lundwall and Terry Humble also generously shared their book-length manuscript of the Spanish period of Santa Rita's history. With all our own work and these more recent studies, we had more rich "veins" of historical "ore" than we would ever need to develop our "mine."

Terry and I devised a strategy to complete the book. We gathered all the historical documents, notes and other records, and photographs and organized them. Then I began to write the text. In January 2008, I lived with the Humbles for two weeks. I composed the first few pages of chapter 1 in Terry's home office on January 22 with my coauthor sitting right next to me. Following that first visit, I devoted some of my time each day during my sabbatical on the Santa Rita project. After every writing session I emailed the latest prose to Terry for his editorial comments. We did this for many months. After completing drafts of the first three chapters, I visited the Humbles again in May, completing chapter 4 during the stay. After mornings behind the computer screen composing, we usually took afternoon breaks and then returned at night to cull Terry's vast collection of photographs. Choosing the pictures was a difficult selection process, but our problem was a good one because we had too many choices. Composing the captions for the illustrations was a more tedious process, but one designed to introduce new information not covered in the text. Terry had written detailed descriptions of the subjects of the photographs. I then wrote captions from his information sheets and we discussed the contents as I composed. It proved to be a joyful and efficient way of completing this part of the project. We were making great progress and our strategy was working quite nicely. Although we would occasionally disagree on the coverage of certain topics or a photograph or maybe a few words placed here and there (e.g., "ungulates" for hoofed animals got cut a couple of times), things went very smoothly. We had not yet completed the initial draft when my fall 2008 semester started, causing a slight delay in the completion of the manuscript. I made one last visit to Bayard and the Humble home in January 2009 to complete most of chapter 5.

In March 2009, we finally finished that first draft. We are indebted to six historians: Jim Bailey, Melissa Burrage, Jolane Culhane, David Hays, Rick Hendricks, and Bob Spude, each of whom read and edited all or most of the manuscript. We then sent it to University Press of Colorado director Darrin Pratt, who had extended a contract to us in September 2007. Terry's intimate knowledge of the facts of Santa Rita's history and my training as an academic historian combined to make this book a thoroughly researched and meticulously edited work. After three additional peer reviews and then substantial revisions, we have completed Santa Rita's story in words and illustrations. We hope that our passion for this special community's history makes this a compelling read with scintillating visuals that give deep insight into one of mining history's most fascinating places. I am eternally grateful to Terry for welcoming me into his world with open arms to examine his personal documentary and photographic archives and for inviting me to stay in his home for a total of eight weeks to complete this project. Our hope is that this community and mining history is appealing to a wide range of folks, from locals who lived this story, to newcomers who want to know more about Grant County and its mining past, to mining historians and other scholars interested in considering *Santa Rita del Cobre* in their own work and in their classrooms. It has been an honor and a pleasure for me to have worked with Terry as well. From my perspective, being a coauthor with a man who worked in the copper mining industry virtually his entire career makes this book a unique collaboration. We hope our partnership makes for an inspirational account of Santa Rita's past.

CHRISTOPHER J. HUGGARD
FAYETTEVILLE, ARKANSAS

Acknowledgments

We thank the following people for their assistance in completing *Santa Rita del Cobre*: Ted Arellano; Jim Bailey; Jackie Becker and Susan Berry of the Silver City Museum; April Brown; Melissa Burrage; Garry E. Bush; Eric Clements; Jolane Culhane; Hannah Desrocher; Jessica d'Arbonne; Dick Etulain; Sydney Flynn; Laura Furney, Dale and Jeanette Giese; Art and Penny Gómez; David Hays of the History Archives, University of Colorado at Boulder;

Rick Hendricks; Kevin Humble; Stephen Hussman, Charles B. Stanford, and Dean W. Wilkey of New Mexico State University, Archives and Special Collections; Andrea Jacquez of Western New Mexico University Library; Greg Kiser; Daniel Kosharek of the Photograph Archives, Palace of the Governors; LaDonna Lowe-Sauerbry; Richard G. Luna; Helen Lundwall; Alexis and Rudy Madrid; Aaron Mahr; Carlos J. Martinez of the Bureau of Land Management, Santa Fe; Guillermo P. Martinez; Tim McGinn; Elmo Mitchell; the late Gerald D. Nash; Genny Olson; Bob Pelham; Richard Peterson of Freeport-McMoRan; Neta Pope; Dan Pratt, Darrin Pratt; Linda Richardson; Brent Robbins; Benito R. Rodriguez; Anthony M. Romero; Duane Smith; Bob Spude; John M. Sully II; Judy Tobler; John Tuthill; Herb Toy; Don Turner; Jerry Vervack; Jose D. Villalobos; and Liping Zhu. We also thank the Social Sciences Department, the Faculty Senate, and President Becky Paneitz of NorthWest Arkansas Community College for Chris Huggard's sabbatical in spring 2008. We extend a heartfelt thanks to our spouses, Micha Humble and Kay Pritchett, for their many years of support for our work on Santa Rita's history.

SANTA RITA DEL COBRE

Introduction

Santa Rita no longer exists. It is not even a ghost town. Locals know it as the "town in space." During the course of the twentieth century, mining companies literally dug up the ground beneath the copper camp. First named El Cobre, or "Copper," by the Spanish for the vast outcroppings of raw native copper, the mining community from its inception centered its activities on digging up the rich, nearly pure nuggets of the red metal in the heart of the Apache

homeland Apachería (see figure 0.1). First the Spanish and then the Mexicans struggled to create an enduring settlement known after 1804 as Santa Rita del Cobre. Violent resistance to their efforts by the Apaches threatened their ability to export the valuable ores and establish a viable camp. By the time the first group of Americans arrived in the 1820s at the locale, the Apaches had already forced three evacuations of the coveted copper mines. Still, the desire for the precious red metal persisted despite financial and diplomatic failures well into the American period after the Mexican-American War. Even after the frontier Americans forced the subjugation of the Apaches to reservations by the 1880s, the town now known as Santa Rita seemed unlikely to be a permanent community. Investors Matthew Hayes, J. P. Whitney, the George

Hearst estate, and the Rockefeller Syndicate, among others, failed to make a successful go of it in the lode-mining era. Like frontier capitalists all over the American West, they grappled with the exigencies of a very isolated spot, limited access to freighting, dwindling ore values, and recalcitrant oxides, among other difficulties.

Large corporate investment in the mines ironically saved the town at least until 1970. In the end, though, Santa Rita succumbed to the successful implementation of massive economies-of-scale mining in the porphyry era—that of open-pit mining of very low-grade ores. One hundred years ago, open-pit mining crews initiated a sixty-year assault on the landscape that led to the demise of the town. Still, from 1910, when the Chino Copper Company began the pit, until 1970, when its successor

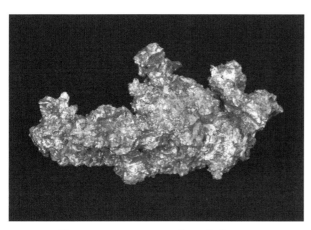

FIGURE 0.1. *This native copper nugget shows the lustrous nature of the raw deposits first discovered thousands of years ago by the Mimbres peoples. "Growing" out of the foothills below an escarpment later named Ben Moore Mountain, the shiny red metal lured successive waves of miners from New Spain, Mexico, and the United States. Mining today still occasionally uncovers this wondrous rich mineral so coveted in modern society.*
COURTESY OF TERRY HUMBLE.

the Kennecott Copper Corporation removed the last of Santa Rita's homes, a distinctive community thrived. New technologies and creative engineering strategies ensured returns on the millions of dollars in investments for the remainder of the town's life. This temporary though extraordinary mining community is the focus of this narrative and photographic history.

This story will introduce the reader to various distinctive themes in mining history. The first two themes are the implementation of frontier mining technology and the conflicts between Euro peoples and Native Americans that came about as a result. The Spanish had scoped out the Sierra del Cobre Virgen (or Virgin Copper Mountains) as early as the 1750s. Not until 1799, however, did they begin claiming and developing mines under Lieutenant Colonel José Manuel Carrasco and the laws of the Spanish Crown. Once these invading Europeans made their way to the coveted raw copper outcroppings as well as some short-lived gold prospects, they employed mining techniques they had inherited from other ventures in Peru and Mexico. This flourish of activity frightened the native peoples, who soon made a concerted effort to remove the intruders. The contact between the Spanish and the Apaches led to a legacy of conflict inherited in 1821 by the Mexicans; violent exchanges led to killings on both sides. The Americans inherited this unfortunate tradition first apparent in 1837 with the Johnson massacre, which motivated the native peoples to forcefully keep the invaders out of their homeland for the next twenty years. The great Chiricahua chief Mangas Coloradas hoped to destroy this legacy as well until his untimely murder in 1863 by US troops. Hence, the application of mining technologies by frontier Spaniards, Mexicans, and Americans forced a nearly century-long conflict with the Native Americans. These two themes dominate the story in the first two chapters.

A third major theme of this volume is the history of the community itself. Although abandoned at times, Santa Rita evolved from a mining-military camp with no more than a few hundred inhabitants, many of whom were forced there against their will, for the first forty or so years into a Victorian Mexican town of about 2,000 by the end of the nineteenth century. Like so many frontier mining inhabitants in New Spain and then the American West, the miners and their families needed protection from the area's natives to carry out their mining endeavors. Hence, Santa Rita, although known as a mining camp administered under Spanish then Mexican law, earned the title of presidio as well. Presidio status afforded the inhabitants enough protection so that they could grow gardens, raise sheep and cattle, and, therefore, build a small community. Prior to the forced evacuation in 1838, the successive mine claimants brought in priests to perform the various religious and civil functions the people so desired as well as soldiers to ensure that mining itself could be carried on. From its beginnings, however, Santa Rita was a closed community. Prior to the American takeover, most of the inhabitants were prisoners who either volunteered or were forced to work the *minas de cobre*. And if they decided to leave, they risked being subjected to capture or a worse fate at the hands of the Apaches, who felt increasingly threatened as the century wore on. Spanish and Mexican authorities punished deserters severely as well if they turned up elsewhere.

This closed-community status remained intact into the twentieth century, when the Chino Copper Company established a company town. In that isolated copper camp, the corporation enticed workers with certain amenities and then constricted them and their families. They lived according to a strict policy of segregation by ethnicity. Mexicans and Mexican Americans lived in "East" Santa Rita and the Anglos in Santa Rita, divided by a valley just east of the "Devil's backbone." Company law enforcers also kept out union organizers whom company officials called "agitators." Until the post–World War II period, company management limited the inhabitants' movements, restricted Hispanics' occupational advancements, coerced them into buying company store products, and literally limited their freedoms in this isolated corner of New Mexico. Despite these limitations, though, the community developed rich family, educational, religious, and social traditions greatly influenced by the Mexican and American cultures. Despite the imposition of welfare capitalism and corporate paternalism, the inhabitants eventually gained freedom from company rule in the post–World War II period. The main impetus for these changes came with the efforts of the local unions, especially the International Union of Mine, Mill, and Smelter Workers (Mine Mill). Under principally Chicano leadership, occupational and social justice became more possible as a result. The evolution of Santa Rita in the twentieth century from company town to independent community before its ultimate demise in 1970 is shared in chapter 4 and in parts of chapter 5.

By the 1880s, Santa Rita exhibited many of the characteristics of most mining camps in the American West in the late nineteenth century. A fourth theme reveals that the copper town experienced a kind of lawlessness that characterized life in most mining camps in the West. Although not as violent and lawless or as male-dominated demographically as many frontier mining towns, Santa Rita witnessed a "wild" period when various corporate investors fell victim to and then perpetrated scams either to make illegitimate claims to the mines or to swindle others for a profit. The famed Santa Rita Mining Association, largely sponsored by the corrupt Santa Fe ring, instigated in the 1860s one of the most notorious efforts to illegally confiscate the valuable ore deposits. This duplicitous attempt to expropriate the richest claims led to a lengthy legal battle won by the Spanish heirs who benefited from the provisions of the Treaty of Guadalupe Hidalgo and the laws of territorial New Mexico. In the 1880s, Boston financier J. P. Whitney attempted to sell the foundering copper mines for $6 million to investors too wise to fall for the trap. In another frontier legal confrontation, the Santa Rita Mining Company, a Rockefeller Syndicate enterprise, tried to "steal" the Pinder/Slip claim from local investors. The result was a lengthy legal and community squabble inherited in 1909 by the Chino Copper Company. This frontier legacy of claim-jumping and contentious litigation places Santa Rita right at the heart of traditional mining history in the late nineteenth-century American West and is covered in chapter 3.

Two additional major themes center on corporate mining and big labor. Like so many mining ventures in the American West, the "Chino enterprise," as famed engineer-historian Thomas A. Rickard called it in the 1920s, blossomed when eastern investors incorporated the Chino Copper Company in 1909. Persistent Boston investor Albert C. Burrage and New York financial giant Hayden, Stone & Company worked out a deal that offered the fiscal backing and engineering know-how to start the open-pit operations. Mining engineer John M. Sully, in fact, proved to the money men, with his successive reports on the New Mexico property from 1906 to 1909, that the economies-of-scale strategy of mass mining would reap tens of millions of dollars in profits. And once Daniel C. Jackling signed on, Santa Rita's fate was sealed. Jackling had already engineered the formation of the world's soon-to-be-largest open pit, the Bingham Canyon Mine near Salt Lake City, Utah. Sully's suggestion that the new copper company implement massive steam-driven shovels to dig the blasted low-grade porphyry ores resonated with Jackling, who first used this strategy in 1906 for the Utah Copper Company. Jackling's endorsement of Sully's propositions and then his willingness to serve as Chino's chairman of the executive committee of the board of directors convinced the investors to go forward with the colossal project. The story of the birth of the Chino Copper Company is covered in chapter 3.

The Chino enterprise proved so successful that it soon became the target of other mining corporations. Profits of nearly $10 million in 1917 alone attracted the attention of the Guggenheims. Already well established in the smelting industry with their American Smelting & Refining Company, the Guggenheims decided to invest in major copper-mining properties as well, first in Chile and then in the United States. To consolidate their multinational properties the Guggenheims incorporated the Kennecott Copper Corporation in 1915, then began acquiring stock in the Utah Copper Company. By the mid-1920s the syndicate controlled the Utah venture, continued to develop the Chuquicamata and El Teniente mines in Chile, and had established two additional mining corporations, Ray Consolidated (Arizona) and Nevada Consolidated. In 1924 and 1926, successively, these two latter outfits purchased the Chino operations, which included the open pit, crushers, and a concentrator. In 1932–1933 the Guggenheims consolidated all four properties—Bingham Canyon, Ray, Nevada, and Chino—under one corporate entity, the Kennecott Copper Corporation.

Kennecott would soon become the top copper corporation in the world. The Bingham Canyon Mine ranked first internationally, and Chino as high as fourth. Kennecott consolidated all four properties in the American West into the Western Mining Division. The corporation's successful domination of the industry translated into exponential profits and reinvestment in each of these mining ventures. After Kennecott acquired Chino in 1933, for example, it soon invested nearly $10 million to upgrade the worn-out and obsolete technologies with newer, more efficient, and larger-scale machinery and equipment. The copper company also constructed a smelter at Chino Mines in the mill town of Hurley just nine miles south of Santa Rita. This rejuvenation program set the pattern for regular upgrading of technologies at the New Mexico enterprise, which included the laying of more than fifty miles of rail lines, the transition from coal power to electricity, the use of the latest drilling and blasting technologies, and the introduction of massive haul trucks, precipitation processes, and innovative smelting techniques. The results were phenomenal. Shovel and haul crews removed tens of millions of tons of ore and waste rock annually, creating the mammoth stairstep benches of the gigantic open pit. Mill men calculated the best strategies to process the ores into concentrates, and smelter men devised new methods to produce increasingly purer copper. By the 1940s, production soared annually to 140 million pounds, rarely dropping below this figure for the remainder of the century.

As early as the 1930s, Santa Rita's fate was apparent to company executives, engineers, and the workers. The steady persistence of the mine operations ate away the ground of Santa Rita. The company began removing homes and then planned for future mine expansion with a clear understanding that the town would eventually disappear. By the 1950s the use of giant haul trucks of 25- to 40-ton capacities expedited the digging of the pit and, simultaneously, the removal of the bedrock foundation of the copper town. Kennecott decided in 1955 to sell the company-owned homes and other buildings to an Ohio real estate firm, initiating the final evacuation over the course of the next fifteen years. Engineering strategies and implementation of massive technologies, such as 85- and 150-ton Lectra Haul trucks, meant that Santa Rita's demise was inevitable. The story of Kennecott's implementation of modern technologies and the consequences for Santa Rita is told in chapter 5.

Another key theme of this volume centers on big labor. For the first forty-two years of the twentieth century the successive corporations blocked unionization at Chino. Corporate hegemony over the workers predominated. John Sully, general manager from 1909 until his death in 1933, crushed all unionization efforts at the New Mexico enterprise in cooperation with his management team, as well as head of security Jim Blair. In 1912, for example, every shovel runner who "sat" on the job in protest of his wages was fired and his name was sent to copper companies throughout the American West. The same fate met a group of men in 1923 after they protested wages and other conditions. After Kennecott acquired the Chino property in 1933, Jackling shut down the operations soon after learning the workers had voted to become certified in the International Union of Mine, Mill, and Smelter Workers. New general manager Rone B. Tempest, in fact, compiled a blacklist of nearly seventy employees who "instigated" unionization. The workers, especially unjustly treated Hispanic "laborers" (a euphemism for Chicano employees), hoped to establish a collective bargaining unit. They believed Mine Mill would assist them in eliminating the dual labor system, the unfair limitations in lines of promotion, unequal pay compared to other divisions in the company, discrimination in the wage scale and occupational advancements, and other traditional injustices perpetrated by company officials. Only after the intervention of the National Labor Relations Board, mainly from 1938 to 1942, and then the US Supreme Court's decision in 1942 to favor union certification did Kennecott have to recognize and begin bargaining with Mine Mill and numerous other unions. The certification crisis is examined in chapter 3. Legal protections and labor activism combined to break down corporate hegemony over workers on a nationwide basis. Chino's story is a microcosm of that transformation in the second half of the twentieth century.

The rise of big labor after World War II brought about steady changes for workers at Chino. The results were astonishing, in terms of both work-related and larger societal reforms. Clearly, Mine Mill and the other unions made gains that altered Kennecott's unfettered domination over the workers at Chino and elsewhere in the American West. Under the leadership of Clinton Jencks, beginning in 1947 Amalgamated Local 890 became the standard-bearer for worker initiatives and demands from this time forward. With his wife, Virginia, "El Palomino," as he was called, vigorously worked to train Chicano labor leaders to fight for their rights at work *and* in the broader community. Albert Muñoz, Joe T. Morales, Juan Chacón, and others soon trusted Jencks. They all worked together to formulate a grassroots labor movement that blossomed beginning in the late 1940s. They learned about the historic injustices against workers in the United States and cleverly placed themselves in that context to fight for their rightful claims to better wages, safer working conditions, greater fairness in promotions, and fringe benefits, such as health insurance and retirement packages. At the same time, leaders like these men and others, such as Art Flores and Cipriano Montoya, inserted broader societal initiatives into their labor strategies. They introduced the rank and file, for example, to political candidates like Henry Wallace in 1948 for president of the United States and David Cargo in the 1960s for governor of New Mexico.

Jencks, Muñoz, Morales, and others also gallantly fought anti-labor and anti-communist efforts of the US Congress to limit unionization and collective bargaining gains. Mine Mill officials of the international office and of Local 890, for example, confronted the unjust anti-communist affidavit provision of the Taft-Hartley Act of 1947. Local 890's first strike in 1948 against Kennecott, in fact, reflected the union's willingness to risk punitive measures by the corporation and threats to its collective bargaining status in opposition to the affidavits, which the US Supreme Court in 1965 finally declared unconstitutional. Local 890's valiant struggle against this unjust law, which also supported the right-to-work doctrine and limited the union's access to National Labor Relations Board arbitration, reflects the civil rights component of the labor movement in post–World War II Grant County. That struggle also included Mine Mill's expulsion from the Congress of Industrial Organizations in 1950 as well as repeated attempts by the United Steel Workers of America (USWA) to raid Mine Mill jurisdictions. Local 890 officials soon realized the need for cross-union cooperation, initiating the formation of the Chino Unity Council in the mid-1950s to unite the industrial and craft unions. The results often translated into wage increases and other benefits for Mine Mill as well as the Metals Trade Council locals (e.g., the Brotherhood of Locomotive Firemen and Enginemen of Local 902, the Brotherhood of Railroad Trainmen of Local 323, and the International Association of Machinists of Local 1563).

These efforts reveal a kind of labor militancy in Grant County not witnessed in the 1950s and later elsewhere in the copper industry. Among the extraordinary achievements of Local 890, after its successes at Kennecott in the late 1940s, was the legendary Salt of the Earth strike of 1950–1952. Targeting the Empire Zinc Company's treatment of its workers, Local 890, whose Chicano leadership worked principally for Chino Mines, won fame in the radical syndicalist movement for its willingness to stand up to multimillion-dollar corporations, as well as government officials, during the height of the Red Scare in the mid-twentieth century. The banned (as communist) and "notorious" *Salt of the Earth* film, starring Local 890 president Juan Chacón as well as labor leader Joe T. Morales, among other locals, still today represents the workers' struggles for occupational as well as societal justice.

Local 890 continued its efforts for labor reforms in the 1950s and 1960s. Numerous strike actions forced Kennecott to more fully address the demands and grievances of the workers. Although Chacón, Severiano "Chano" Merino, and other labor leaders always remained vigilant in fighting for equality at Chino, Kennecott's workers witnessed a dramatic change in their rights on the job and in the wider southwestern community. Chino Mines general manager Frank Woodruff, in fact, came to realize by the 1960s that big labor had made a distinctive mark in labor-management relations. He and officials from corporate headquarters had no choice but to recognize their demands. Even after Mine Mill merged in 1967 into the USWA, the grassroots efforts of Local 890 continued to influence Kennecott's decision-making process concerning its workers. This remarkable success story resulted in Mine Mill and its successor, the USWA, cooperating with Kennecott to set the pattern for the best labor contracts in the copper industry. In essence, Kennecott executives accepted Mine Mill's argument that workers deserved a larger portion of the profits through good wages, formalized grievance procedures, safer working conditions, some of the best health and retirement benefits in all the smokestack industries, tuition support for college students in the county, bonuses for suggestions to improve operations, and other paybacks for service to the company. Labor compromised the traditional corporate hegemony over the workers, and in the end the evolution of labor-management relations facilitated the formulation of an interestingly affluent society. Both Chicano and Anglo workers now had access to the American dream of owning a home, having jobs for generations, and gaining access to the full rights of citizenship. With a standard of living unmatched in New Mexico, Santa Ritans and others of Grant County evolved from a working-class to a working, middle-class community. Labor's role in formulating the new work environment and countywide affluence is examined in chapter 5.

Finally, the reader should be reminded of the environmental costs of mining for two centuries in New Mexico's most prolific and famous mineral district. This theme, peppered throughout this story, has left a legacy that locals cannot ignore. Poor management of resources brought on nature's fury. Floods, for example, inundated the area in the 1890s and early 1900s from the erosion caused by overcutting timber and piling waste and ore dumps in the surrounding mountain ranges. Chino did stave off major flooding and water scarcity problems through its water conservation program. On the other hand, the massive transformation of the landscape throughout the open-pit era in the twentieth century graphically illustrates how mining has affected the local terrain, its human and animal inhabitants, and groundwater sources. The industrial processes of mining, milling, and smelting also left their marks, creating toxic water and air pollution problems. The battle between Kennecott and its workers in the 1970s and later against government officials and grassroots environmentalists attempting to force the copper corporation (and its successors Phelps Dodge and Freeport-McMoRan, in particular) to comply with air and water pollution regulations forewarned of another threat to industrial hegemony similar to the labor movement. In that confrontation Kennecott and its successors eventually realized the societal as well as fiscal benefits of cleaning up their mess, even though recalcitrant at times in formulating a more earth-friendly strategy. Air pollution problems, of course, eventually disappeared with the demolition of the smelter. Water pollution issues remain, es-

pecially in light of the gargantuan needs of the modern operations for this scarce natural resource in the Desert Southwest. Corporate investment in water conservation and cleanup, combined with governmental and environmentalist watch-dogging, has resulted in changes. Whether the current owners will continue to comply with state and federal regulations remains to be seen. Regardless, like the evolution of the labor movement, which required corporate understanding and sympathy for the employees and resulted in changes in workers' conditions, the environmental movement must be taken more seriously for the mining operations to continue to function less toxically in the early twenty-first century.

When Harrison Schmitt, Gilbert Moore, and Ted Arrellano founded the Society for People Born in Space in the mid-1970s to memorialize the former Santa Rita as "space," the parameters for the town's history had been set. It began as a Spanish mining outpost in 1803 and, although at times abandoned, lasted as a community until 1970. This narrative offers deep insight into Santa Rita's mining and communal pasts through words and photographs. We hope the combination of the two gives a complete picture of an exceptional place with a very interesting history.

The visual markers of that place are clear. The Kneeling Nun still sits above the gargantuan 1.5-mile-wide and 1,500-foot-deep pit to remind locals of mining's past and present as well as the Hispanic culture that sustained most of the inhabitants of the former town. The pit itself symbolizes mining life in terms of the profound achievements of modern engineering and technological know-how and provides a stark example of human manipulation of nature. The thousands of visitors who stop at the historical wayside just off Highway 180 on the north rim of the mine learn how the pit was formed and a bit of the history going back to when an Apache man first introduced Carrasco to the rich native copper deposits. They also see the mountain-size waste and leach dumps that can be seen for nearly 100 miles on a clear day in the Southwest. This volume is an attempt to document the mining processes that formulated this scene as well as to remember the people who were responsible for this result. In the end, although Santa Rita no longer exists and is now relegated to a space above the pit, the town's story now has a space in mining and community history.

I

El Cobre

SPANISH AND MEXICAN MINING IN APACHERÍA

I n 1799 José Manuel Carrasco struck virgin copper. The retired lieutenant colonel had rediscovered the richest native copper deposit in North America. Taken to the lustrous outcroppings by a small group of Apaches he had assisted in hard times while serving as captain of the Presidio of Carrizal in northern Chihuahua, the Spaniard thought he had found the mother lode of copper deposits (see figure 1.1). But he knew he and those who followed him to this place would

have to contend with the native peoples. The Spanish officer had spent much of the preceding thirty years pursuing and fighting the nomadic inhabitants of Apachería, the expansive territory of the Apaches. Located in the heart of the Chihenne band's homeland, the *minas de cobre* (copper mines) promised riches to the seasoned Spanish warrior. Or at least that was Carrasco's hope. Unlike Francisco Coronado in the 1540s, who pursued the myth of Quivira and the seven cities of Cíbola, this Spaniard found his treasure. Yet, as this story will reveal, like Coronado, he would experience disappointment and conflict, and the fortune he sought to glorify Spain and to enrich himself would never be realized. Still, six years later in testimony to the Deputacíon de Minería (Mining Bureau) in Chihuahua, he claimed "divine providence"

had intervened to protect him from the risks of journeying into Apache country. God, copper, and glory, he believed, were at hand.[1]

The massive outcropping of this nearly pure copper lay in a parched valley beneath an escarpment in a small range locals today call the Santa Rita Mountains. Like many of the Southwest's rich copper ores, the Santa Rita deposit was in the Mexican Highlands. Part of the northernmost range of the Sierra Madre Occidental originating in Mexico, these mineral-rich uplands exhibit the characteristics of the Basin and Range physiographic province of much of the interior American West. Jutting out of the foothills of what was later named Ben Moore Mountain, just south of the famed Gila wilderness country and southwest of the Black Range, the glistening

FIGURE 1.1. *This Carl Hawk drawing (ca. 1983) depicts the moment in 1799 when an Apache shows Lieutenant Colonel José Manuel Carrasco the outcrop of native copper. Although Carrasco never realized his dreams of making a fortune from the rich deposit, various other miners and mining companies have been extracting ore from the Santa Rita property ever since.*

metallic protrusion was exposed near the confluence of the intermittent Santa Rita and Whitewater Creeks. Part of the northernmost extension of the Chihuahuan Desert, the valley had little vegetation save desert grasses, scrub oak, juniper stands, and various desert wildflowers; Douglas fir and various pine species grew above 5,000 feet in the surrounding mountains to the north and east. Beneath the rugged arid terrain were the richest mineral veins of the future state of New Mexico. Erosion of Tertiary volcanic flows that had uncovered the raw copper had also exposed rich silver, gold, lead, zinc, manganese, and other ores that would by the 1860s begin to entice thousands of hopeful prospectors and miners. Soon thereafter, the Americans would establish the Central Mining District in the newly created Grant County. The Santa Rita copper ores were at the heart of what would become New Mexico's most productive mining district, which also included Pinos Altos, Hanover, Georgetown, Fierro, and many other mining camps[2] (see map 1.1).

Sixty-two years before Carrasco's arrival at the site, Captain Bernardo de Miera y Pacheco, a cartographer from El Paso del Norte, visited the area as a member of Spain's first major military expedition into Apachería. After additional exploratory trips on his own into the region, the skilled soldier-engineer produced in 1758 a crude, hand-drawn map of New Mexico that featured large sections of the Apache homeland. He plotted the copper outcroppings and christened them the "minas de cobre." He named the 8,000-foot-high range to the north the Sierra del Cobre Virgen,[3] or Virgin Copper Mountains, penciling in their jagged features between the upper reaches of the Gila and Mimbres Rivers. Later mapmakers used his map to plot the copper deposits and the mountains named for the coveted metal (see map 1.2).

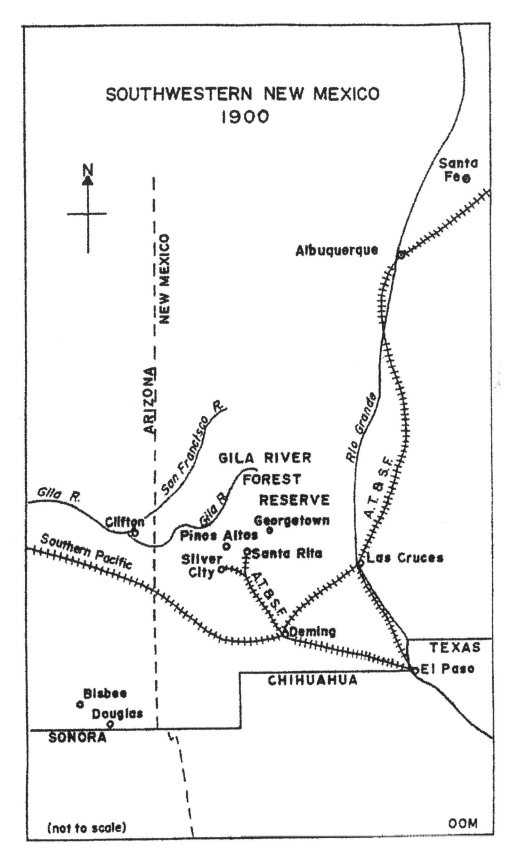

MAP 1.1. *Map of southwestern New Mexico in 1900. Drawn by O. Orland Maxfield.*

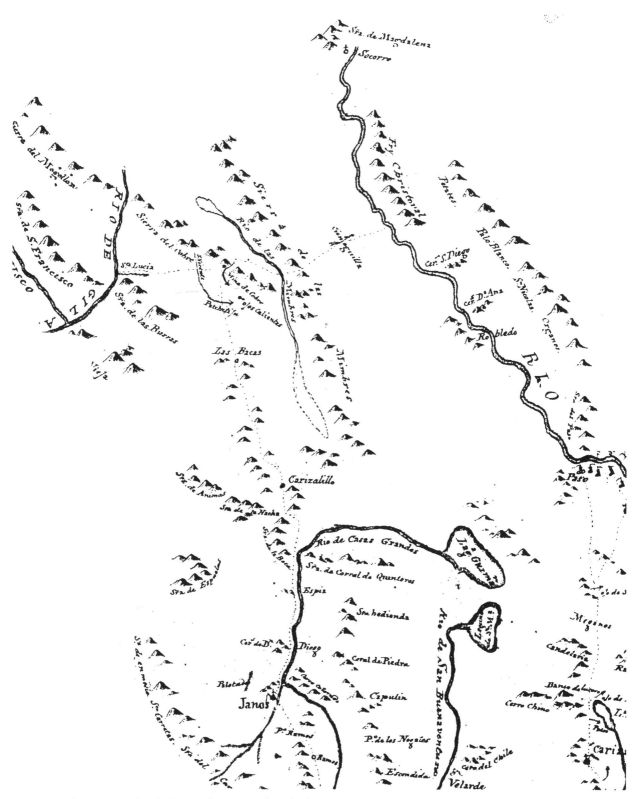

Map 1.2. *Handwritten map (1805) of internal provinces of Coahuala, Nueva Viscaya, and Nueva Mexico by the order of Comandante General Nemecio Salcedo y Salcedo. Ensign Juan Pedro Walker of the Janos Presidio composed this map highlighting the Sierra del Cobre and the "Mina de Cobre" in the north and the Janos Trail (dotted line from Mina de Cobre to Janos) used by Santa Rita miners to freight native copper from the mines to Janos, Chihuahua, and Mexico City.*

In 1777, Miera y Pacheco wrote to King Carlos III with a proposal to settle Apachería. In his letter, he planned for three new hacienda-presidios, one on the Gila River, another at the copper deposits (where he would set up his command post), and a final one on the Mimbres River. Each would be in a new province he named Jila. Like Father Kino in Arizona and missionaries in California, this Spaniard hoped to "manage" the native peoples while simultaneously contributing to the Crown's coffers and his own wealth and glorification.[4] Yet, rather than try to establish a mission at El Cobre, he wanted to create a fortified post, or presidio. A military presence would be necessary, he suggested, especially after the Marqués de Rubí's military inspection of the outposts of northern New Spain in the late 1760s revealed that native peoples, especially the Apaches, had subjected the inhabitants to attacks and other hardships.[5]

Clearly, Miera y Pacheco, along with other officials in northern New Spain, hoped to develop the site, soon to be called El Cobre, to extract mineral riches for the declining empire. The corridor from Mexico City to El Paso to Santa Fe, used by Diego de Vargas in the 1690s to reconquer New Mexico, had not delivered the colonial treasures they had anticipated. Perhaps El Cobre could sate that desire. In the end, however, Spanish authorities rejected the proposal, fearing that the costs of maintaining the outposts in the dangerous Apache country would far exceed the benefits.[6] The Apaches would remain supreme for some time to come.

At the time Carrasco made his first mine claims in 1801 at El Cobre, the Chiricahua Apaches dominated the region. Athapascan peoples originating from Canada, they traversed modern-day southwestern New Mexico, southeastern Arizona, and northern Chihuahua and Sonora, Mexico, beginning in the fourteenth century. Their diets consisted of game, such as turkey and deer, and edible plants like agave and piñon nuts, principally from the San Francisco to the Gila to the Mimbres river valleys of the Southwest. Their nomadic lifestyles precluded the establishment of permanent settlements and instead they constructed *rancherías* consisting of two or three small wickiups, or domed huts, per family that could be easily dismantled and rebuilt or abandoned altogether in times of crisis.

There were four Chiricahua Apache bands: the Chihennes, Bedonkohes, Chokonens, and Nednhis. The Chihennes were the main group in the vicinity of El Cobre. Their territory extended along an east-west axis from the Rio Grande to San Vicente Arroyo (later Silver City), with El Cobre at the center, and north-south from the Black Range to near Janos. Their cousins, the Bedonkohes, ranged from the same arroyo to the future Arizona border from east to west, and then from the Mogollon and San Francisco Mountains in the north to northern Sonora and Chihuahua in the south. The other two bands—the Chokonens, who lived in southeastern Arizona, and the Nednhis, who inhabited northern Mexico—made up the rest of the Chiricahua "extended family."[7]

The Spanish, and later the Mexicans and Americans, interacted mainly with their dominant chiefs. El Fuerte (the "Strong One") was the most famous Chiricahua leader of the nineteenth century. Also known as Mangas Coloradas or Red Sleeves, he lived from about 1790 to 1863. As boys, El Fuerte and other future warrior-diplomats were taught about the importance of certain virtues that came to them, they believed, supernaturally. In their mythology, the deity Usen imparted virtues upon them and taught them to revere the gifts of nature as well as unseen mythical spirits. Mothers told them stories about the mystical Mountain People, who defended them from their enemies and other dangers and prophetically forewarned them of ominous times to come, a reflection of the growing presence of invaders in their territory.

To prepare for these difficult times, their fathers and uncles trained them to endure many hardships, including lack of food and water, having to travel great distances on foot, and bravely suffering through grave injuries. They also learned how to wield bows and arrows, slingshots, and firearms that they acquired in trade or after military victories. The men who would encounter the Euro-Americans also learned to be patient, humble, courageous, and generous. The encroachment of the foreigners, however, taught them the need to be strong physically, mentally, and spiritually. Displays of weakness could result in shame and demotion from chieftainship. Perhaps the two most problematic features of their lifestyles in their relations with the outsiders were their raiding practices and their societal belief in seeking revenge against their enemies. Those warriors who could quietly abscond with domesticated herds of cattle, horses, and sheep, for example, were believed to have been blessed with supernatural powers. On the other hand, seeking vengeance against their enemies was a common part of virtually all cultures and, unfortunately, would lead to countless acts of violence during the nineteenth century between the Chiricahuas and the Spanish, Mexicans, and Americans.[8]

For at least two centuries, the Apaches had engaged in treaties with the Spanish, often agreeing to accept rations to supplement their often meager diets in return for peace. Their traditional dependency on game, however, often created conflicts with the Spanish as the Apaches regularly targeted the invaders' domesticated herds of cattle and sheep for harvest and trade. Considered stealing by the Spanish as well as the Pueblo peoples, Apache raiding regularly threatened the peace and resulted in fractious and often bloody relations.[9] Persistent Spanish encroachment on Chiricahua territory also greatly heightened tensions. Carrasco himself was a principal figure in the testy relationship between the Spanish and the Apaches in Nueva Viscaya and Nuevo Mexico. His friendship with the men who directed him to the virgin copper after nearly thirty years of military pursuit of other Apache men is an example of this complex love-hate relationship. At El Cobre, diplomatic relations between the Spanish and the Chihenne band mirrored this contentious history. Likewise, the growing influx of European and mestizo peoples to this frontier region of northern New Spain encroached on the native peoples' territory, anticipating the borderland conflicts and migratory patterns throughout the colonial Spanish, Mexican, and early American periods.[10] Only when the Apaches were concentrated on reservations in the 1870s did their resistance begin to come to an end.

Carrasco, probably a *criollo*[11] born in 1743 at Julimes, Chihuahua, joined the Spanish colonial army in his mid-twenties, serving for three decades at three separate presidios: Carrizal, Buenaventura, and Janos (see map 1.2). Known for his heroism in battles with the Apaches and for recapturing many a horse herd, he earned a reputation for bravery and rose through the ranks, retiring in 1798 as a lieutenant colonel. In search of a lucrative retirement plan, he asked his Apache friends to take him to the site of the native copper deposits.

Like many prospectors then and later, Carrasco sought financial backing from a wealthy benefactor. In his case, he earned a grubstake from Pedro Ramos de Verea, a successful Chihuahua merchant who had supplied the various presidios of northern New Spain. Ramos de Verea offered Carrasco his mule train to freight the raw copper and unrefined ores from El Cobre

to Chihuahua and on to Mexico City. He also promised the neophyte miner to file mine claims for him at the Deputación de Minería in Chihuahua and to deliver the copper to the royal mint in Mexico City. Further assisting the retired soldier was Pedro de Nava, commandant general of the Interior Provinces. Nava used his authority to put together a ragtag workforce made up of prisoners, a common centuries-long practice among Europeans, from the Chihuahua and Janos jails. Carrasco established a dual-wage system, paying Mexicans one peso a day and Apaches only one-half that.[12] The dangers associated with mining in Apachería made it difficult to entice skilled miners to join Carrasco even at double the normal wage. Nava also sent a contingent of soldiers from the San Buenaventura and Janos presidios in fall 1800 and then again in spring 1801 to reconnoiter the Mimbres Valley and the Sierra del Cobre Virgen. The Spanish officer hoped to frighten away the Mimbreños (the Spanish name for the Chihenne band) in anticipation of Carrasco's arrival at El Cobre in April 1801, when he staked his initial claims and began mining the outcroppings of virgin copper.[13]

By June, Ramos de Verea had registered for Carrasco two copper claims, El Corazón de María and El Corazón de Jesus (the Hearts of Mary and Jesus). Two additional claims, a silver mine, Nuestra Señora de Guadalupe (Our Lady of Guadalupe), and a gold mine, Santísima de Trinidad (Holy Trinity), reveal Carrasco's hopes for multimetal success. According to Spanish mining law, claimants did not actually own the deposits as was the case with land grants. The minerals by law belonged to Crown and country. Regardless, claimants could become wealthy from their mining endeavors. Claims ran about 275 by 500 feet and the government allowed claimants to stake up to three additional claims on a given vein. Development work required a trench of about five feet wide by thirty feet long to legally retain an individual claim. Authorities expected this assessment to be completed within ninety days or the claimants forfeited their rights to the deposit. The Deputación de Minería waived this requirement for Carrasco because of the difficulties of distance, the dangers of Apache depredations, and authorities' hopes for a rich recovery.[14]

By July, Carrasco had sent eighty-four *arrobas*[15] (approximately 2,100 lbs.) of native copper with Ramos de Verea's eight-mule train to Chihuahua on the Janos Trail (see map 1.2). After taxes and Ramo de Verea's share of the first load, which ended up at the royal treasury in Mexico City, Carrasco probably made little profit. Over the next two years, 1802–1803, his lack of mining experience and a series of obstacles—difficulty recruiting laborers, costly freighting charges, high taxes (10 percent for the king's portion and 6 percent for the royal treasury), limited access to refining and smelting technologies (much of the native copper contained a matrix of limestone and granodiorite rock), and costs to hire soldiers to protect El Cobre from Apache attacks—spelled certain failure for the retired officer. In his disappointment, despite a temporary tax waiver, Carrasco decided in late 1803 to sell his rights to the richest copper claims to a well-known Chihuahua entrepreneur and politician, Francisco Manuel Elguea.[16]

Elguea had the means and the connections to introduce more sophisticated techniques at El Cobre. A native Spaniard, he included in his lengthy resumé a lucrative mercantile trade in Chihuahua that allowed him to acquire contracts with the Spanish government to supply frontier presidios as far away as Santa Fe with military accoutrement, farming and smithing tools, food and clothing, and fineries for the elites. He also served on Chihuahua's *ayuntamiento* (or city council), was a subdelegate to the treasury, and completed military service in the militia corps, achieving the rank of lieutenant. The Chihuahua businessman owned a large ranch, Hacienda del Torreón, which provided the horses and mules necessary for his freighting enterprises. These positions, combined with his growing wealth, made him the ideal candidate to develop the minas de cobre.[17]

In accordance with mining regulations, Nemecio Salcedo, commandant general of Chihuahua, appointed Diego Obeso as the superintendent of mines to oversee Elguea's new venture at El Cobre. Obeso, a native of Castile, Spain, and a clerk in Elguea's Chihuahua businesses, soon introduced underground mining to the nascent New Mexico copper industry. Salcedo, who quickly became the greatest advocate of the copper mines in northern New Spain, also gave Obeso the title of *comisario de justicia*. Similar to a justice of the peace, the new comisario had the authority to register mining claims and adjudicate disputes over them as well as to investigate crimes.[18]

Soon after his arrival at El Cobre, Obeso directed the *gambusinos* (miners) to sink a shaft to find the richest veins. The miners soon dug tunnels and carved out stopes, pursuing the veins in an irregular pattern. Using techniques previously implemented at Río Tinto, Spain; Potosí, Peru; and Zacatecas, Mexico, El Cobre miners adopted similar underground strategies.[19] *Barreteros* (pickmen) wielded thirty- to forty-pound iron *barretas* (bars) about six feet in length, wedging them in crevices and cracks to loosen the native copper and other ores. The barreteros also broke away the rich ores with sledge hammers, wedges, and picks. The barreteros or *tenateros* (muckers) then hauled the ore to the surface in *tenatas* (leather ore bags with thongs) or *seronis* (leather shoulder bags). The miners and muckers descended into and ascended out of the shafts on *muescas*, or what came to be known as "chicken ladders," stripped juniper or pine logs notched with steps. They also carried deer or buffalo hide *toneles de agua* (water buckets) down into the mines to quench their thirst and wash their nearly naked bodies (see figures 1.2 and 1.3).[20]

The workers, usually prisoners serving sentences for various crimes or debts, labored under difficult conditions. Shaft entrances often caved in and were so narrow that they were "barely large enough to allow the body of an average-size man to enter hand-over-hand."[21] Loose rock randomly tumbled down the narrow passageways or from unsupported ceilings. The Deputación de Minería addressed these concerns, requiring mineowners to reinforce shaft and adit entrances and tunnels with boxed timbering called crib lathing (see figure 1.4). Sometimes the gambusinos placed overhead beams tied together with rawhide or carved pillars in tunnels and stopes to protect themselves from volatile rock ceilings, hanging walls, or even ore bodies. Muckers often dumped waste rock into worked-out drifts, tunnels, and shafts to reduce their mucking loads and the threat of cave-ins.[22]

The implementation of these extraction strategies, although primitive by later standards, introduced El Cobre's surroundings to the most modern mining technologies of the day. The development of the mines also connected the new copper camp to the Atlantic System, the vast trade network that linked North America to the imperial economy of Spain, the rest of Europe, and their colonial territories. These dual factors—introduction of new technologies and international demand for the copper—also offered the possibility for a long life for the mines and the community, despite the impending conflicts with the Apaches and being so isolated. In the

FIGURE 1.2. *Relics of early Spanish mining discovered in 1915. Chino worker poses at mine portal with rawhide ore bags and notched "chicken" ladder used by the Spanish and Mexican miners from about 1800 to 1838.*

COURTESY OF TERRY HUMBLE.

end, the coveted mineral would pull this remote, Indian-controlled spot into the Western capitalist system.

Because of these dynamics, the Spanish were also initiating the first phase of the impact of the mining culture on the local environment. The dumping of waste materials, for instance, began the gradual reformulation of the local terrain. These rock piles were the first examples in a two-century process of reordering the landscape that eventually culminated in the mountain-size dumps and giant open pit of today. Although the Spanish accounts of the time reveal little about the initial impact of these first dumps, the miners no doubt experienced some flooding problems during the rainy season in late summer and early fall. Still, the environmental impact at this early date was negligible compared to future alterations in the late nineteenth century and throughout the twentieth century. Future miners, in fact, would rework the Spanish dumps, which contained fairly high-grade ores,

well into the twentieth century, suggesting that erosion of the terrain in the vicinity of the mines was limited.[23] Regardless, the Spanish introduced the modern mining culture to the area, putting into motion a new human-nature relationship that over two centuries would place greater stresses on the land, air, and water in the vicinity of the mines.

By December 1803, Obeso had sent Elguea's first shipment of ore to the royal mint in Mexico City. Assayers tested the samples and quickly realized that El Cobre ores were some of the richest to pass through the treasury in a long time. They were of such high quality that one official reported them to be "the cleanest, most malleable, and best suited for coinage" to arrive at the mint in "many years." He recommended a high price of twenty-five pesos per 100 pounds.[24] The mint's endorsement hailed prosperous days ahead for El Cobre. In May 1804, in fact, Chihuahua's commandant general Nemecio

FIGURE 1.3. *Chino worker poses in 1915 with relics of early copper miners at Santa Rita.*

COURTESY OF TERRY HUMBLE.

FIGURE 1.4. *Old Spanish triangle shaft photographed in about 1920 on the east side of the Hearst Pit. Steam shovel operators uncovered this Spanish-era timbered shaft in 1915 during the early phase of open-pit mining. Miners probably constructed this shaft to protect against cave-ins or for ventilation.*

COURTESY OF TERRY HUMBLE.

Salcedo reported to the Ministry of the Indies in Cádiz, Spain, that the new discovery had already resulted in 350,000 pounds of copper production. Salcedo's stamp of approval marked a turning point in the history of the copper mines at what would soon be called Santa Rita. Government authorization would also lead to the establishment of a presidio at the copper mines, so that the ores could make their way to Mexico City to be smelted and then molded into coinage.[25]

In the meantime, Blas Calvo y Muro, a business associate of Elguea, arrived in the sierra in January 1804 with hopes of staking a lucrative gold claim. Carrasco himself had already staked out Gold Gulch about four miles west of El Cobre and still hoped to pan out enough of the lustrous metal to get rich. However, he had done

little or no assessment work on the claim, threatening to end any chance for the desired strike. Calvo y Muro realized Carrasco had not met the requirements of Spanish law to maintain possession and by July had registered his own claim to the gold prospect, naming it the Santa Rita de Cascia. Calvo y Muro's initial success at the Santa Rita caused a short-lived gold rush over the next two or three years and would also introduce Lieutenant Colonel José María de Tovar into the story.

Sometime during this initial phase of mining, the Spanish changed the name of the copper camp to Santa Rita del Cobre. They decided to combine the name of the community's new patron saint, Santa Rita of Cascia, with the site's most distinctive attraction, copper. It is a fitting name because Santa Rita of Cascia (1383–1457), of medieval Italy, was the patron saint of impossible causes. She had spent much of her adult life cloistered, even from the other nuns in the convent, because of an open wound on her forehead. She had allegedly received the injury in the midst of praying after an inspirational sermon on the crown of thorns in the crucifixion story. The oozing sore left her scarred and often bleeding for the rest of her life. The earlier deaths of her husband and two sons, combined with her own reclusive lifestyle, must have resonated with the isolated inhabitants of El Cobre. Soon after 1804, they renamed their mining camp Santa Rita del Cobre in reverence of the suffering saint. The miners themselves often felt abandoned in a difficult cause because of their frontier isolation and their persistent conflicts with the Apaches. Naming a place after a saint was a common practice among Catholic colonists in New Spain.[26]

Tovar first came to the Sierra del Cobre to resolve the dispute between Carrasco and Calvo y Muro concerning the latter's recent claim to the Santa Rita. He ruled in favor of Calvo y Muro, of course, who had Elguea's blessings. But more importantly, Tovar was ordered by Commandant General Salcedo in late 1804 to conduct an inspection of the minas. A Janos presidio officer, Tovar would serve as a key agent in Salcedo's plan to further develop the copper mines. Already commissioned to command 130 soldiers, he and his detachment were very familiar with the terrain because of their previous dealings with Apaches and after having escorted Elguea's mule trains loaded with copper along the Janos Trail (see map 1.2). Salcedo decided Tovar should make an inspection of the Sierra del Cobre copper and gold mines and report what he found.

In December 1804, Tovar reported to Salcedo what was already apparent to Carrasco, Elguea, and the other investors: there were few mines in operation. Mining was dangerous because of cave-ins. The native copper ores were playing out and a much lower-grade ore was being mined. A smelter was needed to refine these complex ores into ingots to greatly reduce the volume of material being freighted by mule. Salcedo took what appeared to be a very negative accounting of the Sierra del Cobre and requested from Spanish authorities permission to make the development of the copper deposits a principal investment of the state of Chihuahua. His appeal granted,

Salcedo soon invested more of Chihuahua's resources in protecting the mining camp and the precious cargo from raiding Apaches, at least until 1810 at the beginning of the Mexican Revolution. Salcedo also announced the establishment of the Real del Cobre (Copper Mining District) and gave superintendent Obeso the authority to register claims and administer Spanish mining laws.[27]

With this new legal, financial, and military support, the Santa Rita del Cobre mining operations began to blossom. Within two years of Tovar's report, the workers implemented a crude smelting process, probably under Obeso's replacement, José Carlos Romero, superintendent and justice commissioner from 1805 to 1807. Soon after the principal deposits of native copper played out, Elguea's men had to devise a way to facilitate shipment of the ores or smelt them on-site to ease the loads of the mules and reduce freighting costs. To do this, they implemented the use of smelting pits. Firing charcoal made from juniper and oak harvested from the local hillsides, the smeltermen broke up the chunks of ore and placed them in clay- and ash-lined pits once the coals were red-hot. Keeping the heat at a maximum temperature with hand bellows, they heated the ore to a molten state, poured it into holes they dug in the earth, skimmed the slag from the top, and then let it settle and cool. In a few hours, the cooled metal hardened into *planchas*, or ingots.[28]

The crude ingots varied by weight and purity. To standardize copper measurements, authorities at the royal mint in Mexico City began in 1807 to require more uniform bars. Smelting the ores in Castilian furnaces, they poured the molten red metal into carved stone molds that they had chipped directly into bedrock. Smelter laborers fed the hearth of the six-foot-high adobe or brick furnace with charcoal, heating it to a desired temperature. They then shoveled in the ore. After some time, the workers tossed flux, probably pine chips or some other quickly burning material, into the firing ore to burn off waste materials (slag) from the matrix (gangue). Workers then skimmed the slag with a hooked ladle. Next, they poured the nearly pure molten copper into metal or stone molds to produce glistening ingots. Standard-size bars had a width of about six inches, a depth of four and a half inches, a length of twenty inches, and a weight of nearly 165 pounds. Two ingots could be tied together and straddled over the back of each mule, whose carrying capacity was usually around 300 pounds.[29]

The environmental impact of cutting timber to produce the charcoal to smelt the ores was, like the dumping, negligible in this time frame. *Carboneros*, or charcoal makers, harvested most of the timber they used for manufacturing charcoal from the Black Range, many miles to the north of the operations. Unlike the massive clearcutting of the late nineteenth century, which laid waste to vast acreages in the Gila River watershed and caused perennial floods from the 1880s to 1910s, the limited downing of juniper and pine stands had little impact on the terrain of the region.[30] The emissions from the crude smelter, although unsafe for the workers themselves, also had an insignificant effect on the local vegetation and wildlife. Perhaps the greatest danger smelter smoke posed was that it warned the Apaches of the presence of the intruders.

Muleteers hauled the refined copper to Mexico City via the Janos Trail. Guarded by soldiers from the presidios of Buenaventura, Carrizal, and Janos throughout the Spanish and much of the Mexican periods, teamsters led their laden mule trains with as few as a dozen and as many as 200 animals through treacherous desert terrain and under constant threat of Apache attack. To facilitate their journey, they beat the trail to precious water sites. They first stopped at Pachitejú (later called Apache Tejo Springs) along today's Whitewater Creek about fifteen miles south of Santa Rita. Then they trekked on to Las Bacas or Ojo de Baca (Cow Springs), another fifteen miles. After that they headed to Carizalillo ("little reeds," which grew in the flowing springwaters) forty miles farther south. Journeying on to additional lesser-known springs, they eventually arrived in Janos for a distance of about 150 miles from Santa Rita (see map 1.2). At Janos they loaded the precious copper cargo on fresh mules for the trip to Chihuahua and Mexico City.[31]

Production figures reflect the increase in copper output with implementation of smelter technology and larger mule trains. In 1805, Santa Rita miners sent 356,000 pounds of refined copper; in 1806, 301,000 pounds; in 1807, just over 100,000 pounds[32] (Apache attacks reduced production for this and the next two years); in 1808, 120,000 pounds of raw metal with 24,000 pounds of ore shipped to Mexico City; in 1809, 110,000 pounds with 43,000 additional pounds of ore. By 1813, Santa Rita miners had produced more than 1.7 million pounds of copper.[33]

This new source of copper helped to feed Spain's insatiable demand for the red metal. Since 1780, King Carlos III's government had established a monopoly on copper production and trade in New Spain to increase access to the red metal to produce armaments. Embroiled in numerous wars in the late eighteenth and early nineteenth centuries, Spain needed copper to manufacture cannons, muskets, pistols, armor, ship fittings, and other military accoutrement. Likewise, Spanish authorities demanded that all Mexican copper be sent to the royal treasury in Mexico City for either transport to Spanish armaments factories or coinage in New Spain. This monopoly put constraints on domestic uses of copper for things as diverse as artillery pieces for the colonial armies, gears and castings for sugarcane-crushing mills in the West Indies, and vats and cauldrons produced by coppersmiths for food storage and preparation. No doubt an illicit trade emerged to provide Mexican craftsmen with copper to make a variety of household items, such as kettles, cups, skillets, candleholders, spits, and other utensils. In addition, goldsmiths, engravers, and silversmiths used copper as an alloy to fabricate jewelry, eating utensils, and other luxury items for the elites. As a result, Santa Rita's raw native copper, which was of a very high grade and extremely easy to smelt, was marketed from the northern frontier of New Spain to Mexico City to Seville and Madrid.[34] Located at the farthest reaches of the Spanish Empire, Santa Rita del Cobre was linked to international wars, households in Europe, and borderland conflicts of the Spaniards and their colonists against Native America.

From 1809 to 1814, a series of factors worked against Santa Rita. In the former year, the viceroy ended the government monopoly on copper because of an abundance of refined metal being received at the mint. The immediate impact was a steep decline in the metal's price and the privatization of the copper industry. Elguea's heirs (Francisco Manuel died in 1806) immediately started a coppersmith business to craft copper products for sale, bells, one of which made its way to St. Louis, probably being the most commonly sought item. Despite their efforts to diversify the demand, however, the price remained low.

Financially strapped and anticipating the impending Mexican Revolution, Commandant General Salcedo decided in 1809 to withdraw troop support from the remote mining camp. He also temporarily terminated the ration tribute paid to the 300 or so Apaches living in the vicinity of Santa Rita. These acts forced Francisco Jáurequi, executor of the Elguea estate and manager of the family businesses, to seek immediate military assistance at the expense of the heirs. The Apaches, to whom Salcedo allocated seed and tools to farm, rebelled by plundering the copper camp herds. Salcedo realized his error and reinstated the ration system a year later, but the damage was done. The Apaches distrusted the Spanish, arguing for additional rations, an expense that fell to the Elguea heirs and others mining there. Despite these difficulties, Apache chief Juan Diego Compá, who grew up at the mission in Janos and spoke both his native tongue and Castilian, began negotiations for peace in 1812 with Santa Rita's new superintendent, José Baca. The results would be a period of general peace from 1814 to 1819.[35]

Spanish authorities began purchasing copper again in 1814, but now with plans to mint new copper coins. The *tlaco* was worth about two cents and the *pilon* one.[36] This new demand, combined with numerous mine closings in southern Mexico as a result of the independence movement, spelled a productive period for the Santa Rita copper mines.

In 1813, Doña María Antonia de Elguea married Pablo Guerra. Guerra assumed control of the deceased Elguea's business interests. Working with Baca, he negotiated new contracts to sell copper all over Mexico. Within four years, muleteers were delivering Santa Rita copper to Durango, San Luis Potosí, Zacatecas, Queretaro, Irapuato, and Arispe.[37] Baca, who served as superintendent from 1810 to 1823, also implemented strategies to monopolize the copper mines, to stabilize the community, and to minimize conflicts with the Apaches, until he left two years after Mexico won its independence from Spain.

Baca's strategy for maintaining control of the Santa Rita mines involved a twofold approach. First, he leased what claims he could not afford to mine principally because of labor costs, so that Guerra and the Elguea family could meet the legal requirement of continual assessment and development spelled out by Spanish mining law. Second, he purchased the rights to as many claims and mines as he could, the acquisition of Carrasco's properties being the most significant. When Carrasco died in 1819, in fact, his son, Carlos, sold his rights to the claims to Baca and Guerra or he simply abandoned them, making them available to the Elguea heirs. Regardless, they had a monopoly on the copper properties at Santa Rita by this time.[38]

The stability of the mining operations in this period allowed for the development of a substantial community at Santa Rita. As early as 1807, Superintendent José Romero had set the pattern for a self-sustaining camp. He hired shepherds to oversee valuable horse and mule herds in the Mimbres and Whitewater Valleys. He worked with Spanish authorities to ensure that a small contingent of soldiers secured the mining site where they had built some breastworks and a stockade for protection. But not until the late 1810s did Santa Rita show signs of permanence, although the Apaches alone would determine that status, as will be seen later in this story. From 1817 to 1819, about 400 people inhabited the mining camp. Many labored in the mines. Others farmed irrigated plots and orchards in the fields near Pachitejú and in the floodplain of the Mimbres River. Still others ranched on the grasslands of the Whitewater and San Vicente Valleys south and west of the mines, respectively. Soldiers, of course, spent time in the camp, now called a presidio, or reconnoitered the Gila, Mimbres, and San Francisco river valleys. Spiritual and communal needs also required Guerra to hire priests to visit the site from time to time to perform marriages, baptisms, and masses.[39]

Three principal factors defined Santa Rita as a controlled or closed society. Each of the superintendents since Obeso, beginning in 1804, had legal authority to enforce Spanish laws concerning mundane affairs, such as squabbles over possessions, and serious matters, such as murder. They also held the authority to jail accused criminals and punish deserters, who, because they were often already serving sentences for debt or more serious crimes, took their chances in Apachería. Just before his death in 1819, Carrasco himself had requested the establishment of a *partido* (district or precinct) with a sub-delegate to the Crown, a request that was denied probably because of the third factor in creating a controlled community, the reality of settling in the heart of Chihenne country. The principal chief, Mangas Coloradas, now led half of the Chiricahua tribe. He had married into the Bedonkohes band, had earned their respect, and had become their chief as well on the death of Mahko. Furthermore, he was bent on removing the aliens from Apache territory. As long as the Spanish (and later the Mexicans) provided military support, the community remained relatively safe, at least at Santa Rita itself. But when fiscal constraints precluded assistance by the military and a continuation of the ration system, the

Apaches ruled, a fact that would lead to the abandonment of the mining camp from 1838 to 1857.[40]

José Baca probably left the superintendency at Santa Rita in 1823 because of the lack of governmental support for the presidio. Independent from Spain only two years, Mexico was in an administrative and fiscal conundrum. As a result, even though Guerra could afford to pay and feed his own workers, the Mexican soldiers lived in impoverished conditions. They lacked the equipment, tools, gunpowder, weapons, and even food to carry out their security duties at the vulnerable copper camp. Apaches, and on occasion Navajos and Comanches, regularly raided the military's horse, cattle, mule, and sheep herds. As a result, Baca departed, probably sensing an impending escalation of Apache intimidations. Although Juan Luis de Hernández replaced Baca that year, the new superintendent left within two years with no record of production for 1824 and 1825.[41]

In 1825, Guerra was on the verge of abandoning the mines altogether. The appearance of a party of American beaver trappers, however, delayed that fatal measure. Sylvester Pattie and especially his son, James Ohio, have lived in southwestern lore for nearly two centuries now. James's account of his tramping and trapping travails is the stuff of legend. For our story, the Patties serve one main purpose: they leased the Santa Rita mines for about two years and paved the way for two future investors, Robert McKnight and Stephen Curcier. They also represent the emerging influx of fur trappers and "explorers" (like Zebulon Pike) as the first wave in the American "invasion" into the frontier of northern Mexico. Although the Mexicans had gained their independence from Spain, the new nation would soon find it nearly impossible to keep out the Americans. The Apaches, however, still remained the biggest deterrent to the Mexican development of the copper mines. Guerra, in fact, could not afford (or was unwilling) to pay for Mexican militiamen to protect Santa Rita. An increase in Apache raids had halted smelting because the carboneros would no longer go into the forests to harvest wood in the Santa Rita Mountains and the Black Range for fear of ambush. Without presidio status, the copper camp was at risk. Regardless, Sylvester Pattie took up Guerra's offer to lease the mines after a miserably poor harvest of beaver furs in late 1825 and early 1826 in the Gila country.[42]

Under Sylvester's management, the leasers produced in 1826 a substantial 184,000 pounds of copper. Guerra was so pleased with the output that he offered the elder Pattie a five-year lease on the mines at an astonishingly generous fee of only $1,000 per year. According to James, his father had solicited the support of other Americans, most notably James Kirker, Nathaniel Pryor, and Richard Laughlin. They were willing to stay to guard the charcoal makers, the farmers and ranchers, and the camp itself. James Pattie also alleged that his father met with the two powerful Apache chiefs, El Fuerte and Mano Mocha, to secure peace while the Americans mined the red metal.[43] Despite these successes, however, the Patties decided to leave Santa Rita in the fall of 1827, probably because they had lost $30,000 in profits from their dual enterprises of mining and trapping when one of Sylvester's partners disappeared with the proceeds. When a personal appeal for a loan by James Pattie to Guerra, who was soon to move María Antonia and the rest of the family to Spain because of the changes brought about by Mexican independence, failed, the Patties decided to seek their fortunes elsewhere.[44]

About the time the Patties left Santa Rita, two investors found out that Guerra was leaving for Spain, so they attempted a takeover of the mines. Robert McKnight, an American, and Stephen Curcier, a Franco-American, soon learned, however, that Guerra did not have to forfeit his claim to the mines even though he was departing Mexico. When the Spaniard's agent, Luis Durasta, threatened a lawsuit if the intruders pursued their plan, they gave up their plot. Taking a more tactful approach, the two men decided in late 1827 to take up Durasta's offer to lease the mines for five years for a nominal fee of one *quintal*, or 100 pounds, of copper a year.[45]

For the next decade, McKnight and Curcier partnered to produce copper at Santa Rita. Little is known of Curcier's previous life, but McKnight had spent the previous thirteen years in Mexico. He served four years in prison as an alleged spy and then six more years as a bond servant. He was freed in 1820 when King Ferdinand VII pardoned all foreign prisoners in Mexico. The seven years before he started mining copper he spent as a family man after he had married a Mexican woman.[46] McKnight served as the man on the ground, managing the mining operations and working with Mexican officials to ensure military support for the soon-to-be fort and penal colony. Curcier remained in Chihuahua City, where he was an established businessman.

With connections in St. Louis, Curcier ordered seventeen freight wagons to supplement the mule trains and to acquire other supplies to restart the mining, smelting, and charcoaling operations. McKnight, in the meantime, put together a workforce, many of whom Mexican authorities had sent from various prisons in Chihuahua and Sonora, and then began the process of reopening the mines, repairing old habitations, and building new housing. Taking advantage of the presence of seasonal residents, mainly American men who trapped and hunted for furs and pelts in winter, McKnight hired the newcomers to protect farmers, charcoal makers, herders, and the community itself. Others worked as miners, smeltermen, and even teamsters, like Christopher "Kit" Carson, who only stayed a few months and later earned fame for other exploits in the Southwest. Santa Rita was emerging as a small way station on the St. Louis–Santa Fe–Mexico trade network materializing in the mid-nineteenth-century southwestern borderlands.[47]

Though no production records have been discovered for the McKnight-Curcier period, the partnership must have been lucrative.[48] Evidence reveals that they sent freight trains of smelted copper and raw ore every three months, suggesting that over the ten-year stretch they must have produced millions of pounds of the red metal. Mexican officials clearly valued their operations as well. In 1832, when the partners agreed to another five-year lease on the property, officials at the Chihuahua treasury decided to mint a new copper coin, the *cuartilla* (about 3 cents).[49] Governor José Calvo refused a little later to reduce the value of this coin when other states had and even issued a law against the circulation in Chihuahua of those states' copper currency. With government support, job opportunities, and growing interest in the region by Americans, Santa Rita grew to 600 or so inhabitants by 1835.

This substantial population posed a new threat to the Apaches. To show their dissatisfaction, the Chiricahuas banded together in the early 1830s to step up their raids, hoping to force the removal of the Mexicans from Santa Rita and the Americans from the Gila and Mimbres watersheds. While the finances were available, Calvo countered these hostile actions with the support of the regular army and regional militias. At one point there were nearly 300 on-duty soldiers at the camp. Yet, Curcier had to assist the government as early as 1831 with food, supplies, and funds to maintain the force, which was necessary because Calvo declared war on the native peoples later that year. Because Calvo was unable to finance the war, he sent Captain José Ignacio Ronquillo to negotiate peace with the Apaches, who would determine whether a permanent community would be established. To thwart the abandonment of Santa Rita, Mexican military officials and the American miners agreed to meet in August 1832 at Santa Rita with Mangas Coloradas, Pisago Cabezón, Juan José Compá, and twenty-six other chiefs. The antagonistic parties agreed to an armistice despite several earlier skirmishes, including a brutal hand-to-hand encounter a few weeks earlier near Santa Lucia Springs. Interestingly, Mangas Coloradas naively thought that the Americans would not treat his people as poorly as had the Mexicans, who frustratingly had ended the Spanish ration system and encouraged the murdering of Apaches, offering bounties for their scalps. By the mid-1830s, Mangas Coloradas's hatred of the Mexicans clouded his view of the Americans, who he believed were more honorable and less likely to indiscriminately kill his people and take over their homeland. His patience and preference for the *norteamericanos*, ironically, allowed for the newcomers to stay at Santa Rita at least until the Johnson massacre in 1837. After the Mexican-American War, the chief initially welcomed the conquerors, who in the end proved to be even more menacing than their predecessors.[50]

The 1832 treaty failed almost immediately when the Mexican government refused to reinstate the ration system. Santa Rita lost a third of its population and the conflict showed no signs of ending, with attacks and counterattacks occurring on a monthly basis. When some Americans met with a group of Apaches at war in 1834 to discuss peace near what is today Clifton, Arizona, and 300 or so of the total group of about 800 Indians soon thereafter died from disease, open hostilities ensued in Apachería.[51]

The response at Santa Rita was to build a fort. Under the direction of Captain Cayetano Justiniani, commander of the Second Division, and with the support of McKnight, the Mexican army built a triangular-shaped fort with two *torreones*, or towers (see figures 1.5 and 1.6). Built of adobe, brick, and stone, the towers, one of which lasted until the early 1950s, were 12 feet high and 18 feet in diameter. Connecting these imposing structures were three walls 200 feet long, 10 feet tall, and 3 feet thick. The workers also built a barracks, cells

FIGURE 1.5. *Remnants of the old Mexican fort at Santa Rita, 1906. For generations local residents believed the Spanish constructed the fort. Records reveal, however, that Mexican authorities cooperated with American operators to build the military outpost in hopes of protecting the copper community from the Apaches.*
COURTESY OF SILVER CITY MUSEUM.

for prisoners, storage rooms, a commissary, and shops of various kinds. All of these structures were supported by *vigas* (ceiling supports) of stripped pine logs from 16 to 20 feet long. Most likely the turrets had platforms inside to house cannons and soldiers with muskets. Behind parapets on top of the thick walls the soldiers could position or mount the three cannons put into commission at Santa Rita. Additional weapons and munitions included two stone mortars, eight cannonballs, seven sacks of grapeshot, fifteen cartridges, one box of gunpowder, a ramrod, and other accoutrement.[52]

This tiny arsenal, however, was not going to deter the Apaches. Convinced that the Mexicans and Americans would be coming in greater numbers, the indigenous peoples geared up for resistance. Aggravating the situation were the trading practices of McKnight and some of the other Americans at Santa Rita and in Apachería generally. Throughout the 1830s they had engaged in illicit trade with the Apaches. Although the documents only record accusations, especially by Sonoran officials, it seems likely that the foreigners were exchanging gunpowder, muskets, and whiskey for horses, mules, and captives that Chihenne and Bedonkohe warriors had kidnapped from the Mexicans. This illegal trade exacerbated an already tenuous relationship between the Chiricahuas and the Mexicans. A peace treaty, signed in 1835 by sixteen Apache chiefs and Justiniani, included provisions for rations, gunpowder (ostensibly for hunting), and other

FIGURE 1.6. *A 1914 photograph of the two remaining towers of the old Mexican fort (the right tower no longer has a roof and is located at the end of the original adobe wall behind the horses). In the twentieth century the successive mining companies used the left tower as a jail first, then as a saloon, and finally as a powder magazine.*
COURTESY OF SILVER CITY MUSEUM.

favors but deteriorated into open hostilities within six months. Though Mangas Coloradas was the prominent chief in the region, Mexican officials instead lavished gifts on Chief Juan José Compá, who they insisted had been elected "general of the chiefs." Given a dress coat, wool military pants with stripes down the legs, epaulettes, a boxed sombrero, and a walking stick with a silver handle, Compá lost standing among some of his tribal members, many of whom afterward looked on him suspiciously.[53]

Despite this internal factionalism, the Apaches still reigned supreme in their homeland. Without question, they wanted to expel the Mexicans and their American friends from Apachería. Soon many of the soldiers and workers (especially those serving prison sentences) began to desert Santa Rita in large numbers. Both the threat of attack and the impoverished circumstances spelled doom for the mine operations and the community. By 1836, McKnight would no longer subsidize the Mexican army, resulting in near-starvation conditions for the troops and their families. The Santa Ritans were desperate. The low point prior to desertion was in the late spring and summer of 1836, when a large group of

Chihennes, Bedonkohes, and Nednhis surrounded Santa Rita. For weeks they threatened daily to attack, sending up smoke signals that terrified the miners. Though the Apaches did not advance, it would only be a matter of time before they forced an abandonment of the copper camp. Authorities in Mexico City had also diverted their military resources to Texas after the Battles of the Alamo and San Jacinto, leaving little doubt in the late 1830s about the fate of the Santa Ritans.[54]

The final blow to renewed peace surrounds a legendary event known as the Johnson massacre. On April 22, 1837, a party of American traders led by John Johnson, a trapper, trader, and miner from Kentucky living in Sonora, happened upon a large encampment of Apaches, about ninety miles south of Santa Rita near a spring in the southern Animas Mountains in what is today the boot heel of New Mexico. Not expecting to meet up with a group led by Chief Juan José Compá, the Americans panicked when they realized they were at a large numerical disadvantage. Deceiving the chief and his companions, Johnson gave them the impression that he had tribute gifts in his wagons. But instead, as they approached, Johnson turned a swivel gun mounted on a mule on the unsuspecting victims. In a few tragic moments, Chief Compá, his brother Juan Diego, Chief Marcello, and perhaps twenty or more other warriors as well as two of Mangas Coloradas's wives were killed. Many others were wounded. The rest scattered. The duplicitous Johnson then took the scalps of the dead to Sonoran officials for a reward. Any chance for a lasting peace in Apachería was lost.[55]

A myth emerged, persisting among some locals today, that the killings took place at Santa Rita. They did not.[56] More important, the bloody treachery ended the first period of Santa Rita del Cobre's history. The massacre led to a total Apache siege of the Santa Rita settlement. As a result, McKnight, the Mexican detachment, the settlers, and even the itinerant trappers and traders abandoned the recently fortified camp in 1838, after months of uncertainty. The Apaches had destroyed the Santa Ritans' gravy train, having whisked away nearly every horse, mule, and sheep in the Sierra del Cobre. Mangas Coloradas and Pisago Cabezón themselves supervised the blockade of the Janos Trail. The coalition succeeded. For the next decade, Mangas Coloradas made it his mission to exact violent revenge against the Mexicans as well as keep them from the coveted copper. His homeland, riddled with that precious red metal that had attracted Spanish and Mexican glory hunters and then the avaricious Americans, would remain under Chiricahua hegemony for another two decades. Despite the twenty-year Apache dominance, however, the Euro-Americans would not forget about El Cobre and the bounty that lay beneath its surface. Even though millions of pounds of copper had already been extracted at Santa Rita, there would be hundreds of millions left for the Americans to mine in the future. Chapter 2 tells the story of this transition.

NOTES

1. Carrasco to Deputación de Minería, Cobre, January 3, 1805, Archivos del Ayuntamiento de Chihuahua 491, reel 154, microfilm, University of Texas, El Paso. The authors thank Helen Lundwall and Terrence M. Humble for allowing us to use their unpublished manuscript, "Santa Rita del Cobre: Copper Mining in Apachería, 1801–1838" (2005), for this chapter.

2. Jerry L. Williams, ed., *New Mexico in Maps* (Albuquerque: University of New Mexico Press, 1986), 23–31; Eugene Carter Anderson, *The Metal Resources of New Mexico and Their Economic Features through 1954* (Socorro: New Mexico Bureau of Mines & Mineral Resources, 1957), 44–45.

3. This mountain range was later named the Black Range.

4. Lundwall and Humble, "Santa Rita," 2–3.

5. David J. Weber, *The Spanish Frontier in North America* (New Haven, CT: Yale University Press, 1992), 211–213.

6. Herbert Eugene Bolton, *Pageant in the Wilderness: The Story of the Escalante Expedition to the Interior Basin, 1776* (Salt Lake City: Utah State Historical Society, 1950), 246–248; Luis Navarro García, *Las Provincias Internas en el Siglo XIX* (Sevilla: Escuela de Estudios Hispano-Americanos de Sevilla, 1965), 88–91.

7. Edwin R. Sweeney, "Mangas Coloradas and Mid-Nineteenth-Century Conflicts," in Richard W. Etulain, ed., *New Mexican Lives: Profiles and Historical Stories* (Albuquerque: University of New Mexico Press, 2002), 139–143.

8. Sweeney, "Mangas Coloradas," 148–152; also see Eve Ball, *In the Days of Victorio: Recollections of a Warm Springs Apache* (Tucson: University of Arizona Press, 1970) and *Indeh: An Apache Odyssey* (Provo, UT: Brigham Young University Press, 1980), for more on Apache lifeways.

9. See Edwin R. Sweeney, *Mangas Coloradas: Chief of the Chiricahua Apaches* (Norman: University of Oklahoma Press, 1998).

10. There is some debate on what to call this area, a frontier of northern New Spain or the Spanish borderlands. They are used here interchangeably with specific reference to New Mexico in the Spanish colonial period. Later in this story, the term *borderlands* is used in reference to the border areas of Mexico and the United States. See Weber, *The Spanish*, 6–10.

11. *Criollos*, or creoles, were Spaniards born in New Spain.

12. Lundwall and Humble, "Santa Rita," 10, 13–14.

13. Carrasco to Tovar, Cobre, November 30, 1804, Archivos del Ayuntamiento de Chihuahua 491, reel 154, microfilm, University of Texas, El Paso.

14. Henry W. Halleck, ed., *A Collection of Mining Laws of Spain and Mexico* (San Francisco: University of California, 1859), 1–8, 220–228.

15. Approximately 25 lbs.

16. Carrasco to Deputación de Minería, Chihuahua, January 3, 1805, Archivos del Ayuntamiento de Chihuahua 491, reel 154, microfilm, University of Texas, El Paso.

17. Francisco R. Almada, *Gobernadores del Estado de Chihuahua* (Mexico City, DF: Imprinta de la H. Camara de Deputados, 1950), 184; Pablo Guerra to Ayuntamiento de Chihuahua, May 17, 1826, Archivos del Ayuntamiento de Chihuahua 491, reel 176, microfilm, University of Texas, El Paso.

18. Salcedo to Tovar, August 10, 1804, Archivos Historicos de Janos, 498, reel 15, microfilm, University of Texas, El Paso.

19. See Otis Young, *Western Mining* (Norman: University of Oklahoma Press, 1970), 62–64; Christopher J. Huggard, "Environmental and Economic Change in the Twentieth-Century American West: The History of the Copper Industry in New Mexico" (PhD dissertation, University of New Mexico, 1994), 21–24.

20. T. A. Rickard, "The Chino Enterprise—I, History of the Region and the Beginning of Mining at Santa Rita," *Engineering and Mining Journal-Press* (November 3, 1923), 758; Young, *Western Mining*, 62–64; Robert L. Spude, "The Santa Rita del Cobre, New Mexico, The Early American Period, 1846–1886," *Mining History Journal* (1996), 8–38; Robert C. West, *The Mining Community in Northern New Spain: The Parral Mining District* (Berkeley: University of California Press, 1949), 18–27; Peter J. Bakewell, *Silver Mining and Society in Colonial Mexico: Zacatecas, 1546–1700* (Cambridge: Cambridge University Press, 1971), 130–134; also see Elinore Barrett, *The Mexican Colonial Copper Industry* (Albuquerque: University of New Mexico Press, 1987), and David J. Weber, *The Mexican Frontier, 1821–1846: The American Southwest under Mexico* (Albuquerque: University of New Mexico Press, 1982).

21. Quoted from Obeso, Cobre, to Salcedo, September 30, 1804, Archivos Historicos de Janos, 498, reel 15, microfilm, University of Texas, El Paso; *The Borderer*, September 21, 1872; *Mining Life*, March 28, 1874; *Grant County Herald*, January 12, 1881; *Silver City Enterprise*, March 23, 1928.

22. Lundwall and Humble, "Santa Rita," 28.

23. For examples of miners reprocessing the Spanish dumps, see Santa Rita Mining Company, Weekly Report No. 155, May 23, 1903; *Silver City Enterprise*, January 29, April 8, 22, 1904; *Silver City Independent*, December 13, 1904.

24. See Rick Hendricks, "Spanish Colonial Mining in Southern New Mexico: A Spanish to English Translation of Documents Relating to El Paso, the Organ Mountains, and Santa Rita del Cobre," *Mining History Journal* 6 (1999): 156–159. Hendricks offers several examples of letters exchanged between Santa Rita's superintendents and officials in Chihuahua from 1804 and 1810.

25. Salcedo, Chihuahua, to Soler, May 9, 1804, Audencia de Guadalajara, 296, reel 59, Guadalajara, Mexico.

26. Herbert Thurston and Donald Attwater, eds., *Butler's Lives of the Saints* (New York: P. J. Kennedy & Sons, 1962), 369–370.

27. Salcedo, Chihuahua, to Soler, December 4, 1804, Audiencia de Guadalajara, 296, reel 72, Guadalajara, Mexico; Tovar, Janos, to Salcedo, December 10, 1804, Archivos Historicos de Janos, 498, reel 15, microfilm, University of Texas, El Paso. Real del Cobre was the first mining district in New Mexico history.

28. Elinore Barrett, *The Mexican*, 63–65, 77–82.

29. Ibid., 89; Spude, "The Santa Rita," 15.

30. See Christopher J. Huggard, "The Impact of Mining on the Environment of Grant County, New Mexico, to 1910," *Annual of the Mining History Association* 1 (1994): 2–8.

31. Max L. Moorhead, "Spanish Transportation in the Southwest, 1540–1846," *New Mexico Historical Review* 32:2 (1957), 108–111.

32. Salcedo, Chihuahua, to Janos, July 1, 1807, Archivos Historicos de Janos, File 18, section 3, 133, Nettie Lee Benson Latin American Collection, University of Texas, Austin. These figures have been rounded.

33. El Cobre to Commander of San Buenaventura, October 25, 1825, Archivos Historicos de Janos, 498, reel 17, microfilm, University of Texas, El Paso. These production figures are probably low because no reports of production have been found for the 1809–1811 period and some shipments may have gone unreported from 1801 to 1813.

34. Barrett, *The Mexican*, 3–4, 42–47.

35. Salcedo, Chihuahua, to Janos, October 18, 1810, Archivos Historicos de Janos, 498, reel 11, microfilm, University of Texas, El Paso; José Baca, Real de Cobre, to Captain José Ronquillo, April 25, 1812, Archivos Historicos de Janos, 498, reel 2, microfilm, University of Texas, El Paso.

36. See David C. Harper, *Coins and Prices: A Guide to U.S., Canadian, and Mexican Coins* (Iola, WI: Krause Publications, 1993), 387, 407.

37. Lundwall and Humble, "Santa Rita," 90.

38. Libro de Registro de Denuncios Mineros, September 15, 1814, Archivos del Ayuntamiento de Chihuahua 491, reel 151, microfilm, University of Texas, El Paso.

39. John M. Sully, "The Santa Rita del Cobre Grant," *Silver City Enterprise*, December 1, 1933; Mateo Sánchez, Chihuahua, to José Miguel Yrigoyen, September 1, 1812, Archivos Historicos de Janos, 498, reel 2, microfilm, University of Texas, El Paso.

40. Spude, "Santa Rita," 11–12.

41. Juan Luis Hernández, Real del Cobre, to Janos Commander, October 24, 1823, Archivos Historicos de Janos, 498, reel 16, microfilm, University of Texas, El Paso; Commander of First Division, El Cobre, to Commandant of Arms, April 30, 1824, Archivos Historicos de Janos, 498, reel 2, microfilm, University of Texas, El Paso; Sweeney, "Mangas Coloradas," 148–149.

42. James O. Pattie, *The Personal Narrative of James O. Pattie* (New York: J. B. Lippincott Co., 1962; reprint of 1831 ed.), 46–49.

43. Ibid., 67–74; Sweeney, *Mangas Coloradas*, 68–87.

44. Pattie, *Personal Narrative*, 121, 249.

45. Letter from William M. Pierson, US Vice-Consul, to Hon. Wm. Hunter, 2d Asst. Sec'y of State, Washington, DC, December 6, 1873, in J. P. Whitney, *The Santa Rita Native Copper Mines in Grant County, New Mexico* (Boston, MA: Alfred Mudge & Son, 1884), 31–32.

46. Frank B. Golley, "James Baird, Early Santa Fe Trader," *Missouri Historical Society Bulletin* 5:3 (1959): 188–190; William Cochran McGraw, *Savage Scene: The Life and Times of James Kirker, Frontier King* (New York: Hastings House, 1972), 43.

47. Lundwall and Humble, "Santa Rita," 124.

48. See Spude, "The Santa Rita," 9; Frederick A. Wislezenus claimed in 1847 that McKnight and Curcier made "a half million dollars" from their copper-mining venture at Santa Rita and they had a "monopoly" on the copper market of Mexico.

49. Notice issued July 21, 1831, Carrizal Collection, 505, reel 3, University of Texas, El Paso; Wilbur T. Meeks, *The Exchange of Media of Colonial Mexico* (New York: Columbia University Press, 1948), 71–73.

50. Sweeney, "Mangas Coloradas," 150–154, contends that even though he was an American, James Kirker probably perpetrated as many bounty-hunter atrocities for profit against the Chiricahuas as anyone—the 1846 massacre of nearly 150 Nednhis and Chokonens in northern Sonora being the most notorious act of brutality. Mangas Coloradas continued to blame the Mexicans because of their scalp-paying policy rather than come to terms with the inevitable aggression of the Americans that would eventually result in the complete takeover of his homeland .

51. Resolution of State of Chihuahua, August 30, 1832, Archivos Historicos de Janos, 498, reel 5; William B. Griffin, "The Compás: A Chiricahua Apache Family of the Late 18th and Early 19th Centuries," *The American Indian Quarterly* 7:2 (1983): 22–45.

52. John Russell Bartlett, *Personal Narrative of Explorations and Incidents in Texas, New Mexico, California, Sonora and Chihuahua* (New York: D. Appleton, 1854), 234–236; Odie B. Faulk, "The Presidio: Fortress or Farce," *Journal of the West* 8:1 (1969): 23–25; Calvo to Justiniani, Chihuahua, May 15, 1835, Archivos Historicos de Janos, 498, reel 26, microfilm, University of Texas, El Paso; Fayette Jones, *Old Mines and Ghost Camps of New Mexico* (Socorro: New Mexico Mines and Minerals, 1904), 35–39; José Hernández, Janos, to Justiniani, June 18, 1835; Calvo, Chihuahua, to Justiniani, May 15, 1835, Archivos Historicos de Janos, 498, reel 28, microfilm, University of Texas, El Paso.

53. "A List of Items" sent from Chihuahua to General Juan José Compá, February 25, 1835, Archivos Historicos de Janos, 498, reel 28; see Weber, *The Spanish*, 227, 229, 231, 233, 253.

54. Rey, Cobre, to Calvo, June 20, 1836, Archivos Historicos de Janos, File 38, section 2, 21, Nettie Lee Benson Latin American Collection, University of Texas, Austin; Sweeney, "Mangas Coloradas," 152–153.

55. See Rex Strickland, "The Birth and Death of a Legend: The Johnson 'Massacre' of 1837," *Arizona and the West* 18:3 (Summer 1976): 257–286; Sweeney, "Mangas Coloradas," 149–154.

56. John C. Cremony, *Life among the Apaches* (Glorieta, NM: Rio Grande Press, 1969), 28–35.

II

Frontier Mining

THE UNDERGROUND YEARS

Santa Rita del Cobre was in the heart of contested territory. The Apaches had the upper hand. The Spanish could not control this borderland region prior to 1821, and the Mexicans found it even more difficult afterward because of the complications of building a new nation-state. The presence of the Americans did mark an interesting turning point. Yet, not until 1848 would the emergent US Empire take possession of the borderlands into which its entrepreneurs,

miners, and trappers had made inroads. Even then the Apaches would still reign supreme in the region for at least another ten years. Indicative of the importance of the precious red metal, the Chihenne and Bedonkohe bands would soon be known as the "Coppermine" Apaches to US military personnel, gold rushers, and boundary commissioners who briefly visited the coveted mining site in the interim period 1838 to 1857.[1] This metaphor for the Indians also reveals the American belief that Santa Rita was part of their "manifest destiny" to open a new mining frontier. This imperialist ideology prompted the takeover of Mexican territory and the pushing aside of the indigenous peoples. With their aggressive actions, the norteamericanos opened the "new" Southwest to modern mining and a taste of

gilded age industrialism during the second half of the nineteenth century.

The first reappearance of Americans took place in 1846 in anticipation of the Mexican-American War. The Army of the West, under General Stephen Watts Kearney, visited the former copper camp.[2] Kearney's column was on the march to attack Mexican forces in California and to occupy major settlements between the Rio Grande and the Pacific Ocean as part of the military strategy to take the Southwest. On their arrival to the abandoned presidio, Kit Carson led a group of officers on a tour of the ghost town, blaming the Apaches for its demise and lamenting that the men who made "fortunes" were forced to leave. Among the sojourners was Lieutenant William H. Emory, who later recorded his impressions of Carson's tour. "There are the remains of some twenty or thirty adobe houses, and ten or fifteen shafts sinking into the earth. The entire surface of the hill into which they are sunk is covered with iron pyrites [fool's gold] and the red oxide of copper."[3] Captain Henry Smith Turner, also on the guided excursion, noted in his diary the ruins of previous mining efforts and the rich ore piles that still remained. He concluded that the camp had been hastily abandoned. "The rock breaks easily and the pick appears to be the only tool used formerly. Occasional veins of pure copper, very yellow from the quantity of gold it contains, traverse the whole mass . . . water had filled many of the abandoned chambers . . . The fort which was built to defend the mines was built in [the] shape of an equilateral triangle with round towers at the corners . . . [It] was built of adobe [and] was still in tolerable preservation; some remains of the furnaces were left, and piles of cinders . . . Charcoal in quantities was used."[4] The copper mines clearly had potential for greater development because both observers made a point to record the abundance of pastureland, timber, and water that could support a renewed effort at mining and community building. But they had far greater concerns during the Mexican-American War, so they did nothing to realize this potential during their short visit that year.

The Americans succeeded in winning the war. The newly won lands, the so-called Mexican Cession, included what is the heart of the Southwest today with Santa Rita as part of that conquered territory. About the time of the 1848 signing of the Treaty of Guadalupe Hidalgo, which ended the war, the Americans discovered gold in California. As part of that frenzied rush, several hundred Forty-niners stopped at the old mining camp, which was located just north of the principal southern route to the Pacific coast. Even the pathfinder himself, John C. Frémont, passed through the camp. Their presence mainly alienated the Apaches, who under Mangas Coloradas's leadership, exacted supplies as tribute from some of the sojourners and ambushed others.[5]

Under these conditions, Francisco Elguea (son of Francisco Manuel) decided in 1849 to make an appeal to the acting governor of New Mexico, Donaciano Vigil, to set up military defenses for his family's mine claims. He correctly ascertained that he retained rights to the Santa Rita deposits under the provisions of the Treaty of Guadalupe Hidalgo. Unknown to Elguea, the Americans

had already sent Brevet Major Enoch Steen and a detachment of fifty soldiers to seek a peace agreement with the Chiricahuas. The encounter had turned out badly, though, with Steen later claiming to have defeated a band of warriors at Santa Rita. Peace seemed increasingly improbable. Steen's suggestion that the Americans establish a fort at the site fell on deaf ears. Regardless, Elguea seems to have succeeded that year in leasing the property to James Magoffin, an American trader who lived in El Paso on the Mexican side of the Rio Grande. No record of production under Magoffin, however, has been uncovered.[6]

The following year, New Mexico became a territory and the United States began negotiations with Mexico to survey the boundary between the two young republics. In part because Santa Rita had the largest number of inhabitable structures between the Rio Grande and Tucson and US boundary commissioner John Russell Bartlett was curious about the mines, the border negotiators made their headquarters at Santa Rita. Bartlett and Third Infantry commander Lieutenant Colonel Lewis S. Craig renamed the old presidio Cantonment Dawson sometime during their one-year stay. They did not renew mining activities, however. While there, Craig hoped to buy peace from Mangas Coloradas and other Apache inhabitants while finalizing the boundary between the United States and Mexico.[7]

The most salient contributions of the boundary commission for this story are two woodcut illustrations of Santa Rita (see figures 2.1 and 2.2). These illustrations, which appeared in Bartlett's *Personal Narrative of Explorations* (1854), are the earliest known images of the once-thriving copper camp. *Valley of the Copper Mines from the South*, a wide-angle vista of the ghost town, shows several features: namely, the Mexican fort, several of the adobe huts, the stockade and perhaps a livery stable, and oxen, wagons, and tents that they brought to the site. The two prominent creeks, Santa Rita (*center*) and Whitewater (*center right*) are also visible in the bucolic scene. The second drawing, *Presidio at the Copper Mines*, offers a closer look at the fort, the bluff known as La Bufa to the Spaniards and Mexicans, adobe buildings, tents, wagons, and livestock. These two depictions of Santa Rita help to substantiate the extent of the community during the Spanish and Mexican periods.[8]

Emory renamed the escarpment in the background of *Presidio at the Copper Mines* after his good friend

FIGURE 2.1. Valley of the Copper Mines from the South *from John Russell Barlett's* A Personal Narrative of Exploration and Incidents in Texas, New Mexico, California, and Chihuahua, *1854. This woodcut is the earliest illustration of Santa Rita and reveals the Mexican fort, a corral, habitations, and the landscape in the mid-nineteenth century.*
COURTESY OF NATIONAL ARCHIVES.

Captain Ben Moore of the First Dragoons. On the north edge of Ben Moore Mountain stands to this day an obelisk-like stone tower the first Spanish-speaking settlers called La Aguja (The Needle). Soon after the Bartlett visit, Hispanics of the area renamed it the Kneeling Jesus or the Kneeling Saint (Santa Teresa). By the early 1870s, it became known as the Kneeling Nun, perhaps in recognition of the copper camp's patron saint, Santa Rita (see figure 2.3). Countless myths were told about the devoted, unwavering nun's role in protecting the miners and their families. When violent earthquakes nearly toppled the spiritual monument in the 1880s, locals feared that their meditating supplicant would be destroyed and leave them vulnerable to the perils of mining. Yet it survived. Today modern technology threatens to bring her

down. Porphyry deposits lie directly beneath the pious stone. Many devotees of the Kneeling Nun, some of whom sought national monument status in the 1990s for the landmark, hold out hope that the current owners will not allow the growing open pit to swallow up this historic signpost.[9]

The Americans attempted in 1852 to establish a lasting garrison called Fort Webster at Santa Rita soon after the Bartlett crew left (see figure 2.4). But this endeavor failed because of Apache intimidation. Mangas Coloradas and Chiefs Delgadito, Coletto Amarillo, and Ponce negotiated in the 1840s and 1850s for a treaty securing ownership of their homeland with the copper mines at its heart.[10] Perhaps as many as 900 Chiricahuas lived in the vicinity of the old camp and in the Mimbres

Presidio at the Copper Mines.

FIGURE 2.2. Presidio at the Copper Mines *from Bartlett's* A
Personal Narrative, *1854. The United States–Mexican boundary
commission established its headquarters in 1851 at Santa Rita.*
COURTESY OF NATIONAL ARCHIVES.

Valley. Despite the soldiers' attempts to assuage the
Apaches with gifts, tensions remained palpable with
little chance for a resolution. Soon after Brevet Major
Israel Richardson of the Third United States Infantry established a new fort in January 1852, the Apaches mobi-
lized to force a withdrawal. After a short six weeks, the
soldiers began clamoring for a retreat. Corporal James
Bennett, who arrived with sixty additional soldiers on re-
prieve from their pursuit of Apaches elsewhere, recorded
in his diary for February 20 that "there are 50 men here,
all frightened out of their wits. They have old wagons,
logs, barrels, rocks, and other articles too numerous to
mention, piled around their fort, making it impossible

to get to it . . . They expect momentarily to be attacked
by the Indians." Within six months, they abandoned the
garrison and established Fort Webster II to the east on
the Mimbres River.[11]

The Apaches did not realize their dreams for a protected homeland in south-central New Mexico Territory
either. Although Indian Agent Michael Steck and Territorial Governor David Meriwether agreed to such a res-
ervation in a deal worked out with Mangas Coloradas
in 1855, US senator Thomas Rusk from Texas nixed it
in Congress to protect the South's interests in laying a
southern route for a transcontinental railway.[12] Against
the wishes of the indigenous people, mining would

FIGURE 2.3. *Photograph of the Kneeling Nun (unknown date). The most popular legend surrounding this monolithic rock formation tells of a Santa Rita nun falling in love with a Spanish soldier she had nursed back to health. Having compromised her vows of chastity, she climbed to the bluff to kneel in prayer for forgiveness only to turn to stone after being struck by lightning.*
COURTESY OF TERRY HUMBLE.

be renewed within two years. The Americans clearly viewed the Indians as impediments to mining, agriculture, railways, and other economic developments. Native American claims to the land and their people's fortunes were swept aside in favor of conquest. In the near future, the aging chief, who by this time was farming along Mangas Creek near Santa Lucia Springs, regrettably would have to abandon his new agrarian lifestyle to fight once again. The Americans proved to be even more covetous of his homeland than the previous invaders.[13]

The westward movement of Texans during the gold rush placed Santa Rita back on the mining map. As early as 1854, San Antonio cattle runner John James learned

of the rich deposits at Santa Rita, probably while watering his herd at Ojo de Baca on his way to California. He set up camp on the Copper Mine Road, the old Janos Trail. By 1857, he and his brother-in-law, James R. Sweet, the mayor of San Antonio, and another businessman of that town, Jean Batiste LaCoste, decided to secure a lease of the mines. Sweet was a successful merchant and trader with the means to provision a mining venture and LaCoste was a self-taught engineer with similar investment interests. Another backer, Simeon Hart, enticed his father-in-law, Leanardo Siqueros, and his son, Francisco, of Parral, Chihuahua, to supervise the new operations. The Texans[14] brought supplies and

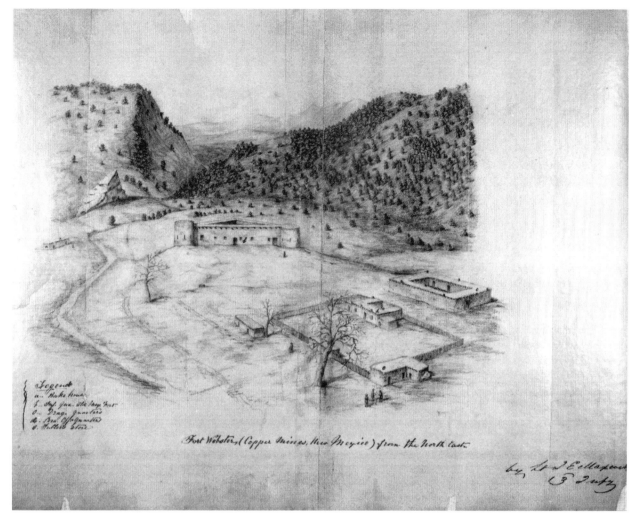

FIGURE 2.4. Fort Webster (Copper Mines, New Mexico) *from the northeast, 1852.*
Drawn by infantry officer Lieutenant Joseph Edward Maxwell, this illustration depicts
Santa Rita and the reconstructed military post, known to the Americans as Fort Webster.
COURTESY OF THE NATIONAL ARCHIVES.

connections to markets back East to the borderlands partnership and the Mexicans contributed their mining expertise.[15]

After a fall 1857 examination of the minas de cobre, Siqueros negotiated with Elguea's agent, Juan Ibern y Mandri, for a seven-year lease to begin the following April. The lease called for the partners to pay 600 pesos a year after the first two years and to remove the Apaches, or the contract would be voided (a high likelihood after a shooting incident left a Chihenne woman dead on that preliminary visit). On his arrival the next summer with miners and family members in tow, Siqueros began shipping native copper to Chihuahua City. When Mangas

Coloradas and his band began farming that same summer along Mangas Creek west of Santa Rita and seemed content with rations of corn and beef from American authorities, Siqueros and his partners had a sense of security in their new undertaking. The operation was in full swing within a year.

The Mexican superintendent soon oversaw the reopening of once-worked shafts, the firing of an adobe smelter, and the shipping of ingot bars to international markets. After initially using the same mining and refining techniques implemented by the Spaniards and Mexicans before him, Siqueros soon brought in a Chihuahuan furnace expert and a boilermaker to operate the new steam-powered bellows of a more modern smelter. German metallurgical smelting techniques, combined with Mexican mining strategies, resulted in millions of pounds of copper output. By 1860, Siqueros was shipping nicely smelted 150-pound ingots by wagon to San Antonio. From there, Sweet made sure the precious cargo made its way to the port of La Vaca to be loaded onto schooners headed for harbors in New York City and Europe. The Sweet-LaCoste-Siqueros venture brought handsome dividends. The director of the US Census Bureau reported in 1862 that Santa Rita ranked second in copper output nationally, with only the prolific Michigan mines outproducing the remote New Mexico operations.[16]

The community also exhibited qualities that made it unique in frontier mining history. Rather than the rough-and-tumble, male-dominated, and violent camp scene indicative of most Rocky Mountain mining towns, Santa Rita was settled by families. According to the 1860 census, many of the miners came with their wives and children. No doubt this demographic feature brought stability to the camp with 214 miners and their families. Census takers also recorded that at least ten women worked as mine laborers at this time, possibly as ore sorters aboveground. Interestingly, Mexican and Mexican American miners, perhaps with the exception of the 1870s to 1890s period, generally lived with their families throughout the history of copper mining in southwestern New Mexico. Migration of miners and laborers across the border would be common well into the twentieth century. As mining historian Robert L. Spude contends, "Santa Rita [in the 1860s] reflected more an extension of the Mexican mining experience than the mining frontier of the American West."[17]

A close examination of the 1860 census also reveals that virtually all the miners were born in either Mexico or New Mexico in an era when many mining districts in the West banned Hispanics from working underground. Conversely, other factors do suggest that Santa Rita exhibited qualities that were synonymous with frontier conditions elsewhere in the American West. High alpine isolation, expensive operational costs, need for better transportation like the railroad, incessant conflicts with indigenous peoples that would lead to the natives' subjugation, hardships brought on by inclement weather, difficulties with increasingly complex ores, and dangers from bandits all characterized this frontier mining camp

and nearly all others from the Mexican to Canadian borders. Further typifying western frontier mining was a gold rush to the area in 1860 when prospectors discovered rich placer deposits at Pinos Altos, only fifteen miles northwest of Santa Rita. Still another frontier theme would emerge in Santa Rita's history soon thereafter: a complicated legal battle over control of the richest copper claims.[18]

In 1861–1862, several factors worked to force a shutdown of all operations at Santa Rita. Sweet and LaCoste terminated their partnership with Siqueros after he accumulated a large debt that his replacement, Mariano Barela, struggled to overcome. They also grappled with inclement weather and the difficulties of smelting more complex ores. Production plummeted. By then they also had to contend with the outbreak of the Civil War. The southerners now risked a high chance of complete failure at this point despite high copper prices.[19] Sweet tried to salvage the mining venture by sending Barela fifteen wagons loaded with supplies in hopes of seeing them returned to San Antonio filled with copper. Mostly, Sweet feared that wartime conditions would destroy their mining enterprise. "The frontier is breaking up and taking all things, [like debt, transportation costs, and the war], into consideration, we are in a sad state."[20]

By spring 1862, Santa Rita was again under threat of becoming a ghost town. Civil War politics had led to the establishment of the Confederate Territory of Arizona, and Sweet and LaCoste had supported the creation of this slave territory. Still, the southern firm abandoned the site altogether two months after the Union won the Battle of Glorieta Pass in March 1862. A few weeks later, General James H. Carleton's California Column marched into Santa Rita and commandeered the company's supplies, copper ingots, a steam engine, and a boiler. Carleton made the decision to shut down the operations to keep the precious alloy out of the hands of the Confederate sympathizers. Soon after Carleton's departure, the Coppermine Apache chief Mangas Coloradas, now supported by his son-in-law Cochise's warriors, began a reoccupation of the copper camp. They began to use the site as headquarters for assaults on Pinos Altos (with more than 800 inhabitants) and elsewhere. Santa Rita again entered a dormant mining period.

The Indian removal of the miners at Santa Rita, however, came at a high cost for the Apaches. Tired of the fighting, especially after participating in the Battle of

Apache Pass in July 1862 in the Confederate Territory of Arizona,[21] the septuagenarian Mangas Coloradas and a small group of warriors went to Pinos Altos on January 17, 1863, to seek peace. The townspeople, angry over recent depredations, arrested the aged chief, and a military escort the following day took him to Fort McLane, about twelve miles south of Santa Rita. Brigadier General Joseph West, under orders from General Carleton to capture or kill the powerful Apache headman to break the will of the Chiricahua resistance, accused Mangas of unprovoked acts of brutality. The Chihenne warrior claimed he had acted in self-defense. That evening West gave a tacit order to kill the celebrated Indian chief. When the sentries began that night to torture the bound Mangas Coloradas by poking his uncovered legs and arms with their red-hot bayonets, the tough old statesman sat up and exclaimed in Spanish that "he was not a child to be played with." With these final words, several of the soldiers shot him, one execution-style in the back of the head. They then maimed his body and beheaded him, an unimaginable insult. Carleton later reported that he was killed while attempting an escape, "a frontier euphemism for murder."[22] El Fuerte's death symbolized the beginning of the end of the Chiricahua's domination of the southwestern section of New Mexico Territory. Though Apache chief Victorio and the famed warrior Geronimo would continue into the 1880s to resist the takeover of their homelands, the Americans were bent on subduing the indigenous peoples as part of their belief in their God-given right to conquer Apachería and the rest of the Southwest.[23]

The ubiquity and determination of the Americans in the Southwest precluded a long-term abandonment of Santa Rita this time. A renewed interest surfaced in 1864 after geologist Richard E. Owen and his assistant E. T. Cox of Indiana State University inspected the deserted mine works. They reported "large quantities of Native copper, as pure as that of Lake Superior."[24] Once in the hands of the Union government in New Mexico Territory, the report motivated General Carleton, Governor Robert Mitchell (also a former Union general), Attorney General Charles Clever, US Marshal John Pratt, and other spoils men to claim the coveted mines in June 1866. To carry out this takeover, they established the controversial Santa Rita Mining Association (SRMA). The conspirators then paid several miners during the winter of 1866–1867 to reopen some of the mines to be-

gin efforts to patent the claims under the new Mining Law of 1866. They also enlisted the protection of troops stationed at the recently established Fort Bayard, only five miles west of Santa Rita. Taking advantage of postwar conditions, the influential claimants ignored the legitimate ownership and possessory rights of the Elguea heirs and their lessees Sweet and LaCoste.

In January 1867, Alexander Brand kicked out the SRMA's hired operatives. Several years earlier, he had replaced Barela as Sweet and LaCoste's superintendent. A land surveyor, medical doctor, and southern sympathizer, the Frenchman Brand had engaged in earlier business ventures with LaCoste and he resented the intervention of the corrupt Santa Feans. He and other anti-unionists led a movement to annex the western portion of Doña Ana County to Arizona Territory. Though the secession effort failed, he and others in 1868 successfully established a new county carved out of the western portion of Doña Ana named, ironically, Grant County.[25] Brand simultaneously worked with other miners in the new county to revamp local mining laws under the jurisdiction of the new Pinos Altos Mining District. Complaining of the dangers of traversing Apache country to file claims in Mesilla on the Rio Grande, they made Central City (also known as Santa Clara), five miles west of Santa Rita, the county seat to serve as the new mining district headquarters. Brand himself then administered the codes after being selected as the first county clerk. He registered the most recent claims and kept the land records for Santa Rita, an apparent conflict of interest in his battle with the SRMA and the Elguea heirs for legal control of the copper mines. Taking advantage of his new authority, Brand partnered in 1869 with local miner James Fresh, who as the new superintendent continued development and mining work on the Santa Rita claims.[26]

By this time, the SRMA had countered with their own fraudulent efforts to patent the claims under the national Mining Law of 1866. They dug a twenty-foot shaft, renovated some of the buildings, and called in mine surveyor R. B. Willison to meet the criteria for patent rights (see map 2.1). María Antonia Elguea de Medina (widow of Elguea and Guerra), in the meantime, hired Chihuahua attorney M. H. MacWillie to represent the Elguea heirs. With the blessings of María Antonia, the Mexican barrister filed an objection to the Santa Fe mining association's bogus claim to the Santa Rita mines. She hoped to regain full ownership rights and then to sell the property.

In April 1870, US land commissioner Joseph Wilson handed down his ruling in favor of the Elguea heirs. Basing his judgment on the 1865 New Mexico territorial mining law, Wilson cited Willison's survey, requested, ironically, by the SRMA, as evidence that the heirs met the law's requirements to retain ownership. "The Santa Rita del Cobre is . . . the kind of property which . . . the territorial mining act of . . . 1865 . . . classed as '*mine and mineral ground heretofore occupied in this territory*' and is subject to relocation only after mining has ceased . . . for a period of ten years or more."[27] Since Brand and Fresh were leasing from the heirs (they inherited the lease from Sweet and LaCoste), and the SRMA had never staked a legal claim to the copper mines, both groups' claims were shaky at best. At this juncture, the deceitful Santa Feans and covetous Texans had lost the legal battle to the Spaniards. This frontier squabble also caused the shutdown of mining operations and the desertion of the camp.

Circumstances at Santa Rita, and in Grant County as a whole, changed dramatically over the next two years. After the Civil War, transportation, capital, and market barriers slowly disappeared and the growing military presence offered more protection to immigrants looking for work in the mines. Symptomatic of these changes, the new boomtown of Silver City sprang up overnight in 1870, refocusing frontier mining in southwestern New Mexico on precious metals (see figure 2.5). The Elguea heirs soon sold their interest in the Santa Rita claims to New Yorker Martin "Matt" Hayes. Hayes had come west during Denver's gold rush in the 1860s and had cultivated relationships with powerful mining men Jerome Chaffee (future US senator) and David Moffat, who made fortunes at Georgetown, Colorado, and in other Rocky Mountain ventures. In 1872 the two magnates sent Hayes to New Mexico Territory in search of new prospects with knowledge that the Denver & Rio Grande Railroad planned to lay a line south into Mexico. Within a year Hayes learned about the dynamics of the Santa Rita legal morass. Making connections with the Elguea heirs' newest attorney, John S. Watts, himself part of the famed Santa Fe ring, Hayes eventually purchased the Santa Rita copper mines for $15,000 from Pablo Guerra, son of the late María Antonia de Medina of Bilbao, Spain, and Dolores Elguea, her granddaughter of Chihuahua.[28]

Hayes's effort to gain a monopoly over the Santa Rita mines was complicated by the flood of newcomers to

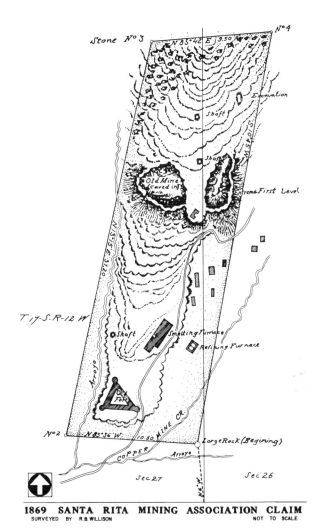

1869 SANTA RITA MINING ASSOCIATION CLAIM
SURVEYED BY R.B. WILLISON   NOT TO SCALE

MAP 2.1. *Sketch of survey of Santa Rita del Cobre Mineral Claim, 1869. The Santa Rita Mining Association commissioned this hand-drawn survey map in an attempt to claim the Santa Rita copper mines under the Mining Law of 1866. The provisions of the Treaty of Guadalupe Hidalgo, however, ensured that the Elguea heirs would retain control of the coveted mineral property.*
COURTESY OF THE HISTORIC AMERICAN BUILDINGS SURVEY, NATIONAL PARK SERVICE.

Grant County. The frenzied rush in the 1870s had enticed prospectors from all over the territory to stake "new" claims under the federal Mining Law of 1872. That year, James Fresh, with new partner John Magruder, for example, staked the Chino[29] claim, attempting to preempt the Elguea heirs' legal right to the Santa Rita property. Soon, prospectors invaded the site, claiming nearly every inch of the old copper camp. Rampant frontier speculation characterized this period of individual and corporate greed that targeted mineral lands of the public domain.

FIGURE 2.5. *Downtown Silver City, New Mexico, 1911. Beginning in the 1860s, miners descended on Grant County in search of gold and silver, introducing southwestern New Mexico to the American mining frontier.*
COURTESY OF SILVER CITY MUSEUM.

Despite the complex legal circumstances brought on by this claim grabbing as well as transportation and technology difficulties, however, Hayes gradually began to develop Romero Hill. Miners probably implemented single- and double-jacking techniques (see figure 2.6) to free the ores for ascent to the jaw crushers above. Hayes also introduced reverberatory smelter technology adopted from Swansea, Wales, to produce copper matte for his first shipment by wagon in 1875 to rail connections in south-central Colorado. Trains freighted the unrefined copper to the Baltimore Copper Works. The Hayes outfit sent about 1 million pounds of copper that year to the Maryland refinery. Clearly, Santa Rita was now connected to the major investment markets in the northeastern United States. Eventually, New York and Boston financiers would put up the capital and find the stockholders to develop the New Mexico properties, culminating in 1909 with the incorporation of the Chino Copper Company.

In the meantime, local proprietors opened a saloon, a restaurant, and a corral in hopes of permanence at Santa Rita. Close by, Fresh and Magruder's Mimbres Mining & Reduction Company was also producing at copper claims in Santa Rita and silver mines in Georgetown. Their smelter on the Mimbres River produced more than 200,000 pounds of copper in 1875–1876. The nationwide financial stresses of the crash of 1873, though, halted operations altogether by 1877. Santa Rita was abandoned again, with only a caretaker and perhaps a couple of other families living in the copper-laden hills.[30]

In April 1881, Hayes and his partners decided to sell their rights to the Santa Rita properties after purchasing the Chino, Guadalupe, and Yosemite claims from Fresh and Magruder earlier that year. The Coloradans were able to convince Joel Parker Whitney of Boston to buy them out for $350,000. He took the risk, banking the success of the enterprise on the arrival a month earlier of the Southern Pacific and Atchison, Topeka, & Santa Fe

FIGURE 2.6. *Underground miners in Georgetown, New Mexico, posed for this 1880s photograph. Note the candlesticks used for lighting and the sledgehammers and drills used in single- and double-jacking techniques. Santa Rita's underground miners worked under very similar conditions, although no photographic record exists to illustrate their techniques.*

COURTESY OF SILVER CITY MUSEUM.

railroads to Deming, just fifty miles south of Santa Rita. Plans for a line to Silver City were already in the works, persuading Whitney to believe another spur would soon be laid to Santa Rita, eliminating a key obstacle to profitable mining at the copper camp. Just a year earlier, Chief Victorio had been killed, and "concentration" on reservations was on the horizon for his subjugated Chiricahua tribe. This dramatic change convinced Whitney of the viability of his new venture even though several Grant County miners lost their lives to the Apaches during the early 1880s.[31]

Despite his own scare with Mimbres warriors while on a hunting trip east of Santa Rita a couple of years earlier, Whitney took the risk that the copper mines would bring large profits. The Bostonian had invited his own

expert, John Slawson, former mine superintendent in the Lake Superior district, to examine the properties. The engineer gave a glowing report of their potential. Just after the purchase, Whitney confidently reported to the Boston press that Santa Rita's mines might even "eclipse" the value of the prolific Michigan properties. The partners predicted that the dumps alone at Santa Rita were worth $2 million and million-dollar-a-year profits could be expected. The promotion worked and Whitney soon raised $250,000 from New England investors after incorporating the Santa Rita Copper & Iron Company (SRCIC). Using funds procured from his other Colorado and New Mexico firms—the Bonanza Development Company and the New Mexico Land and Cattle Company—he bought nineteen full and eleven fractional claims spanning 509 acres. Further strengthening his position, Whitney incorporated the Carrasco Copper Company, which purchased nearly all the remaining full and fractional claims in Santa Rita. He bought an additional fifteen claims in 1882 for $1,000 from Eugene Connor. Santa Rita's mineral properties were completely monopolized the following year when the Secretary of the Interior gave title to sixty more claims covering 1,200 acres to Hayes, who immediately turned over the properties to the SRCIC (see map 2.2).[32]

With this cache of claims, Whitney hired Slawson to supervise a collection of Cornish miners, who soon deepened the shafts of the Romero workings. The company also invested in steam-driven hoists to replace the horse-drawn whims and windlasses that were substitutes for the old chicken ladders. A hoist man could more efficiently lower and raise miners and deliver ore to the surface in cages. Millmen sent the crude material through forty Cornish stamp mills and jigs in the Southwest's first concentrator (or mill), built by contractors Ben Harrover and C. H. Wilkie. The concentrates from the mill then went through blast furnaces at the smelter. Though cordwood was still the principal fuel, coal brought in by rail could now be used to produce hotter temperatures in the smelters and, consequently, create higher purities of copper matte. Slawson soon advertised in the *New Southwest* the need to purchase 200,000 adobe and red bricks, 25,000 bushels of charcoal, and 5,000 cords of wood. He also hired carpenters and sheepherders to sustain the burgeoning multicompany enterprise (see figure 2.7).[33] The human impact on the local environment changed dramatically with these new industrial developments that sustained Santa Rita's copper operations as well as many others in Grant County. The miners and their employers would pay a heavy price in the near future.

The sibling copper companies soon began substantial production despite having to operate intermittently because of unpredictable water shortages, high freighting costs, fickle copper prices, and uncooperative complex ores. Despite these obstacles, the mill produced four to eight tons of copper concentrates a day during the next two years after they scrapped the stamp mills in favor of roll crushers, Evans slime tables, and Collom jigs, all adopted after mill engineer Theodore Schwartz (an MIT graduate) recommended these changes. The company smelted some of the new concentrates but ended up shipping most of it—about forty tons a week—in

sacks to Detroit. They freighted the native copper in barrels. The company's production for 1883 only reached 658,000 pounds of pure copper and 866,000 pounds of 76 percent concentrate during the first four months of the year. Water shortages and low copper prices precluded year-round operations. Perhaps as much as 2 million pounds of copper was produced that year, not nearly enough to justify the $400,000 already invested in the Santa Rita operations. Whitney got cold feet by the end of the year. Even before his new superintendent, P. I. Mitchell, could blow in the new water jacket furnaces at the smelter, he had ordered operations halted. Within a month, he began offering the properties to eastern and European investors through a promotional flyer titled "Native Copper Mines of Santa Rita." He claimed that the mines would garner a net profit of nearly $1.4 million a year. When the Bostonian gave a tour of Santa Rita to one group of British investors who had recently acquired copper properties for $1.2 million at Clifton, Arizona, they were astonished at his whopping $6 million price tag. They turned him down. Not even the endorsement of English mining engineer Theodore Whyte, representative of the Earl of Dunraven, who had investments in Colorado, could convince the New Englanders to take the risk on Santa Rita.[34]

Despite the shutdown in 1883, Santa Rita began to take on characteristics of a frontier mining town. The previous year, Grant County commissioners granted Santa Rita's petition for its own precinct, "starting at the Kneeling Jesus and going northwest to the division between Hanover Canyon and the branch of Shingle Canyon, then due west two miles. Then [its boundaries ran] southwest to the forks of the road running down Whitewater Creek and towards Fort Bayard, below Henry Gatton's house, then due south two miles and from that point straight back to the place of beginning."[35] Most of the new inhabitants were men (about 400) who worked for the Santa Rita Copper & Iron and Carrasco Copper companies and independent contractors. Eight saloons popped up along with three stores, a stage line, a sawmill, and a post office. Sigmund Lindauer (future postmaster) operated the Santa Rita Copper & Iron Company Store, which sold liquor and housed a butcher shop. In the spirit of free enterprise, M. H. Marks opened a private mercantile with a saloon called the Opposition Store. Benno Rosenfeld served as Santa Rita's first postmaster, housing a mail room at his store. Another store, run by Dr.

J. B. Warren and Joe Cline, also sold spirits and sported a popular dance hall. Robert Gamble's saloon welcomed Spanish-speaking guests because he had procured a *baile* (dance) license. Joseph Andrews offered hotel and restaurant services at the Broadway. Ed Marriage's stage ran coaches from Silver City to Georgetown with a stop in Santa Rita. When Marks's Opposition Store and Warren and Cline's saloon burned down in May 1883, Santa Rita was symbolically warned of hard times ahead. Whitney, in fact, had just closed his operations. During such lulls, local entrepreneurs stayed flexible and packed up their goods and headed throughout the 1880s and 1890s to the "boom towns" of the moment, like Georgetown and Shakespeare.[36]

Copper price fluctuations in the early 1880s foiled Whitney's speculative venture. He could not find a buyer so he began leasing the properties. Not until 1897 did the George Hearst estate purchase an option on the extensive mineral deposit. When mining engineer Arthur Wendt evaluated the properties in 1886 to assist Whitney in unloading the claims, he found little mining going on. "The Santa Rita mines are entirely idle; and unless developments in depth on the iron-ore outcrop should expose a richer and different character of ore [that is easier to concentrate] from that treated in the stamp-mill, the property is likely to remain idle for a considerable time."[37] Whitney had lost his patience. The long-awaited rail line had not come, even though he had

MAP 2.2. *Map showing the property of the Santa Rita Copper & Iron Company and the Carrasco Copper Company, Grant County, New Mexico, 1881. By the early 1880s, J. P. Whitney had consolidated much of the rich Santa Rita mineral claims into one property.*
COURTESY OF GRANT COUNTY CLERK'S OFFICE.

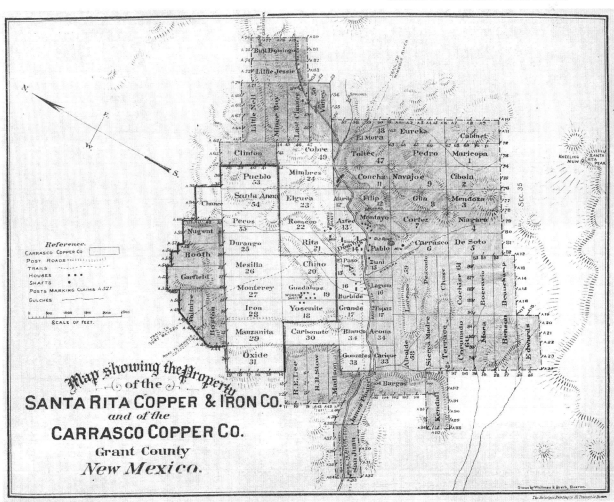

FIGURE 2.7. *Santa Rita, 1882. Note the Romero Mine and Mill to the right and workers' homes to the left. Downtown was located behind the mine works.*

COURTESY OF PALACE OF THE GOVERNORS PHOTO
ARCHIVES (NMHM/DCA), NEG. NO. 144629.

been instrumental in bringing the Silver City, Deming & Pacific Railroad to Silver City in 1883 amid great celebration and a railhead was only eighteen miles south of Santa Rita at Whitewater. Even the installation of a telephone in 1883 could not inspire a revival of the mining operations at Santa Rita.[38]

From 1884 to 1909, leasing outfits dominated the Santa Rita mining landscape. Slumping copper prices and poor management by Whitney resulted in low output from 1884 to 1886. Then the crash of 1893 stifled production again for two years. Another lull occurred from 1902 to 1905. During the earlier, more prosperous, twenty-five-year period, more than 200 miners leased the Santa Rita copper properties. Local mine contractors, such as Herbert and Arthur Dawson, D. B. Gillette, James Gilchrist, Ashton and Roach, Mariano Islas, the Peterson Brothers, J. W. Bible, E. W. Clark, McLaughlin and Curtis, and a host of smaller operators, took advantage of more favorable industrial conditions, such as stable copper prices and the arrival in 1891 of a rail line to Hanover and another spur in 1898 all the way to

Santa Rita. Contractors also participated in the royalty system of the Santa Rita Copper & Iron Company and then, after 1899, of the Santa Rita Mining Company. The emergence of local smelters and mills in Silver City and Hanover also buoyed small-time development projects.

Dr. Franklin A. Mahone, who served as SRCIC's superintendent from 1886 until his death in 1897, supervised the leasing of Whitney's properties. His replacement, Arthur Dawson, who had been instrumental in developing copper, iron, and zinc properties at Hanover (he had served as superintendent of Southwest Coal & Iron Company and the Hanover Improvement Company), leased to more than 150 miners, covering seventy-seven claims and 1,500 acres in Santa Rita. Driscoll and company, in partnership with James Gilchrist, was the main leaser in Santa Rita in 1897–1898 just after the George Hearst estate took an option on the properties. During that two-year period they employed more than 100 miners and put to work fifteen six-horse wagon teams to haul the ores to the Silver City Reduction Works. This nearby smelter processed the majority of Santa Rita copper ores until 1902, when it burned down. Subsequently, most ores were shipped to the Kansas City Consolidated smelter in El Paso and some made their way to the Comanche Mining & Milling Company in Silver City until 1907, when it was reduced to ashes. After that, the Chino Copper Company sent Santa Rita copper concentrates to the American Smelting & Refining Company furnaces in El Paso.

The Romero shaft by 1900 had numerous drifts at the 100-, 200-, and 300-foot levels. During the busiest times, it was worked by five different contractors, each working anywhere from three to fifty miners. By that time there were nearly two miles of tunnels just in the Romero workings. Contractors discovered massive sheets of native copper as well during this development period. Most of the raw copper "nuggets" weighed from five to eighty pounds. But on occasion a massive "chunk" was unearthed, as was the case in 1899 when three lessees working the Montoya brought up a ten-thousand-pounder. Such finds brought wanted attention to the district searching for a more visible identity among corporate giants, such as General Electric, the Rockefellers, and the Guggenheims.[39]

The development of operations at Hanover also enhanced opportunities for lessees. After the SRCIC mill burned on July 4, 1895, for example, miners were able to send their ores to the New Mexico & Arizona Smelting Company, a subsidiary of American Zinc & Lead Company. The milling and smelting business treated ores from the Chino, Carrasco, Yosemite, Robert E. Lee, Oxide, Guadalupe, Romero, Rita, and other mines in Santa Rita, as well as ores from Pinos Altos, Mogollon, and White Signal. General Electric developed the Ivanhoe Mine on the outskirts of Santa Rita, further providing visibility and resources for copper development in the Central Mining District. The Chino Copper Company eventually purchased this property in 1918 to bring it into the corporate fold. The SRCIC also benefited from a regional iron smelter in Pueblo, Colorado. During 1897–1898 the company sent iron ores to the Rockefellers' Colorado Fuel & Iron Company works, which treated about 3,000 pounds of concentrate a month from the Booth, Nugent, Garfield, Boston, and Chance mines. By 1903, mills and smelters from Tyrone to Silver City to Hanover and Santa Rita served the prolific district, making it by far the most productive mining area in New Mexico Territory.[40]

The frontier mining era was coming to an end at Santa Rita. Availability of outside capital and modern technologies meant that corporate mining was on the horizon. Then, in 1897, two events foretold of bigger things to come. First, Whitney's long-awaited hope for a rail spur to Santa Rita came to fruition. In December, investors from Chicago, Illinois, and Santa Fe and Las Vegas, New Mexico, incorporated the Santa Rita Railroad Company, bringing in $50,000 from the sale of 500 shares. By June 1898, the first iron horse steamed into the camp's modest depot. When the Atchison, Topeka, & Santa Fe purchased the line in 1900, freighting was guaranteed for the remote copper town. Second, the George Hearst estate of San Francisco decided to take a two-year option on the Santa Rita properties. Familiar with the district after having purchased the Silver City Reduction Works and the Pinos Altos Gold Mining Company, the Hearst interests hoped to cash in on the underdeveloped properties in southwestern New Mexico. They bought the Santa Rita Company Store as well and then offered the mines to sub-lessees. Well-known metallurgical expert Benjamin Thayer soon took the reins as manager, moving his wife and three children from California. George Stevenson, son of the famous Adlai, came aboard as superintendent. Santa Rita was poised to take advantage of the escalating copper demands in the United States to contribute to the nation's growing appetite for electricity, automobiles, and munitions at the dawn of the "Copper Century."[41]

The huge demand for the red metal at the end of the nineteenth century caused a surge in its value in mid-1899 to 17 cents a pound. This high price, combined with the arrival of the railroad, made Santa Rita more palatable to corporate investors. Though the General Electric Company had toyed with the idea of investing in the mines, it would be the Rockefeller Syndicate that would make the first effort to implement economies-of-scale investment in Santa Rita. Hoping to buy up most of the lucrative copper deposits in the United States, the syndicate's Amalgamated Copper Company, predecessor of the Anaconda Copper Corporation, focused its attention in the summer of 1899 on the rich mineral deposits of Grant County. Under the guidance of Arthur Dawson, the business collective paid for a thorough inspection of the Romero works, sinking several thousand dollars into dewatering efforts to unveil the veins below. Their representative, William H. Burrage of Boston, accompanied Dawson down into the Romero shaft. He surfaced convinced that profits could be made. After some negotiations, the Hearst estate agreed to relinquish its option for $200,000. Amalgamated then paid Whitney and the SRCIC an additional $1.2 million for the seventy-nine claims.

The copper conglomerate also purchased several additional nearby claims and Julius Welgehausen's ranch to procure water rights. The new boss in town also gave contractors a sixty-day notice of the termination of their leases in August 1899 and warned "squatters" to evacuate their premises. Santa Rita's century-old works now became an entity of one of the largest multinational copper corporations in the world. Notable among the giants in the syndicate were William D. Rockefeller, Albert C. Burrage, Marcus Daly (of Butte, Montana, and Anaconda Copper fame), Henry Huddleston Rogers, Thomas W. Lawson, and even Whitney, who retained a minor interest in the properties. The Amalgamated Copper Company, a subsidiary of the Standard Oil Company, hoped to monopolize the copper industry, as its parent corporation and the Carnegie Steel Company had in the oil and steel businesses, respectively.[42]

In many ways, this transaction forewarned the end of an era at Santa Rita. Rampant speculation, claim grabbing, technological inefficiencies, and exorbitant freighting costs, although not completely gone, were basically obstructions of the past. J. P. Whitney's rush to invest in modern equipment before fully understanding the technological needs for milling the complex ores characterizes the fly-by-night nature of frontier mining throughout the American West. Yet, in the end, the proud Bostonian earned a profit from his risky if poorly thought-out venture. He never realized, however, his $6 million dream.[43] His successors would reap most of that additional profit when the Chino Copper Company bought them out a decade later and then introduced the economies-of-scale strategies of open-pit mining and flotation milling.

By fall 1899 the Rockefeller Syndicate incorporated the Santa Rita Mining Company (SRMC) with a working capital of $5 million. Shares sold for twenty-five dollars apiece. Arthur Dawson temporarily stayed on as superintendent and William H. Burrage performed the duties of agent and manager. Engineer Benjamin Thayer soon thereafter took on both of these positions until 1903, when the parent company, Amalgamated Copper, wooed him away. On Thayer's departure, mine foreman Mike Riney was promoted to superintendent, and book-keeper/stenographer John Deegan earned the title of general manager.[44] Under these men's leadership, the SRMC continued development work and soon began leasing sections of the extensive subterranean works.

For the next six years the company developed the underground tunnel system, carving out thousands of feet of new drifts and stopes. Company miners pounded sledgehammers on their steel bits to loosen the ores that muckers handloaded into railcars for transport in the tunnels. Steam-driven hoists lifted the men and ores to the surface in carefully fitted buckets or skips. Black smoke billowed from tall stacks on Romero Hill, providing the best example of the advent of the industrial revolution in the quaint New Mexico town. Housed in the soot-stained buildings were hardworking boilers that drove the machinery that separated the precious mineral from the unwanted waste materials while spewing welcome emissions into the air. Surrounded by growing heaps of ore rock and mill tailings, the manufacturing complex rose above the horizon like a behemoth from below (see figure 2.8). Locals preferred industrial "progress" over the potential health risks of the smoky operations.

Beneath this distinctive monument to industrialism miners cut through bedrock at a rate of eighty feet a month in the numerous tunnels and shafts crisscrossing this dark underworld. Soon cross-country drifts connected the Romero, Hearst, Carrasco, Chino, Santa Rita, and other claims, especially at the 200- and 300-foot levels. Miners completed the tunnel that linked the Romero and Carrasco Mines in October 1901, opening new veins to be worked by either the company or its lessees. A year earlier the Hearst interests had dug a similar drift from the Carrasco to the Hearst. The Santa Rita Mining Company discovered that the drift was slightly angled downward rather than level. This engineering flaw caused "bad" air to accumulate in the shafts, especially near the 300-foot level. To remove this threat to the miners, the company sank another shaft for ventilation about 900 feet east of the Hearst Mine. Known as the Hearst Air Shaft, the outlet created a vacuum effect that drew noxious gases out of the deep workings. This technique revealed the growing need for safety measures to catch up with the dangers of handling the increasingly powerful technologies of the day.[45]

Prior to the emergence in 1910 of open-pit mining, the peak year of production in Santa Rita was 1901. That year, Grant County produced nearly 10 million pounds

FIGURE 2.8. *This 1909 photograph shows spewing emissions from the Romero concentrator, massive piles of cordwood to fuel the mill, and the emerging residential landscape of the industrial community. Note in the upper right the miners' boardinghouse, which eventually had 100 rooms and burned in 1928.*

of copper, most of which came from the efforts of the Santa Rita Mining Company and its lessees. Nearly 600 miners were scattered in the vast underground catacombs that resulted in the shipment of about fifty railcars of ore a month to the Silver City Reduction Works. Among the heavier producers were Crawford and Fritter on the Rita; Bayne and Company on the Chino; and Burns and Company on the Chino, Lee, and Romero. Second-tier shippers were Spafford and Company on the Lee; Herbert Dawson on the Aztec, Chino, and Cortez; and F. G. Burns on the Aztec and Rita. Other prominent lessees in this period included Felipe Casares, Portwood and Company, McGregor and Company (who previously ran a mill in Georgetown), and Islas and Ernest. Charlie Sun, who took a lease on a section of the Romero, was the first Chinese mine contractor in Santa Rita. Their successes increased everyone's odds for making profits. Costs, in fact, dropped with higher yields so that by the

end of 1901, per pound expenses fell from about ten to five cents, nearly doubling profits.[46]

Most of the lessees mined using simple single- and double-jacking technology. After drilling their blast holes, they filled them with dynamite for end-of-shift blasts. When the dust settled, muckers collected the chunks of ore and, if fortunate, nuggets of native copper. They then loaded them into modern ore cars or less costly burlap sacks. Many of the lessees used horse-powered whims and windlasses to bring the miners and their ores to the surface in buckets. Animal power usually cost far less than steam. Likewise, independent miners saved on expenses, using their skills to construct wood head frames, hoist wheels, and pulley supports from timber they cut in the Gila Forest Reserve (see figure 2.9). As late as 1908, more than 100 lessees supervised their own operations, hiring as many men as their share of the royalties could sustain.

FIGURE 2.9. *Ruelas claim and horse-drawn whim, 1910. Many Santa Rita miners leased copper properties from the Chino Copper Company. Here leaser Miguel Ruelas (fourth from left in the center group) and his workers (probably relatives) operate a horse-drawn whim to raise and lower miners and ore from the Chino claim.*
COURTESY OF BENITO R. RODRIGUEZ.

As long as the price of copper remained above twelve cents a pound and freighting costs were reasonable, miners took ore to the company mill. They sent concentrates to Silver City until 1902 and then to El Paso after that. As with miners' wages, the company based smelter payments on the purity of ores and the price of copper. This sliding scale could benefit company miners, as was the case in the summer of 1903, when they received a 50-cent-a-day raise. During layoffs, like the one just a few months after the pay increase, married men were kept on over single ones. For lessees, pay worked differently. During the SRMC's first two years, 1899–1900, the lessees split the proceeds fifty-fifty with the company, unless the ore assayed below 10 percent purity. In those cases the lessee got nothing. This penalty caused many of the

contractors to quit. So, beginning in 1901, the company equally divided profits on ores below 10 percent and paid lessees 75 percent on returns above 10 percent. From 1905 to 1909, another change occurred so that lessees could earn, at most, 70 percent on returns on prices below 15 cents. If the price rose above that, they got 65 percent. Good prices from 1905 to fall 1907 favored the company, though the lessees were guaranteed 75 percent. When the price dipped below 15 cents over the next four years, the lessees earned 70 percent, a trend that continued until 1911, when the Chino Copper Company terminated all leasing contracts.[47]

Another strategy focused on abandoned dumps. A two- or three-man team could earn a good living by jigging the old ore or waste piles, some of which were left by the Spanish, on the Carrasco, Hearst, and other claims. Essentially, the miners built a large frame over ditches they dug adjacent to the dumps. They then placed screens in the frames that could be jigged, or shaken, to separate crushed gangue or dirt from the raw pieces of native copper (see figure 2.10). Use of these handmade jigs may have been the most lucrative way a small operator could "mine." There was so much virgin copper still left in the waste heaps in the early 1900s that overall ore returns came in at a high 22 percent purity.[48]

During 1902 the SRMC changed its mill leasing policy. Horace Moses, who would play a key role in Chino's future operations, served as general manager of the mill. Previously under Moses's supervision, the company operated the concentrator for six months and then leased it to Crawford and Fritter the following six months. This six-month lease arrangement, however, proved to be inefficient because each party discovered that when they returned to the mill, the machinery and equipment had worn out. It often took weeks to repair damaged slime tables, settling tanks, steam boilers, and other apparatuses. Consequently, both parties agreed to alternate running the mill every other month. This strategy reduced repair as well as production costs. That same year the company also had to renegotiate its smelting contract when copper prices dropped. Instead of working out a contract with the new Comanche company in Silver (the Silver City Reduction Works burned in 1902), the SRMC and its lessees began sending most of its ore cars to the El Paso branch of the Kansas City Consolidated smelting company. Unfortunately for the miners, freighting costs escalated as a result and output dropped in 1903, not to fully recover before 1909, when the Chino Copper Company purchased the entire Santa Rita copper field.[49]

Miners developed incredible skills during the underground era of copper mining at Santa Rita. They also took risks that sometimes turned out tragically. The US Census reported in 1900 that Santa Rita had nearly 2,000 inhabitants.[50] Townspeople had a sense of community and they began to celebrate holidays, especially

FIGURE 2.10. *Hand-jigging the old Santa Rita dumps, ca. 1910. These men made a living leasing the waste dumps from the Chino Copper Company.*
COURTESY OF TERRY HUMBLE.

the Fourth of July. Independence Day brought the multi-ethnic residents together to enjoy picnics, play baseball, and set off fireworks. The highlight of the holiday was the drilling contest. Single- and double-jackers lined up to hammer their steel bits into solid rock. In 1901 many of the 2,500 visitors that day gathered around the sweaty drillers as they vied for first prize. Each team or individual was allotted fifteen minutes to bore a hole into the igneous rock taken from just below the Kneeling Nun on Ben Moore Mountain. Despite the appearance of some Cornish drillers, or "Cousin Jacks," the Barela brothers, a local Santa Rita team, took the double-jacking purse after their concerted effort. Single-jacking champion Tranquilo Morales of Silver drilled all the way through his nineteen-inch boulder in a mere fourteen minutes, contradicting his first name, "Tranquilo," which means "calm down" in Spanish. The Hispanics' success did not go unnoticed by Superintendent Thayer, who wrote a few months later that one of his foremen "had secured a very good crew of Mexican contractors on the Lee and I am in hope that they will be able to do more than the worthless class of white men which he has been working."[51]

But stereotypes poorly typify the miners' laboring qualities, as the following year's drilling contest reveals. Con O'Neil, a Butte, Montana, transplant, and George Husband, a Santa Ritan, drilled three-fourths of an inch deeper than Valentine Guiterrez and Alfred Barrios, who worked in Pinos Altos but resided in Santa Rita. Juan Avalos won the single-jacking contest over L. W. Ligon by just half an inch. The first-place double-jackers won $100 and the second-placers $25; the single-jackers took home $50 and $25, respectively. Clearly, the skilled miners of Santa Rita and other mining sites in Grant County came from varied ethnic backgrounds. Two weeks later, double-jackers Guiterrez and Barrios challenged O'Neil and Husband to a rematch, with each squad throwing $100 into the winner-take-all pot. The former team got their revenge, making $50 apiece for their quarter hour of exertion. It was "a sight worth travelling many miles to see," reported the *Silver City Independent*. The winners drilled nearly three inches deeper into the granite. The hammer had slipped in O'Neil's hand, causing their steel to chip on three separate blows. Barrios and new partner Anastacio Ponce won the two-man contest and Tranquilo Morales the single-jacking one in 1907.[52]

These matches drew the community together during the joyous occasion of celebrating the nation's inde-pendence. But deaths and serious injuries in the mines galvanized the town even more. Falling rock, deadly gases, and torturous falls claimed most of the victims. In January 1899, for example, Manuel Lucero, a trammer working for Gilchrist and Clark, was killed by falling rock in the Cortez Mine. Tragedy struck just a couple of weeks later when Ramón Martinez died of asphyxiation from deadly gases that nearly took the lives of Tirzo Martínez and Francisco Lucero. In June of that year, well-known contractor Arthur Dawson needed several weeks to recuperate after he nearly died in a rockfall on the Romero. The following year, a rockfall claimed the life of Aurelio Delgado in the Mother Lode Mine. Rockfalls took the lives of several more miners over the next few years: Francisco Alvillar (1901), Viterbo Cordova (1902), Narcisso Lascaro (1904), and Juan Durán (1907).

Perhaps the saddest fatalities involved two young brothers, José and Calilsto Jimenez, who perished two days before Christmas 1902 in a cave-in. Superintendent Thayer wrote in that month's company report that "the cave of ground amounted to practically nothing, but the rock which fell upon the two little fellows was suffi-ciently large to crush out their lives." The SRMC official immediately sent notice to all the lessees, "prohibiting the working of boys in the properties of the company, in any capacity [even as sorters aboveground] whatsoever." This admonition must not have made too much of an impression because Jesus Valencia, only ten at the time, almost lost his life the following March. Working on the Crawford and Pronger lease at the Chino, Valencia slipped and tumbled 108 feet to the bottom of the shaft. Expecting to find the grisly remains of the young lad, the rescuers were amazed to find the tyke still conscious and sitting in an upright position. His landing had been soft-ened by a sack of candles, a financial loss that everyone celebrated, especially after the company doctor's exami-nation failed to uncover a single broken bone.[53]

One day after Francisco Alvillar was killed, another accident took G. H. "Paddy" Welch's life. He, along with Valentine Guiterrez, P. H. Williams, and Roy Whittlesey, had ascended to the surface after blasting a stope deep in the Hearst shaft. After taking a supper break, they descended back into the dust-settled chamber only to find that five rounds had not detonated. "Welch and Guiterrez," Thayer later recorded, "had cleaned four of the five holes. Welch [then] inserted [into the final hole] a three-quarter inch pipe with a half-inch rod running through the pipe called a 'gun' for removing mud and water from the holes. An explosion occurred immediately. The fuse had been removed but the cap must have stayed in the hole and was set off by the iron of the gun. Welch was blown almost literally to pieces, and Guiterrez had one eye blown out, although he had been blind in this eye all of his life. He was fortunate in not having his good eye injured and he is doing well in the Company hospital." Guiterrez, Williams, and Whittlesey all made full recoveries.[54]

Another grave danger was falling down shafts in the pitch darkness of the underground. On November 30, 1902, Amado Roblero, while working for lessees Crawford and Pronger, forgot to shut the trapdoor at the collar before descending in the cage of the Chino shaft. When the miner reached the entrance to one of the levels, he was struck by an ore car that had just been emptied and had rolled into the unguarded cage, knocking him into the shaft and a fatal fall of several hundred feet. Dioncio Portillo lost his life in a similar accident in December 1904. In this case, the miner forgot to ring the bell for the hoist man to return the shaft platform so that he could push his ore car into the ascending cage. When he rolled the full car into the open shaft, the force pulled him into the manmade cavern and a 300-foot drop to his abrupt death. Jay Hiler lost his life at age nineteen when he slipped and smashed his head on some equipment in the Santa Rita concentrator. In August 1907, Commodore Perry Crawford experienced a fatal fall of more than 145 feet in the Ivanhoe Mine. A well-known contractor, Crawford had made a small fortune in various business ventures after moving to Grant County in 1873. Cosponsoring the construction of the Deming & Silver City Railroad and supervising the digging of the cross-country tunnel that connected the Santa Rita and Hearst shafts were just two of his many contributions to the mining district.[55] These tragedies were symptomatic of minimal precautions in an era when few mining districts and states had protective laws and safety experts to improve company and miner awareness. They also illustrate the inherent risks of mining underground.

The dangers of mining, however, did not preclude community building in Santa Rita. The often-abandoned and sometimes uninhabited copper camp finally seemed poised for a permanent settlement. Eager to establish a company town, the Santa Rita Mining Company contracted its first couple of years in town to build hundreds of homes as well as a company store. By the end of summer 1899, the corporation had already constructed a set of general offices that would be used for the next fifty years about 100 feet north of the old Mexican fort. By spring of the next year numerous officials' residences had also been erected. William Burrage moved with his new wife into one of the eight-room, thirty-four- by thirty-eight-foot frame homes. Thayer, at the time Burrage's assistant, was assigned an adobe house that his wife, four children, and he lived in for the next three years. Later, they moved into the general manager's house (see figure 2.11). During the year 1900, the company contracted with P. J. Bratley to erect a substantial twenty-eight-room boardinghouse, a mess hall, and a saloon (operated by Bill Ernest, former owner of Ernest's Roadhouse) to room, feed, and "lubricate" the mostly single miners. Carpenters renovated the old mess hall, transforming it into the town hall and a residence. The company annually built additional modest cottages to accommodate its growing workforce and even rented spaces for tents until it sold out to the Chino Copper Company. One of the towers of the old Mexican fort was also transformed into a temporary jail. Carpenters lined the walls of the adobe turret with two-inch planks so the prisoners could not burrow out. Within a year, a modest jail replaced the makeshift dungeon.[56]

The company chose the Booth Mine for its water supply. Not particularly rich in ore, the shaft, which gushed 216,000 gallons a day, replaced an artesian well near Santa Rita Creek as the town's main source of water. St. George Robinson laid a four-inch pipeline from the shaft in fall 1901 to each of the businesses as well as the "leading" residences. Any new businesses could request running water from the Booth. The company lagged a bit in offering electricity, which finally came to town in November 1907 when a power plant was constructed. And, again, the prominent citizens were the first to get this service.[57]

Santa Rita's 2,000 residents also needed a school and a hospital. Bratley constructed the healthcare facility in late 1900. There, Dr. Charles F. Beeson attended to patients with a nurse assistant. All SRMC employees paid a dollar a month for expenses, a charge later set at one dollar for single men and two for families. Health officer Dr. James Johnson also tended to community medical needs.

Bratley's company also built a new schoolhouse to replace the two-room adobe school in the old Mexican

FIGURE 2.11. *Downtown Santa Rita, ca. 1911. The Santa Rita Mining Company constructed many of the buildings in the downtown scene in this photograph. Note the general manager's house with the peaked roof in the center. The Chino Copper Company inherited these structures, which served the corporation's officials and other employees until the 1950s.*
COURTESY OF SILVER CITY MUSEUM.

fort. The "elegant" building was paid for in part with donations collected at two separate dances—the Anglos held a "Grand Ball" one weekend in November 1900 and a week later the Chicanos hosted their own "Baile." These philanthropic socials brought in seventy-eight dollars. The SRMC also donated $100 for the educational facility. Designed to accommodate 100 students, the facility saw 230 students enroll in January 1901, putting a heavy burden on teachers Janie Trevarrow and Mary Riney. To their relief, about 100 fewer pupils enrolled for the following spring semester. Regardless, the increase from only four children attending classes two years earlier reveals the unprecedented population growth in the copper town. Ben D. Moses was selected as principal in 1902 for the new Santa Rita school district. Grace C. Osmer

served as his assistant principal. The local press complimented their efforts in the face of exponential growth. "Principal Moses has proven himself a splendid disciplinarian as well as instructor," the *Silver City Enterprise* reported in March 1903. "And Misses Grace Osmer and Kate Crawford [teacher] are doing good work with the smaller pupils. The town hall is being used by Miss Crawford as a schoolroom, the school being too crowded."[58]

Crowding continued to be a problem for the Santa Rita School during the next few years. When enrollment topped 350 in fall 1907, the district decided to build a new schoolhouse. Henry Brixner won the contract to design and oversee the building's construction until an accident required Walter Reubles, a master carpenter, to complete the job. Costing a reasonable $700, the new school was one story and made of adobe bricks laid in lime mortar. Opened in November 1908, the structure contained four classrooms of varying sizes within its overall dimension of 100 square feet. As a safety precaution, each room's doors opened to the outside so "there will be no jamming of children against closed doors" in case of fire. Spacious windows brought ample light into the rooms and, when open, offered excellent ventilation. Lillian Jackson of Pinos Altos taught the primary students, Inez Huff of Santa Rita and Ella Smith of Las Vegas the intermediate grades, and Principal James Higdon of Silver City the advanced pupils. "The citizens of Santa Rita feel elated and are grateful to the Santa Rita Mining Company through whose courtesy and the interest manifested by its genial manager, John Deegan, the bulk of the money that the building cost was secured." The copper camp was subtly experiencing the transition to a company town.[59]

The most prominent symbol of the emerging company town was the Santa Rita Company Store. Built in spring 1900, the store became the symbol of corporate power (see figure 2.12). A large structure of nearly 5,000 square feet, it was made primarily of adobe. There, patrons could purchase a multitude of items, from coal to cradles (see figure 2.13). By summer, the company notified other mercantile owners to close shop, which later led to a lawsuit on the Pinder claim. Ferdinand "Fred" Risque took over management duties and Clarence Bayne, a popular Silver City youth, came aboard as one of the clerks. To encourage employees and their families to shop at the company store, the SRMC printed scrip or *boletas* in coupon booklets. On a purchase, the store manager or a clerk tore out the boletas equal to the amount of the sale. Another version of the scrip booklet had to be punched. Regardless, the company kept close track of clients' expenditures, though it also extended credit (promise-to-pay) to ensure that its workers spent their hard-earned dollars in the company's mercantile. Store profits paralleled the company's and lessees' level of production.[60]

The company's desire to establish a monopoly in the grocery business motivated its officials to begin to intimidate other mercantile establishments. Using strong-arm tactics, Manager John Deegan threatened to fire employees and to evict renters who traded with non-company businesses. This strategy to enforce company-town rule soon immersed the SRMC in a complex, drawn-out lawsuit over fair competition as well as control over the mine

claims in Santa Rita. Known as the Pinder/Slip contest, the lawsuit pitted independent businessmen against the Santa Rita Mining Company. First claimed in 1881 by Benjamin Harrover and William McBreen, the James Pinder[61] claim came to J. P. Whitney the following year as part of a transaction with Eugene Connor. Like many mining speculators of the day, Whitney hoped to consolidate as many claims as he could to ensure a monopoly over the copper field. Such control, however, was contingent upon carrying out annual assessment work on each of the properties. When the Hearst estate purchased its option on all of Whitney's claims in 1897, the responsibility for meeting the criteria for possessory rights fell to the California business group. If the legal requirements were not met, the Pinder would revert to the public domain.

Ayrus Hamilton discovered in 1898 that the Hearst interests had not completed the required work in the previous year. So he staked the neglected claim, putting out markers on April 16, 1898, and named it the Slip. An ambitious businessman, Hamilton saw the potential for non-mining development at the site. The principal highway ran parallel to Santa Rita Creek and passed through the property. Soon a rail line from Hanover Junction (San Jose) would be laid to Santa Rita. Without fanfare, he built a home there and then opened a saloon by midsummer of that year. Yet, when Hamilton shot and killed Juan Biscarra, after the Mexican man drew a knife on him in the saloon on July 18, 1898, the SRCIC and Hearst group paid little attention to the site even though they had secured a right-of-way across the property a few months earlier. Acquitted by a coroner's jury because of self-defense, Hamilton decided a month later to sell the Pinder/Slip for $500 to John M. Storz. The fresh occupant soon contracted for construction of a new saloon and gave Benno Rosenfeld permission to erect his own general store. Storz then negotiated with the Deming law firm of Baker and Wamel to build a butchershop and another store with a barbershop in it. The collection of buildings soon became known as Storzville, which was no more than a rail stop where the claimant and his associates hoped to profit from traffickers passing from San Jose to Santa Rita.[62]

The Santa Rita Copper & Iron Company first attempted to evict Storz and his friends in 1899 through a lawsuit filed in the Third District Court in Silver City. The company wanted the occupants to abandon the property and to pay $5,000 in damages. But the court dismissed the case a few weeks later. The corporate attempt to monopolize the Santa Rita copper field was foiled for now. In the meantime, Storz began running a gambling operation. The mixture of booze and gaming soon resulted in an unfortunate accidental killing one night when a drunken George Stevenson shot at a man who had ordered him to holster his pistol as he wildly waved it in the air. The act of intimidation cost local bartender Billy Woods his life when the stray bullet entered his head while he was fast asleep in a nearby building. This sordid event, which later included the saga of Stevenson's escape from one jail in 1900 and a foray into Mexico before being recaptured and then himself murdered a year later in another prison, angered the new mineowners in town, the Santa Rita Mining Company. Yet the company did nothing at this time to acquire the Slip. Over the next two years several partners purchased partial interests in

the claim and others established mercantile operations, the G. L. Turner & Son General Merchandise store being the most prominent (see figure 2.14).[63]

When valley-wide floods really slowed mining operations in 1904, the SRMC began to feel the financial pinch and decided to make an attempt to force out the Turners, whom the company viewed as a threat to their own store's profit potential. John Deegan soon started telling his employees that they would be fired if they traded with the Turners, who, later court testimony revealed, offered better prices. In the meantime, the company filed for a patent on the claim, forcing the hand of the current owner, James N. Upton, who acquired the property in 1901. Learning of the company's takeover attempt, Upton filed suit against the SRMC in state court for obstructing his rights to possess the site. He also sued for $5,000 in damages. The following March, the court

FIGURE 2.12. *Santa Rita Store Company, ca. 1915. Built by the Santa Rita Mining Company, this mercantile became the company store for the Chino Copper Company. Company employees used* boletas, *or coupons, printed by Chino to make purchases in the store. A symbol of company authority, the store offered a variety of goods, from coal to cradles to caskets.*

COURTESY OF TERRY HUMBLE.

FIGURE 2.13. *Inside the mercantile, ca. 1912. In this photograph of either the Santa Rita or the Hurley Company Store, patrons pose with the store manager and the clerks. Chino offered diverse items, from canned beans and fresh meat to shovels and jewelry. Note the young girl in the lower right corner.*
COURTESY OF SILVER CITY MUSEUM.

ruled in Upton's favor. The SRMC appealed the case to the New Mexico Supreme Court, which heard arguments in January 1906 and then upheld the lower court's decision. In their anger, SRMC officials continued to intimidate employees who chose to shop at the Turner store. Further enraging them was the fact that they had to enter this very store to do business at the Santa Rita Post Office inside.[64]

The SRMC's boycott backfired. In July 1905, the Turners filed a restraint-of-trade suit against the company once they learned of Deegan's bullying tactics. Asking for $50,000 in compensation, the store owners were pleased that their request for a venue change was granted and the proceedings took place in Socorro. The jury ruled in their favor as well by awarding them $25,000. The company appealed the decision, of course, but eventually lost

again in 1907 in the United States District Court in Las Cruces. The jury once more sided with the store owners, ruling that the mining company was in restraint of trade under the Sherman Anti-Trust Act of 1890. The judge dampened the Turners' celebration, though, when he decided to fine the mining company and its store $1,000 apiece and to rescind the earlier $25,000 award.

Later, Deegan convinced his friend Al Owen to offer to pay $15,000 for the Pinder/Slip. Upton agreed to the price, not knowing that Owen was serving as the company's agent. And even though the SRMC did in 1908 gain possession of the property, the company filed another lawsuit the following year in an attempt to remove the Turners, who were renting the lot from a Mr. Wamel, the newest owner. In the end, the suit was settled in 1910 after the Chino Copper Company had

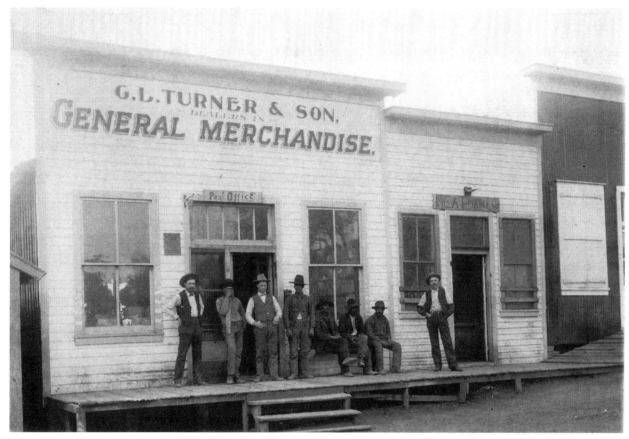

FIGURE 2.14. *Turner & Son General Store in Santa Rita, 1906. George Turner's store sat on the controversial Pinder/Slip mining claim. Prior to Chino's near-monopoly on retail trade, the Turner business served as a store, meat market, and post office. Embroiled in a five-year lawsuit, Turner eventually sold out to Chino and the company moved the building to downtown Santa Rita and converted it into the assay office.* COURTESY OF DON TURNER.

purchased the copper field from the SRMC. Chino paid Wamel for his lot and then temporarily leased it to the Turners, who had to leave by December 1, 1910.[65] The new corporation had won, marking a clear transition to company-town status for Santa Rita. The individualistic claim grabbing of the frontier period had come to a close and now the new boss, the Chino Copper Company, was in town. Eventually, Chino used the claim for a water source known as the Pinder Well. It also became the site of the Pinder shops in the open-pit era.

The final example of the end of frontier mining at Santa Rita involves water and timber management. For more than three decades independent miners and mining companies had acted irresponsibly concerning the environment. This short-sighted, cut-and-run attitude that permeated western American thinking and prac-

tices motivated the frontier industrialists' treatment of the nation's vast forest- and grasslands. Managing resources like water and timber rarely entered into their plans for mining, logging, and grazing except to accumulate these natural resources for profit whenever they could. This laissez-faire worldview celebrated the machine in the garden. In the dry Southwest, miners, loggers, and ranchers exploited the land and its natural gifts as thoroughly as the Mining Law of 1872, their finances, and technology would allow. The results often created obstacles to their expected outcomes: lost profits and unwanted expenses. In the end, their destructive tendencies contributed to the punitive response of the poorly treated environment.[66]

By the early twentieth century, overcutting of timber and overgrazing of grasslands formulated a dangerous recipe for massive flooding. The unbelievable appetites of mining, milling, and smelting operations for wood to fuel steam boilers and to timber deep mines resulted in clear-cutting of vast sections of the Gila wilderness in the Black Range to the north and in the surrounding hills of Santa Rita. The removal of the timber eliminated an important stabilizer of the soils in the dry countryside. Overgrazing further loosened the volatile topsoils of the arid landscape. Mills and mines required thousands of board-feet of cribbing timber and boiler fuel (see figure 2.8). During the first two months alone in 1903, the SRMC contracted for 150,000 running feet of timber, equivalent to 20,000 eight-foot heavy posts. Timber consumption was staggering. More than 200,000 cattle and sheep regularly grazed the same woodlands. Signs of the impending crisis for these excesses arrived as early as the late 1880s in Grant County's seat, Silver City. Without the glue of tree and grass roots to hold the dirt in place, rains began washing the valuable earth right through downtown. Ore dumps and tailings piles also transformed the now vulnerable mountainsides into channels for gulley washers. Under these conditions, floods regularly began eroding away Main Street so that by 1902 nothing remained of it. Known as the "Big Ditch," the giant arroyo funneled away rich ores, mine equipment, valuable livestock, commercial buildings, and town lots with monotonous regularity. By 1910, the mining community had built water breaks and gated dams to deal with their sins rather than admonish loggers and grazers to reduce their destructive practices. They also tried to lay blame for the floods at the feet of Hispanic sheepherders, easy scapegoats for the entire community's abuses. Not until the US Forest Service set aside the Gila National Forest and then enforced new timber-cutting and cattle-grazing policies in the 1910s and 1920s did conservation take root and remediation begin.[67]

Santa Rita experienced a similar outcome. Floods took out rail beds in 1891, 1895, and 1896 on the Hanover line and then in 1902 and 1904 along the Santa Rita spur. The two most destructive examples occurred during these latter two years. In both years, freighting of ores came to a standstill for months after floods took out large sections of the newly installed Santa Rita Railway. In 1902, a massive torrent washed half of Ernest's Roadhouse down the usually trickling Whitewater Draw. Sam Allen's orchard went with it. The San Jose rail bridge and about three miles of the line washed out between Whitewater

and Hudson just south of Santa Rita. Superintendent Thayer reported on August 30 to his New York bosses: "I have never seen such a volume of water go through this canyon, it washed away the entire dump of the Montoya and filled the deep Montoya shaft to its collar. For a time it threatened the Aztec shaft, and many of the homes of the people were threatened. I took the entire crew from the mines and sent them to their homes to look after their families."[68]

The most damaging flood raged through Santa Rita in 1904. A ten-day downpour in October cost the mining community tens of thousands of dollars. Witnesses recalled seeing a fifteen-foot-high wall of mud slurry come crashing down Santa Rita Canyon, destroying everything in its path. Ore dumps, mine machinery, workshops, iron rails, boxcars, grazing cattle, and company homes disappeared in the mucky mess. Sadly, Santa Ritan Guadalupe Gonzales drowned while saving her children, whom she managed to pull to safety before fatigue loosened her tenuous grip on a tree branch in the raging floodwaters. She was later found a mile downstream. Ernest's Roadhouse was gone for good this time. A mile and a half of the Santa Rita spur washed out too. The costs were staggering. Already having to suffer through three weeks of no rail line in August because of an earlier flood, the mining community lost an additional five months of service after the October deluge. Despite a rise in copper prices in late 1904, the SRMC and its lessees could not take advantage of the increase. In early November, John Deegan considered freighting ores by wagon, but at $2.50 per ton the expense was prohibitive. In addition, other costs mounted as a result of the flood. Dewatering and re-timbering the inundated shafts required more wood and coal purchases to fuel the steam-driven duplex pumps and sawmill operations. Ore and supply transport by wagon from Hanover rather than the cheaper rail service further strained the miners' pocketbooks. Successive winter snowstorms over the next five months further exacerbated their misery and frustration and brought still more wet conditions. Their frontier sins had come home to roost.[69]

Though washes would occasionally revisit Santa Rita and elsewhere in Grant County, they would never top the 1904 flood. Old-timers fifty years later would recall the devastating torrent that probably influenced the Santa Rita Mining Company's decision to sell its vast copper properties to the Chino Copper Company. Ironically, that sale in 1909 marked several interesting shifts in the mining history of Santa Rita. Chino would implement conservation practices in its business strategies that would ameliorate the frontier abuses. Like many other corporations in America in the Progressive era, they began to carefully manage water, timber, coal, and other natural resources. Chino's construction of a massive water-diversion project serves as one example of the concerted effort and financial investment in controlling the precious resource water. The company also eliminated the industry's dependency on timber by replacing this fuel source with coal. The federal government facilitated this transition to conservation of natural resources as well with more vigilant management of logging and grazing in the nearby Gila National Forest. Conditions greatly improved and uncontrollable floods rarely revisited the area.

Other factors also meant an end to the frontier mining period. The corporation created a company town. Miners were recruited and then expected to abide by company rule or be escorted out of town. Union men were blacklisted. The company was *the* law. Free enterprise was replaced by the company store and a monopoly on trade. On the mining front, iffy investments like those of the Santa Rita Copper & Iron and Santa Rita Mining companies were replaced with a stable corporation, the Chino Copper Company. Its vast new fiscal resources soon translated into economies-of-scale investment in new steam shovels, giant crushers, flotation milling, and haulage machinery. The porphyry copper era was at hand.

Likewise, there was no longer an ongoing contest over the copper territory. The Apaches had been displaced a full generation earlier. The era of small-time mine operators at Santa Rita would soon be a distant memory. The frontier had come to an end. The arrival of the Chino Copper Company would guarantee the development of a substantial community at Santa Rita. The introduction of open-pit mining ensured that probability for the first time in the copper camp's history. Ironically, the digging of the massive pit would also be the beginning of the end of that same town. The rise of corporate mining and the company town and then their eventual demise are core themes of the next two chapters.

NOTES

1. Robert L. Spude, "The Santa Rita del Cobre, New Mexico, The Early American Period, 1846–1886," *Mining History Journal* (1996), 10; Morris E. Opler, "Chiricahua Apaches," in Alfonso Ortiz, ed., *Handbook of North American Indians*, vol. 10 (Washington, DC: Smithsonian Institution, 1983), 399–421; Edwin R. Sweeney, *Mangas Coloradas: Chief of the Chiricahua Apaches* (Norman: University of Oklahoma Press, 1998), 2–11, 66–81.

2. Kearney agreed to peace with Mangas Coloradas, who hoped the Americans would defeat the Mexicans in the war. And although the Americans would not abide by the peace treaty, Mangas Coloradas welcomed Kearney and his soldiers because he hated the Mexicans and wanted them to be defeated. One Apache chief, perhaps Mangas Coloradas himself, was alleged to have said to Kearney: "You have taken New Mexico and will soon take California, go then and take Chihuahua, Durango, and Sonora. We will help you . . . The Mexicans are rascals; we hate and kill all of them" (Edwin R. Sweeney, "Mangas Coloradas," 132).

3. Ross Calvin, ed., *Lieutenant Emory Reports: A Reprint of Lieutenant W. H. Emory's Notes of a Military Reconnaissance* (Albuquerque: University of New Mexico Press, 1951; reprint of 1848 edition), 97–99.

4. James M. Cutts, *The Conquest of California and New Mexico* (Albuquerque, NM: Horn & Wallace, 1965), 183–184.

5. Spude, "The Santa Rita," 10–11.

6. A. B. Bender, "Frontier Defense in the Territory of New Mexico, 1846–1853," *New Mexico Historical Review* 9 (1934): 255–258; Spude, "The Santa Rita," 11.

7. John Russell Bartlett, *Personal Narrative of Explorations and Incidents in Texas, New Mexico, California, Sonora and Chihuahua, Connected with the United States and Mexican Boundary Commission, During the Years 1850, '51, '52, and '53* (New York: D. Appleton, 1854), 225–232; Spude, "The Santa Rita," 11. Bartlett admired Mangas Coloradas, writing that he was a leader of "strong common sense and discriminating judgement," and in his communications may have convinced the Apache chief that he and his men would remain peaceful and had no plans to reopen the mines; see Sweeney, "Mangas Coloradas," 155.

8. Ironically, Bartlett and his counterpart, General Pedro Garcia Condé, originally made an error in drawing the border between Mexico and the United States, resulting in a later need, in large part because of southern Americans' desire for a southern railroad route to the Pacific Ocean, to initiate the Gadsden Purchase of 1854. To assuage tensions with the Mimbres Apaches, Bartlett also recounted how he had given the tribe beads, cotton cloth, and shirts. Mangas Coloradas received a special gift of blue broadcloth, a frock coat lined with scarlet and gilt buttons, a pair of pants with a red stripe, a pair of shoes, a white shirt, and a red silk sash. Lieutenant Colonel Craig also gave him an officer's uniform with epaulettes. Clearly, the Americans understood the importance of tribute gifts, a tradition first used by the Spanish in their relations with the Apaches.

9. See Neal W. Ackerly, *A History of the Kneeling Nun: Evidence from Documentary Sources* (Silver City, NM: Dos Rios Con-

sultants, 1999); Ross Calvin, ed., *Lieutenant Emory Reports*, 88, 97–98; *Mining Life*, September 20, 1873; Rossiter W. Raymond, *Statistics of Mines and Mining in the States and Territories West of the Rocky Mountains* (Washington, DC: Government Printing Office, 1874), 337; *Weekly New Mexican*, January 30, 1877; *Grant County Herald*, May 28, 1881; *Silver City Enterprise*, June 26, 1885, May 6, 1887; Paul I. Wellman, "Above a Great Copper Mine the Kneeling Nun Still Prays for Mercy," *Kansas City Star*, April 18, 1937; Kennecott Copper Corporation, *Chino World*, August 16, 1976; Ricardo Muñoz, *The Kneeling Nun: My Eternal Mother* (Las Cruces, NM: Del Valle Printing & Graphics, 1983); and Vic Topmiller Jr., "Making the Kneeling Nun a National Monument in Name of Mother Teresa" (unpublished flyer, September 12, 1997) are just a few of the sources of information on the Kneeling Nun.

10. Undoubtedly, the Apaches understood the value of the copper mines by this time and may have already planned to use them for leverage in treaty, trade, and other negotiations.

11. James A. Bennett, *Forts and Forays: A Dragoon in New Mexico* (Albuquerque: University of New Mexico Press, 1948), 34–35.

12. Spude, "The Santa Rita," 11–12; W. Turrentine Jackson, *Wagon Roads West: A Study of Federal Road Surveys and Construction in the Trans-Mississippi West, 1846–1869* (New Haven, CT: Yale University Press, 1964), 35–44, 110–112, 215–240; Dan L. Thrapp, *Victorio and the Mimbres Apaches* (Norman: University of Oklahoma Press, 1974), 44–46; also see Kathleen P. Chamberlain, *Victorio: Apache Warrior and Chief* (Norman: University of Oklahoma Press, 2007).

13. Sweeney, "Mangas Coloradas," 156.

14. Sweet was actually Canadian and LaCoste French, but they had lived in Texas long enough to be considered Texans.

15. Spude, "The Santa Rita," 13.

16. Ibid., 14–16, 8.

17. *Eighth Census of the United States, 1860*, Doña Ana County, New Mexico, microfilm copies at Silver City Public Library, Silver City, New Mexico; Spude, "The Santa Rita," 15.

18. See Duane A. Smith, *Rocky Mountain Mining Camps: The Urban Frontier* (Lincoln: University of Nebraska Press, 1967); Paula Petrik, *No Step Backward: Women and Family on the Rocky Mountain Mining Frontier* (Helena: Montana Historical Society Press, 1987); and Liping Zhu, *A Chinaman's Chance: The Chinese on the Rocky Mountain Mining Frontier* (Niwot: University Press of Colorado, 2000), for comparisons of frontier mining life.

19. Sylvester Mowry, *Arizona and Sonora: The Geography, History, and Resources of the Silver Region of North America* (New York: Harper & Bros., 1866), 22–25.

20. Quoted from Spude, "The Santa Rita," 17, italics in the original.

21. Seven months later, in February 1863, the Union created Arizona Territory, changing the borders of the Confederate Territory of Arizona from an east-west boundary to a north-south border as it exists today.

22. This is a partial quote from Sweeney, "Mangas Coloradas," 158; also see pages 155–159.

23. Spude, "The Santa Rita," 18; also see Darlis Miller, *The California Column in New Mexico* (Albuquerque: University of New Mexico Press, 1982).

24. Quoted from Spude, "The Santa Rita," 19.

25. Spude, "The Santa Rita," 20; see Conrad Keller Naegle, "The History of Silver City, New Mexico, 1870–1886" (master's thesis, Western New Mexico University, 1943), for a discussion of the Grant County secession movement in the 1860s.

26. Spude, "The Santa Rita," 20–21; *Santa Fe Weekly Gazette*, December 8, 1866.

27. Quoted from Spude, "The Santa Rita," 21.

28. Spude, "The Santa Rita," 22–26; Hayes did not pay the Elguea heirs until 1880 or 1881 after a trip to Spain had failed to turn up a Spanish land grant for the Santa Rita del Cobre. To avoid legal complications he paid the Elgueas to gain full title to the property by 1881.

29. Chino, which is the Spanish word for "Chinese," was used as a slang term here, meaning iron pyrite or fool's gold. There is no record of a Chinese person's effort to claim this mine; see Spude, "The Santa Rita," 22.

30. *Grant County Herald*, May 23, June 20, August 29, September 26, November 28, 1875, February 17, 1877; John M. Sully, "Chino Copper Company: The Story of the Santa Rita del Cobre Grant and Its Development from Discovery to the Present Date" (Reprint from Official State Book for Distribution at Panama-California Exposition at San Diego, 1915), 150–153; Spude, "The Santa Rita," 24–26; *Grant County Herald*, May 23, 1875.

31. Grant County Deed Book 7, 250–254, Grant County Administration Building, Silver City, New Mexico; Richard Miller, *Fortune Built by Gun: The Joel Parker Whitney Story* (Walnut Grove, CA: Mansion Pub., 1969), xvii; *New Southwest*, June 4, June 28, August 13, 1881; also see Melissa D. Burrage, "Albert Cameron Burrage: An Allegiance to Boston's Elite, 1859–1931" (master's thesis, Harvard University, 2004), 240; *Silver City Enterprise*, July 23, 1880; *Grant County Herald*, March 29, 1879; Spude, "The Santa Rita," 26; Chamberlain, *Victorio*, 205–207.

32. Spude, "The Santa Rita," 26; *Southwest Sentinel*, August 8, 1883; also see J. P. Whitney, *The Santa Rita Native Copper Mines in Grant County, New Mexico* (Boston: Alfred Mudge & Son, 1884). After his resignation from the Santa Rita Copper & Iron Company, John Slawson was killed by Apaches near Clifton, Arizona Territory, as was the company attorney, John P. Risque. John Magruder was the only survivor of the attack; see *New Southwest*, November 5, 1881, April 29, May 6, 1882.

33. Miller, *Fortune Built by Gun*, 104–107; Grant County Deed Book 8, 90, 765–769; Spude, "The Santa Rita," 27–29; *New Southwest*, September 10, November 26, 1881, March 11,

1882. For a discussion of how stamp mills worked, see Duane A. Smith, "How It Worked: The Stamp Mill," *Montana: The Magazine of Western History* 58:1 (Spring 2008): 63–65.

34. Spude, "The Santa Rita," 29–30; Huggard, "Environmental and Economic," 44; *New Southwest*, April 8 and October 28, 1882; *Silver City Enterprise*, December 28, 1882, March 23, 1883; Whitney, *The Santa Rita*, 5–9, 15, 22–25.

35. Grant County Commissioner's Record Book 1, 157, Grant County Court House, Silver City, New Mexico.

36. Spude, "The Santa Rita," 29; *New Southwest*, March 4, August 5, 26, October 14, December 23, 1882; *Silver City Enterprise*, November 16, 1882, February 22, 1883; Grant County Commissioner's Book 1, 226, Grant County Court House, Silver City, New Mexico; *Silver City Enterprise*, May 11, 16, 1883.

37. Arthur F. Wendt, "The Copper Ores of the Southwest," *Transactions of the American Institute of Mining Engineers* (New York: American Institute of Mining Engineers, 1887), 25–77.

38. *Silver City Enterprise*, September 14, 1883; *Southwest Sentinel*, April 7, 14, and August 4, 1883.

39. *Silver City Enterprise*, July 10, 1891, October 4, 1895, March 12, April 27, September 21, November 19, December 17, 24, 1897, January 7, 28, February 11, July 22, October 28, 1898; *Silver City Independent*, December 14, 1897, January 17, February 28, March 10, May 16, 1899; *Silver City Enterprise*, January 27, February 3, 10, 17, March 10, 31, April 14, May 19, 1899.

40. *Silver City Eagle*, July 10, 1895; *Silver City Enterprise*, July 12, 1895, February 21, April 3, 1896, January 27, April 28, May 5, 16, 1899; Huggard, "Environmental and Economic," 98; *Mineral Resources of the U.S.* (1906), 304; *Silver City Enterprise*, February 14, 1902.

41. *Silver City Enterprise*, October 1, 15, 1897, November 19, December 31, 1897, April 8, May 13, 1898, January 13, 1900; Miller, *Fortune Built by Gun*, 111; see Michael P. Malone, "The Close of the Copper Century," *Montana: The Magazine of Western History* 35 (Spring 1985): 69–72.

42. T. A. Rickard, "The Chino Enterprise—II: Later History and Formation of Present Company," *Engineering and Mining Journal Press* (November 10, 1923): 803; Arthur B. Parsons, *The Porphyry Coppers* (New York: Arthur B. Parsons, 1933), 207–209; *Mineral Resources of the United States* (1906); Spude, "The Santa Rita," 30; *Silver City Independent*, May 16, 30, July 4, 1899; *Silver City Enterprise*, March 17, May 12, 19, June 9, 30, 1899; also see Burrage, "Albert Cameron Burrage," 128–220, 250–309; and F. Ernest Richter, "The Amalgamated Copper Company: A Closed Chapter in Corporate Finance," *The Quarterly Journal of Economics* 30:2 (February 1916): 387–407, who argues that the Amalgamated Copper Company made a play to pool all the major copper properties in the United States to monopolize the industry. These authors also show how A. C. Burrage, the Rockefellers, and H. H. Rogers involved their family members in the copper industry in Michigan, Montana, New Mexico, Chile, and elsewhere. Urban Broughton, president of the Santa Rita Mining Company (SRMC), for example, was H. H. Roger's son-in-law; William H. Burrage, manager of the SRMC, was A. C. Burrage's first cousin; and William G. Rockefeller, SRMC financier, was William A. Rockefeller's son.

43. Spude, "The Santa Rita," 30–33, contends that Whitney failed because he invested in technologies that were not effective enough to generate a profit, especially since the Santa Rita ores were so complex and knowledge of how best to treat them was limited.

44. Santa Rita Mining Company, Weekly Report No. 185, December 19, 1903 (John Deegan to SRMC president Urban H. Broughton in New York); *Silver City Enterprise*, October 2, December 11, 18, 1903; *Silver City Independent*, October 6, 17, 1903.

45. Santa Rita Mining Company, Weekly Report No. 71, October 12, 1901, No. 80, December 14, 1901.

46. Santa Rita Mining Company, Weekly Report No. 31, January 5, 1901; *Silver City Enterprise*, March 15, May 3, 1901; *Silver City Independent*, April 2, June 18, October 22, November 19, 1901; Santa Rita Mining Company, Weekly Reports No. 52, June 1, 1901, No. 135, January 3, 1903, No. 157, June 6, 1903; *Silver City Independent*, December 8, 1903; Santa Rita Mining Company Smelter Returns for October 15, December 26, 1906; *Silver City Enterprise*, June 21, 1912.

47. Santa Rita Smelter Returns, 1900–1909; Chino Copper Company Smelter Returns, 1909–1911.

48. Santa Rita Mining Company, Weekly Report No. 155, May 23, 1903; *Silver City Enterprise*, January 29, April 8, 22, 1904; *Silver City Independent*, December 13, 1904.

49. Santa Rita Mining Company, Weekly Reports No. 92, March 8, No. 94, March 22, No. 96, April 5, 1902.

50. *Twelfth US Census*, Grant County, New Mexico, Precinct 13, Santa Rita.

51. *Silver City Enterprise*, July 5, 1901; *Silver City Independent*, July 9, 1901; Santa Rita Mining Company, Weekly Report No. 93, March 15, 1902.

52. *Silver City Enterprise*, July 10, 1903; *Silver City Independent*, July 7, 14, 21, 1903, July 9, 30, 1907.

53. *Silver City Enterprise*, February 3, June 23, 1899; Grant County Probate Minutes, Book 5, 360, 541, 590, and Book 6, 137, 296; Santa Rita Mining Company, Weekly Reports No. 132, December 13, 1902, No. 146, March 21, No. 147, March 28, 1903; *Silver City Enterprise*, July 6, 1900, May 24, 1901, December 12, 1902, March 27, 1903, August 23, 1907; *Silver City Independent*, July 29, December 2, 23, 1902, March 24, 1903, August 27, 1907.

54. *Silver City Enterprise*, May 24, 28, 1901; Grant County Probate Minutes, Book 5, 613.

55. Santa Rita Mining Company, Weekly Reports No. 131, December 6, 1902, No. 238, December 24, 1904; *Silver City Independent*, December 2, 1902, December 27, 1904, August 27, 1907; *Silver City Enterprise*, December 12, 1902, January 15, 1904,

August 23, 1907; C. P. Crawford Biographical File, Silver City Public Library, Silver City, New Mexico; and correspondence between Terry Humble and Hugh L. Kelley, December 1991.

56. Santa Rita Mining Company, Weekly Reports No. 6, July 14, No. 13, September 1, No. 22, November 3, No. 28, December 15, 1900; *Silver City Enterprise*, August 18, 1899, February 23, April 13, June 1, June 15, July 27, November 30, 1900, April 5, May 3, 1901, December 21, 1906, September 13, 1907; *Silver City Independent*, November 14, 1899, March 6, June 15, November 13, 1900, February 5, November 19, 1901, December 18, 1906, September 17, 1907.

57. *Silver City Enterprise*, October 5, 1900, November 13, 1908; Santa Rita Mining Company, Weekly Report No. 17, September 29, 1900.

58. Santa Rita Mining Company, Weekly Reports No. 7, July 21, No. 13, September 1, No. 22, November 3, 1900; *Silver City Enterprise*, October 5, November 30, 1900, January 18, 1901, May 31, 1902, March 3, 1903; *Silver City Independent*, September 13, October 18, November 1, December 13, 1898, January 17, October 3, 1899, August 19, 1902.

59. *Silver City Independent*, July 16, 1907, September 15, 1908; *Silver City Enterprise*, May 29, July 24, October 2, 1908.

60. Santa Rita Mining Company, Weekly Reports No. 5, July 7, No. 6, July 14, 1900; *Silver City Enterprise*, April 13, July 27, September 21, December 7, 1900; *Silver City Independent*, March 6, June 16, July 17, August 7, 1900.

61. When Harrover and McBreen first staked the claim they named it the Jim Pender. They soon renamed it the James Pender only to finally agree on the James Pinder in September 1881; see Terrence M. Humble, "The Pinder-Slip Mining Claim Dispute of Santa Rita, New Mexico, 1881–1912," *Mining History Journal* (1996): 93.

62. Humble, "The Pinder-Slip," 94; Huggard, "Environmental and Economic Change," 48; Grant County Mining Deeds, Book 36, May 19, 1899, 568–569, May 20, 1899, 573–574, Book 40, September 30, 1902, 434; *Silver City Enterprise*, May 26, 1899; *Silver City Independent*, May 16, 23, 1899; Grant County Proof of Labor Deeds, Book 36, February 2, 1899, 486–487.

63. Humble, "The Pinder-Slip," 94–95.

64. Ibid., 96.

65. Ibid., 98–99; *George L. Turner & Son vs. Santa Rita Mining Company and Santa Rita Store Company*, Civil Docket Book F, Case no. 3997, 509, Santa Fe Archives; *Silver City Independent*, December 25, 1905.

66. See Christopher J. Huggard, "The Impact of Mining on the Environment of Grant County, New Mexico, to 1910," *Annual of the Mining History Association* (1994): 2–8.

67. Huggard, "The Impact," 4–5.

68. *Silver City Independent*, September 2, 1901; *Silver City Enterprise*, July 3, 1891, July 12, August 9, 1895, July 24, August 7, 28, 1896, August 29, 1902, October 14, 1904; Santa Rita Mining Company, Weekly Reports No. 117, August 30, 1902, No. 136, January 10, No. 142, February 21, 1903.

69. Santa Rita Mining Company, Weekly Reports No. 227, October 8, No. 228, October 15, No. 232, November 12, 1904, No. 246, February 18, 1905; *Silver City Independent*, October 11, 1904; *Silver City Enterprise*, May 6, October 14, 1904; Santa Rita Mining Company, *Annual Report* (1904), 16.

III

The Chino Years

THE OPEN PIT, THE MEN, AND THEIR METHODS

The Chino Copper Company initiated open-pit mining on September 23, 1910. That day the new corporation's Marion 91 steam shovel Number 2 scooped its first bucket of low-grade porphyry copper (see figure 3.1). Thus began the stairstep descent in the quest to remove hundreds of millions of tons of earth peppered with minute flakes of copper deep in the bowels of the mountainous terrain. A new era had dawned at Santa Rita. Using the latest

economies-of-scale technology, the newly incorporated mining company joined the international open-pit copper family that would dominate the industry throughout the twentieth century. No longer would underground high-grade mining feed America's frenzied demand for the red metal. Rather, aboveground, low-grade digging would characterize the extraction of the most popular base metal. Chino was part of the growing American domination of the industry that by 1916 led the world with nearly 60 percent of overall copper production. This exploding North American supply provided copper for the expanding military arsenals of the globe, and the highly conductible metal played a key role in the rise of consumer demands in the 1910s and later for electricity, automobiles, and appliances.[1] This worldwide need,

combined with the application of new open-pit technology, changed Santa Rita's dormant potential from one of little hope to one with the promise of a prosperous future.

Soon five steam-shovel crews of nine men each dotted the hilly horizon below Romero Hill (see figure 3.2). Locomotive engineers inched their ember-belching engines along so ore cars could accept each new bucket of precious rock. The billowing black smoke from the modern machinery's boilers and engines translated into exponential earnings for the nascent copper corporation. Within four short years the enterprise employed 2,000 workers and produced more refined copper than in the previous 100 years, despite the incredibly low grade ores of about 2 percent purity. Company executives

FIGURE 3.1. *Marion Model 91, No. 1 steam shovel loads the ore cars on locomotive No. 1 below the Romero workings, 1910.*

COURTESY OF PALACE OF THE GOVERNORS PHOTO
ARCHIVES (NMHM/DCA), NEG. NO. 047639.

witnessed the precipitous rise of their opening-day five-dollar stock of 1909 to sixty-two dollars by 1917. Profits soared to nearly $10 million in this latter year as well. The Chino Mine, as it became known, soon ranked as the fifth largest copper producer in United States and the seventh most prolific pit internationally, positions that would only improve with time.[2] This assured productivity made Chino a prize property that would be purchased successively by Guggenheim-controlled companies Ray Consolidated Copper Company in 1924 and then Nevada Consolidated in 1926. The parent firm, the Kennecott Copper Corporation, finally acquired the mine in 1933, making it a key sibling in its expanding open-pit family.

Chino's founding superintendent John Murchison Sully deserves much of the credit for the explosive rise of the company to preeminence in copper-mining history (see figure 3.3). Like many of the prominent engineers of the American West, Sully was from New England. Born in 1863 in Dedham, Massachusetts, he earned a mining engineering degree in 1888 from MIT. He began his career in Tennessee for the Chickamauga Coal and Iron Company, which he worked for in the 1890s both

as an engineer and as superintendent. His ambitions led him in about 1902 to the Hite's Cove Mine in Mariposa, California, and then soon after that to Anniston, Alabama. There he served as general manager of the Woodstock Iron Works, a stint he later claimed introduced him to steam-shovel mining techniques. While in California he became friends with General Electric mining engineer Donald Palmer. Palmer lured the up-and-coming professional in 1904 to work for the Hermosa Copper Company at its Ivanhoe Mine and other GE properties in Grant County, New Mexico Territory. In search of a steady flow of copper for GE's appliance manufactures, Hermosa directed Sully in his first year to complete an extensive examination of the Santa Rita Mining Company's nearby property. He was charged with evaluating whether to purchase the foundering operation, which despite some successes was at a crossroads because of complex and declining grades of ores.[3]

Sometime in late 1905, Hermosa superintendent John W. Bible received a telegram from GE's mining head, Denis Riordan, instructing him to enlist Sully to investigate the Santa Rita copper deposits. From

ROMERO 6310 BENCH
Dec. 17-10

FIGURE 3.2. *Four steam shovels in operation below the Romero works on December 17, 1910. Note the track gang (left center).*
COURTESY OF TERRY HUMBLE.

December 1905 to September 1906 Sully supervised several churn drill crews that sank sample holes to chart the ore bodies of the property (see figure 3.4). Establishing map coordinates, called "Chino coordinates" to this day, along east-west and north-south axes at 100-foot intervals, Sully chose the Chino claim as his center point of reference, thus explaining the soon-to-be name of the open pit. He and his crews took 4,300 samples of ore to be assayed for purity and as measures of "overburden," or waste rock, to determine removal costs. Sully took samples every three feet, and "records of the materials were carefully kept," as he wrote in a revised version of the report. "When the facts collected in prospecting and sampling the ore bodies was [sic] supplemented by geologic and general mining cost data, the whole body of information systematically gathered was used in estimating the tonnage of rock rich enough to be classified as ore."[4]

Like many engineers of the day, Sully took a scientific approach in calculating the potential values of the porphyry deposits. Technological efficiency, cost-effective management, low wages, and resource conservation characterized this emerging Progressive-era worldview. Even in collecting the numerous samples, the engineer had hired his churn drill crews with an eye to efficiency concerning costs and time. Spanish-speaking, technologically savvy Anglo drillers, for example, were assisted by three "Mexican" laborers on each of the three drill crews in operation. Mechanically knowledgeable bilingual "bosses" brought more efficient results, using cheap labor for most of the hard work. Though his superiors at GE would cringe at the overall cost of the sampling process, this "scientific," if unfair, labor system would characterize the Chino operations for much of its history.[5]

Before taking the reins at Chino, though, Sully still had to be convinced of the efficacy of moving forward

FIGURE 3.3. *John Murchison Sully, ca. 1910. Sully, who served as general manager of Chino from 1909 until his death in 1933, was known as the "father" of the company town of Santa Rita.*
COURTESY OF JOHN M. SULLY II.

with plans to develop the New Mexico property. His mining engineering training instilled in him a need to focus on the task at hand. So he worked in 1906 with Hermosa mill superintendent N. C. Titus to test the porphyry samples at a 100-ton pilot mill. The experiment proved quite disappointing initially, with only 47 percent recovery of the copper. A quick adjustment at the mill with the addition of another jig, however, soon showed a recovery of about 70 percent. Sully began to realize the Chino ores were more complex than others, such as those at Bingham Canyon, Utah, but did not let the relatively small difference in the returns discourage his effort. Regardless, he believed that at least 9 million tons of 2.5 percent ore sat beneath the ground in the Santa Rita hills. Using a conservative estimate based on mining engineering calculations then in vogue, Sully also predicted handsome profits of about $17 million, especially if the

company implemented "mass mining" using steam shovels to dig a giant open pit.

He also estimated a twenty-eight-year lifespan for the proposed pit based on production from a 1,000-ton-a-day mill. Earlier speculators like José Manuel Carrasco, Matt Hayes, and J. P. Whitney would have been flabbergasted at the young engineer's calculations. Still, General Electric balked at the opportunity. A distraught Sully resigned in November 1906 only a month after turning in the report. Later, Sully wrote of his vindication in a 1915 publication he prepared for the Panama-California international exposition: "It was left for an unknown engineer, by chance assigned to this search for truth [about the Santa Rita potential] to prove, by painstaking detail work, investigation and careful study of the correlated facts . . . [that] this deposit was valuable beyond the wildest dreams of all who had passed before him."[6]

Soon after his resignation from Hermosa in late 1906, Sully ventured to Mexico in search of new prospects. While there, he made acquaintance with Standard Oil attorney W. D. Pearce, who was instrumental in financing the Chihuahua al Pacifico Railroad in Mexico and who found Sully's pitch for the Chino property convincing. Pearce encouraged Sully to pursue his aspirations for developing the property. Soon after returning to the United States in April 1907, Sully requested permission from GE's mining chief, Riordan, to distribute the report. Riordan agreed but instructed Sully to contact the Santa Rita Mining Company (SRMC) as well for consent. Several months later he wrote John Deegan, superintendent at Santa Rita, to request that he contact the company president, Albert C. Burrage (known as "A.C."), and seek authorization to circulate the report. Deegan soon responded, informing Sully that vice president Urban H. Broughton agreed to share the report and that Burrage, who by now knew of the prospectus, had sent engineer Walter H. Wiley of Colorado to evaluate Sully's conclusions. Seeing an opportunity to tout his findings, Sully headed to Santa Rita in May 1908 to meet with Wiley to discuss possibilities. Wiley gave the affirmation that Sully hoped for and then he contacted Burrage with his positive outlook. Soon afterward, perhaps in the late summer or fall, Burrage invited Sully to Boston to discuss his proposition.[7]

Agreeing to $25 a day plus expenses, Sully took the rail trip from Santa Rita to Boston, not yet realizing that the company president also had hopes for open-pit mining

FIGURE 3.4. *Chino churn drill and crew, ca. 1910. A crew of five drillers and their helpers pose next to a steam-driven Cyclone model churn drill. Sully used samples from an earlier crew to predict the size of the ore body to promote the property back East. Engineers used the samples the drill crews generated to determine blasting and mining strategies for the operations.*

COURTESY OF TERRY HUMBLE.

at Santa Rita. Burrage had just returned from Europe, where he had been studying copper-mining technology, geology, and metallurgy. To facilitate his understanding of copper he had constructed an experimental village in southeastern Massachusetts several years earlier to try out unproven steam-shovel technology as well as complex low-grade ore-milling techniques in an effort to learn how to deal with the exigencies of the open-pit phenomenon. Burrage's adventuresome nature had led him to invest in copper mines in the Upper Peninsula of Michigan and elsewhere in the American West. He was instrumental at the turn of the century in attempting to monopolize the copper industry as a key member of the Rockefeller Syndicate's Amalgamated Copper Company. His aggressive approach eventually inspired him in the 1910s to incorporate the Chile Copper Company, which developed the prolific Chuquicamata Mine. When he sold the Chile property for $25 million in 1912, he was able to recover from major losses in some of his previous copper-mining investments. Five years earlier he already had big plans for the Santa Rita properties.[8]

On his return to the United States in late 1907 he learned of Sully's report. He soon procured a copy from Charles A. Coffin, president of General Electric Company, explaining why Wiley was already in New Mexico when Deegan contacted Sully in May 1908. Burrage "was so impressed by what he read [in Sully's 1906 report]," T. A. Rickard tells us, that he chose to ignore engineer C. C. Burger's earlier negative accounting. At their initial meeting in Boston, Burrage "cross-examined" Sully on the "smelting and marketing [of] copper . . . Sully's weak point[s]."[9] Probably satisfied more with Sully's predictions for Santa Rita than with his answers on metallurgy, the financier turned self-taught engineer introduced the young professional to Thomas Lawson, another partner in the SRMC. Lawson showed a "keen interest" in Santa Rita's prospects, and they all three agreed that Sully should return to New Mexico and update his now two-year-old report.[10]

On his return to New Mexico Territory in June 1908, Sully immediately began additional assessment work on the Santa Rita property. During the course of taking 6,000 fresh ore samples, his new acquaintance, A. C. Burrage, stopped by on July 7 to see how things were going. They were developing an important business friendship that would eventually pay off. "A lawyer by profession, a capitalist by circumstance, a keen man of business, with an intensely active mind,"[11] Burrage became "impressed with the truthfulness of [Sully's] findings."[12] In his updated report, Sully would raise his ore body estimate to 12 million tons at 1.63 percent copper and $18.5 million in possible profit, a figure that would be more than tripled in the first ten years. Three months later the financier telegraphed Sully, instructing him to return to Boston. The two put together an abstract and took it to the well-known New York financial firm of Hayden, Stone & Company, which had ties to the Utah Copper and Ray, Arizona, porphyry copper properties. Wiley soon joined the two advocates, who met with Galen Stone and Charles Hayden to make their pitch for financial backing.

Burrage and Sully's strategy resulted in the introduction of another engineer, A. Chester Beatty, who, already in the West, was instructed to take the rail to Santa Rita. Soon Sully, Wiley, and Beatty met at the properties to discuss the values in the ores both from the collected samples and from new assays of some of the dumps. After two days "on the ground," Beatty and Sully returned east to meet with Burrage and their potential investors. In what would be one of several twists to the story, described by Sully as "one full of pathos and human interest," Beatty proved uncooperative, to the "disgust" of Burrage and his associate Thomas Lawson, who was now on board. Hayden and Stone "declined to underwrite the bonds. The business was near a fiasco." Burrage was perturbed because his option on the property would soon play out. But he refused to give in. Convinced of the efficacy of mass mining, he sent Sully to Bingham Canyon, Utah, to examine Daniel C. Jackling's efforts at steam-shovel techniques there. The Bostonian then directed him after that to assess some coal properties near Socorro, New Mexico, anticipating the fuel needs of the shovels and locomotives necessary for the economies-of-scale strategy. Sully then traveled back to Santa Rita by November to continue his sampling work.[13]

By early 1909 Sully and SRMC superintendent Deegan were testing the ores "with no particular success." Burrage still persisted, ordering Deegan to send him $30,000 from the corporate coffers, most of which came from the profits of the Santa Rita Company Store, to finance the promotion of the property. He also offered Sully the superintendency of the SRMC, which the younger man refused, only willing to accept in February the title of "superintendent engineer in charge." In the meantime, Lawson had used the funds to produce a prospectus of the Santa Rita deposits and then distributed 65,000 copies in hopes of enticing stockholders. His effort flopped and the meager funds collected were "returned to the few simpletons that had been caught by his bait." Plans to float the proposal to European investors were abandoned. Burrage persevered and was able to enlist the support of his longtime friend and financier H. H. Rogers and other members of the Rockefeller Syndicate, who with the Bostonian had established the Amalgamated Copper Mining Company and who still held 50,000 of the 75,000 total shares in the SRMC. Burrage and Lawson controlled the other third. Sully continued to try to convince Beatty and Hayden & Stone to take a risk on Santa Rita.[14]

Nothing changed until about May 17, when Burrage informed Sully while they were in New York City that Rogers had agreed to underwrite the project. At the last moment, though, Rogers postponed the signing of the deal because of the opposition of his business partners—Rockefeller, Lewisohn, and Broughton, in particular—who had been instrumental in the formation of the SRMC and who no longer believed the Santa Rita property would pay off. Former SRMC superintendent Benjamin Thayer also "had no faith in the optimistic estimates" of Sully, an irony not lost on T. A. Rickard, who had visited Santa Rita and Hurley for a week in 1922 and then wrote a five-part series of articles in 1923–1924 on the "Chino Enterprise." He harshly criticized the future president of the Anaconda Copper Company in his *Engineering and Mining Journal-Press* account for failing to realize the potential of the gargantuan Santa Rita ore body Thayer himself had managed since 1902.[15]

In a sudden reversal, however, Rogers changed his mind and decided against his associates' wishes, informing Burrage and Sully on May 18 that they would meet the next day to seal the deal. The two partners were ecstatic when they left the Belmont Hotel together early on the morning of the nineteenth after having celebrated their long-awaited goal. Fate, though, took a cruel turn that morning. Only thirty minutes after parting company with Sully, Burrage telephoned his partner to give him the grim news that Rogers had died of a stroke early that morning.[16] All looked lost. But the Boston financier would not give up. As Sully recalled later, "this disappointment happened at the eleventh hour, but President

Burrage, who had now become determined to see the property come into its own, while considering his old associates first, had not relied on their assistance entirely, turned immediately to a firm of powerful bankers" from New York for financing.[17] Burrage knew he had to act quickly as well, because although Rockefeller and Broughton had promised him they would sell their stocks to him, that opportunity for a majority option on the property would expire in five days. He "would have put his own money into the venture," Rickard recorded, "for he had faith in it, but just then he was not in a financial position to do so."[18] But the deal was "saved in the nick of time." At the last hour Hayden & Stone agreed to finance the project, signing the deal the following Monday in Boston. Unfortunately for Burrage, however, with the incorporation agreement, which resulted in the initial sale of 750,000 five-dollar shares, he and Lawson "lost control" of the venture, only able to procure 125,000 shares (see figure 3.5).[19] Perhaps the Rockefeller Syndicate hoped to take a controlling interest under Amalgamated Copper's umbrella.

Regardless of who won domination of the new Chino Copper Company, all agreed that the steam-shovel technology Sully first proposed in his 1906 report would be implemented at the Santa Rita property. Sully then returned to New Mexico, but now as superintendent of Chino. Soon after that Hayden asked Daniel Jackling, architect of the Bingham Canyon open-pit project, to visit Santa Rita and assess its potential. Jackling gave a positive review of the New Mexico prospect and then agreed to serve as chairman of the executive committee of the board of directors of the new corporation. The investors and managers held modest expectations for the company, which soon had $3.5 million at its disposal to purchase the mining, crushing, and milling machinery to start the operation. Sully and his team began preparing to produce copper. Still, nobody involved anticipated the unprecedented profits that would be made from the Grant County enterprise.[20]

Chino soon had a corporate office in New York City with numerous highly placed officials in other copper enterprises sitting on its board of directors. Along with Jackling, they included well-known financiers Charles M. MacNeill, Charles Hayden, Spencer Penrose, Berthold Hochechild, and even A. Chester Beatty, who may have secretly had faith in the risk all along. John Sully, general manager, headed the list of on-site officials that also

FIGURE 3.5. *Chino Copper Company stock certificate, 1920. Chino stock sold for $5 a share initially in 1909 and quickly rose to $62 by 1917.*

COURTESY OF TERRY HUMBLE.

included Horace Moses as superintendent of mines, William H. Janney as superintendent of mills, and many other engineers, technicians, and experts (see figure 3.6). A coterie of consultants filled out the professional ranks: R. C. Gemmell as mine engineer, Frank G. Janney as mill engineer, and George O. Bradley as mechanical engineer. Each had offices in Salt Lake City, where they kept up with developments at Utah Copper's Bingham Mine. Sully, in fact, hosted Jackling, other board members, and additional investors, among them Sherwood Aldrich, president of the Ray Consolidated and Gila copper companies, to join him in November 1909 on a tour of Santa Rita. Arriving in his lavish private railcar, Cypress, with this distinguished contingent, Jackling symbolically welcomed Chino into the growing porphyry copper empire that within a few years would consist of Bingham Canyon, Nevada Consolidated, Ray and Ajo in Arizona, and Braden and Chuquicamata in Chile, among others. Soon afterward, Sully began sending monthly reports on the operation directly to Jackling.[21]

During the next five years Chino officials purchased as much land as they could to initiate their plans to map out a controlled industrial as well as communal landscape in southwestern New Mexico. By early 1914 the company owned in Santa Rita 2,645 acres of mine property, of which about 2,400 was patented. In total it acquired nearly 20,000 acres by 1920, much of which A. F. Kuehn, secretly posing as a rancher, purchased for a mill site, water rights, dumping lands, and a buffer zone (see map 3.1). Chino officials chose a flat cow pasture nine miles south of Santa Rita called Hurley siding, along the Atchison, Topeka, & Santa Fe line and near a water source, the Brahm or B Ranch Spring, as their mill site. In 1911 the Hurley mill began producing concentrates to send to the American Smelting & Refining Company's furnaces in El Paso. Chino also built a power plant at Hurley equipped with three Nordberg Corliss 100-revolutions-per-minute engines and Allis-Chalmers three-phase, sixty-cycle generators that produced 24,000 volts of electricity. The company also constructed a 3-million-gallon concrete reservoir to store water for the eight 500-horsepower Heine water-tube boilers that were cooled by two towers that reintroduced the used water into the reservoir.[22] With the first three years' profits the company invested in additional machinery—totaling ten steam shovels, twenty-one locomotives, and fifty six-yard, fifty twelve-yard, and twenty-four twenty-yard ore cars—to mine the pit and haul the ores to the primary crusher in Santa Rita and the concentrator in Hurley.

The demands for water of the steam-powered machinery and the milling processes forced the Chino Copper Company to devise a land-based strategy to acquire millions of gallons of water a day. In addition to owning 20,000 acres of land, the corporation acquired rights to the precious resource on 300,000 acres in Grant County. These acquisitions ensured steady supplies from Apache Tejo (the Apaches' former campsite of Pachitejú), Cameron, Twin Sisters, and Whiskey Creeks and the

FIGURE 3.6. *Santa Rita staff, February 1911. Front row,* left to right: *unknown; Hugo Miller, chemist; Fred Ely, sample chief; George Webster, auditor; Ira Brown, Sully's secretary; Horace Moses, mine superintendent; Mr. Brown, supply clerk; Walter Pattison, shovel foreman; and E. Perry Crawford, engineer. Middle row,* left to right: *Mr. Snell, engineer; Mr. Watt, engineer; Jim Wroth (one step above middle row), engineer; L. E. Foster, chief engineer; Morris McGrath, assistant engineer; unknown; Joe Heistich, stenographer; Mr. Gray (in front of post), store clerk; and George Cameron, chief chemist. Back row,* left to right: *R. E. McConnell, engineer; H. E. Treichler, engineer; Charles Morrell, team boss; Red Moseley, drill foreman; unknown; Deacon Pearce, accountant; and Mr. Palmer, mess-hall manager.*
COURTESY OF TERRY HUMBLE.

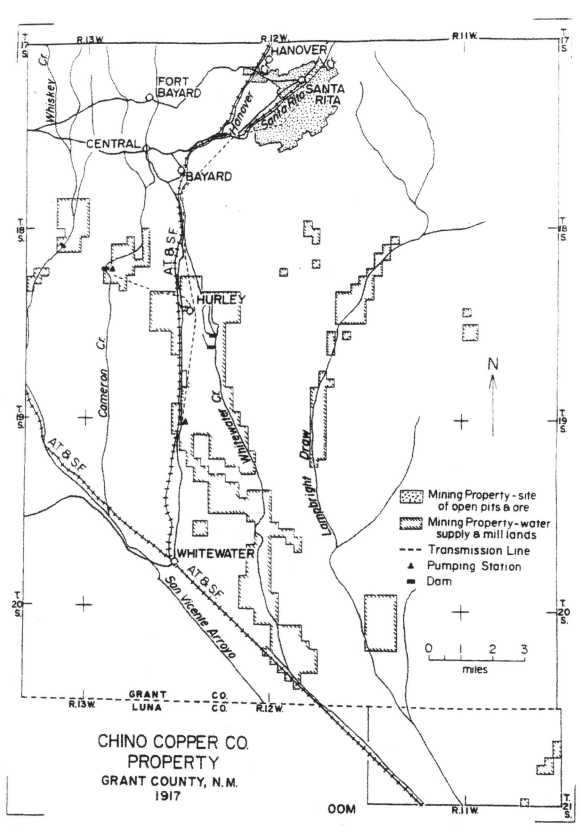

MAP 3.1. *Chino Copper Company property map, 1917, showing thousands of acres of mine and mill lands.*

B Ranch as well as the water rights to the lands of the Victorio Land and Cattle Company. Like late nineteenth-century miners who filed far more mine claims than they needed to gain access to fuelwood, Chino acquired far more land than was necessary for the operations to acquire water rights. Water would remain an essential natural resource commodity for the copper company.[23]

The water pollution complaint of Mattie N. Dowell in 1912 illustrates the company's obsession with controlling the water and land rights in the vicinity of the mining and milling operations. In early 1912, Dowell, with several children to provide for, had claimed a 320-acre homestead four miles south of Hurley under the Desert Land Act. By July, soon after staking her claim, she realized that Chino was dumping raw tailings from its Hurley mill into Whitewater Draw, which bordered her homestead. Knowing the federal law required her to irrigate, she soon figured out that Whitewater Creek was so contaminated with a toxic sludge that she could not use the deplorable slime for irrigation purposes, so she filed a lawsuit in October with the Sixth Judicial District Court of New Mexico. She argued that the mining company's dumping system obstructed farming because the tailings were "poisoning and making [the] said lands dangerous, unsafe, unfit, non-productive and valueless for agricultural, grazing or any other purposes." She also complained that Chino's blasting hindered her farming efforts.[24]

Chino's response was typical of mining companies throughout the American West. In their motion, company attorneys argued, first, that Dowell had not proven that dumping and blasting were occurring at all; second, that her complaint threatened one of the state's most productive economic ventures (the economic hardship argument that the industry would use in response to water and air pollution laws especially in the 1960s and later, as well as against labor demands throughout the twentieth century); and third, that Dowell's farm was undeveloped and valueless and, therefore, less significant to society than the mine operations. Chino then filed a demurrer hoping to get the complaint thrown out. On March 7, 1913, however, the district judge ruled against the corporation and then gave company attorneys fifteen days to make a response. Sully and his lawyers replied a week later, reiterating their claim that there was no pollution damage and that Dowell's land was too parched to farm. Whitewater's creek bed was too "steep and precipitous,"

they contended, to allow spillage onto her "absolutely worthless, unproductive, and useless" land. They further asserted that the "tailings did not, do not and will not destroy or in any manner injure" Dowell's property. But then the company tacitly admitted their environmental sins, reporting that workers were at that very moment constructing dams, dikes, ponds, and an impounding reservoir to stop the flow of the yellow sludge: the "works . . . [had] progressed to a point where they entirely hold up, store and prevent the moving off the lands . . . any . . . tailings from the . . . mill." In other words, the company understood the threat to lands downstream from the mill wastes but was unwilling to admit it outright. The judge realized this offense and ruled in Dowell's favor, and so the Chino Copper Company offered to buy her homestead, which she eventually agreed to sell.[25]

The Dowell case reveals how Chino and other companies in the American West often thwarted attempts to stop toxic dumping in the vicinity of their mining-milling complexes.[26] To protect itself Chino purchased as much land as possible (initially about 20,000 acres) to create a buffer zone between neighboring settlers and the industrial operations. If farmers or ranchers were still affected by the pollution, despite the buffer, the company then worked to acquire those individuals' lands, as was the case with Dowell. This land-grabbing scheme ensured that Chino could do what it had admitted to in the lawsuit, contaminate the surrounding environment. Often mining companies created subsidiary land and cattle companies to disguise their intent to buy ranchlands for water rights and dumping as Chino had done with A. F. Kuehn. In the end, the strategy of operating exclusively on privately owned lands allowed the copper company to continue to pour pollutants into Whitewater Draw and exempt it (as well as future owners) from federal regulations that protected only the public lands. Private ownership of mine and mill property also protected the company from costly lawsuits, which for mine officials translated into higher profit margins. Purchasing Dowell's land also reflects the industry's polluter-pays approach to dealing with other landowners who complained of water and/or air pollution. The industry-wide stand was either pay for the right to continue to pollute or pay for the land itself.[27] This ploy worked for corporate polluters until federal and state laws in the 1960s and later forced the mining companies to change their practices or face multimillion-dollar fines imposed by the Environmental Protection Agency and state environment divisions.

In the end Sully and his team of engineers used their technological know-how and modern machinery to forge a clear hegemonic control over the land, water, air, and other natural resources. The environment was at the mercy of the mining men. Yet unlike the frontier miners of the late nineteenth century, whose laissez-faire attitudes led to massive destruction of forestlands, watersheds, and landscapes, without much concern for the future, corporate authorities implemented conservation measures. These strategies ensured longevity for the operations even though the principal goal of conserving the natural resources—from water to coal to natural gas and limestone used in the operations—was to maximize profits for the dividend holders rather than because of a genuine concern for the environment. The company's acquisition of vast acreages of land, for instance, to acquire water rights is one such strategy. They then used drilling, piping, and hauling technologies to ensure a steady supply of the precious liquid to run its steam-powered equipment, to transport ores as slurry, and to offer domestic reserves for the townspeople. In their efforts they also devised elaborate strategies to reconfigure the terrain to funnel floodwaters away from the emerging pit, the town of Santa Rita, and other resources at the operations. The best example of this manipulation of the natural landscape was the construction in the mid-1920s of a water diversion channel around the open pit. Some measure of water control and its conservation was essential in the arid Southwest.

Still, the company's hegemonic power also meant that polluting the water was acceptable. This attitude pervaded Sully and his engineer's thinking, as exemplified in the way the company addressed Mattie Dowell's water pollution complaint. In the end, though, the corporation's fiscal resources allowed officials to "buy off" Dowell and continue dumping contaminated waste materials. As will be seen, corporate hegemony of the environment would remain intact until the 1970s.

Chino also purchased large acreages of land to construct two company towns. With these tracts the corporation expanded Santa Rita and then built Hurley from the ground up. Soon each of the industrial towns was dotted with readymade family-size homes, single miner's boarding rooms, a hospital, clubs, libraries, and, of course, the company stores.[28] This planned residential-industrial landscape quickly paid huge dividends for the stockholders, company officials, and its managers. Under Sully's fatherly guidance the company formulated a system of welfare capitalism that even allotted the general laborers, many of whom migrated to the area from Mexico, a decent wage as long as they followed company rules, stayed on "their" side of the towns, and refused to unionize. Under these circumstances Chino's production soared over the next decade. From 1912 to 1920 the company produced an average of 57 million pounds of copper, peaking in this period at about 80 million in 1917, with a low of 27 million the first full year of output in 1912.[29]

To achieve this success Sully believed that he had a right, even a duty, to impose his authority over the workers. In effect, Chino's superintendent placed corporate interests and company profits above fair wages and collective bargaining, especially for Hispanic employees. Prior to World War II, in fact, corporate hegemony at Santa Rita meant that workers, like the environment, were at the mercy of company authority. In the case of workers, the general manager viewed himself as a father figure who was to be obeyed and respected under the guise of welfare capitalism. As long as Sully, for example, believed that he was providing good-paying jobs, housing or access to lots to construct housing, a school, a hospital, and other amenities, the employees were expected to abide by *company* rule at the *company* works in the *company* town. This perspective on the relationship of workers to their employer, Sully and the board of directors insisted, precluded the need for unions for Chino's workforce, whether skilled or unskilled. As a result, corporate hegemony over the workers characterized employment conditions at the operations especially prior to the 1940s.

The company may have justified its paternalistic treatment of workers because of the makeup of Santa Rita's population. Sully, no doubt, espoused the view that the immigrants were "fortunate" to have reasonable wages at the mine works and would not complain of strict company rule for fear of deportation, as had been the case at Phelps Dodge properties in 1917 in southeastern Arizona. The 1920 US Census of Santa Rita reveals that of the 3,565 people counted, exactly one-third of them, 1,190, were born in Mexico. Of 561 heads of household with Spanish surnames, 423, or 75 percent, were born in Mexico. Of those heads, 120 were born in New Mexico. By the 1930s, when Santa Rita's workers began to seek union certification and greater workplace rights, fewer inhabitants had been born in Mexico. Of 3,889 residents, the 1930 census recorded that 2,420, or 62 percent, were Hispanic. Of that total, only 905, or 23 percent, were born in Mexico. Enumeration of heads of household also reflected a gradual shift to fewer Mexicans, with 313 of 459, or about 68 percent, down from 75 percent ten years earlier. New Mexico–born Hispanics numbered 132, or about 29 percent. Though census records for households are not available since the 1930 census, fewer and fewer of Santa Rita's inhabitants were born in Mexico. Generation after generation of New Mexico–born Mexican Americans dominated the demographic landscape in the copper town as well as in the employment pool at Chino.[30]

Corporate domination of the environment and the workers allowed company officials to forge ahead with investments in the best science and technology of the day. To realize the mine's potential, for example, Sully and his team first had to understand the geologic makeup of the porphyry ore body to determine the economies-of-scale strategy to implement. At Santa Rita, the low-grade copper ore had been infused into limestone in the geologic past principally because of volcanism. The deep underground superheating of the limestone bedrock mass by magmatic flows caused mineralization of various types of copper on different occasions over millions of years. Various forms of copper mineralization resulted, from the formation of the nearly pure lustrous native copper and glistening purple cuprite to malachite, azurite, and chrysocolla. All of these types of copper still appeared in the Chino ore bodies mainly along fractures and fissures. But they could no longer be mined profitably using underground lode strategies, even though the company did very briefly employ miners to use block-caving methods in the Hearst-Carrasco section of the workings. The bulk of the low-grade copper ores had been infused in the limestone by a porphyry intrusive of which monzonite-porphyry contained the bulk of the disseminated oxide and sulfide copper targeted for mining: "the deposition of copper from ascending solutions during periods of thermal activity following in the wake of active volcanism, was supplemented . . . by the deposition of secondary copper precipitated from descending solutions."[31] The Chino Copper Company targeted the ore body containing microscopic specks of copper Mexican

workers called *mosqueado*,[32] which was occasionally infused with native copper peppered with exquisite spiked cuprite specimens. Eventually, plate tectonics uplifted the ore body, making it far more accessible to humans in very recent geologic times. The initial dimensions of the Chino ore body were about 4,000 square feet at the surface and 600 feet deep, figures that would increase with new prospect drilling and the growth of the pit.[33]

Churn-drill crews continued to map out the ever-growing ore body after steam-shovel operations began in September 1910. Under the supervision of drill superintendent W. A. "Red" Moseley and his assistant Walter Belford, Chino implemented two kinds of churn-drilling methods. First, prospect or scout drill crews pounded six- or eight-inch steel bits deep into the porphyry mass an average of 500 feet (the deepest was 1,635 feet) to continually update the measurements of the ore bodies to determine future mining strategies and to calculate profit potential. Within four years of scout drilling more than 300,000 feet of holes, engineers increased the size of the ore body to 90 million tons and increased the mine's life expectancy to forty years. This prospect drilling revealed four distinct ore bodies. The north and northwest sections together were crescent-shaped and the south and southwest sections mirrored them. Collectively these coalescing ore bodies formed an oblong formation running northwest to southeast with an additional southwest-facing point on the ore grid (see map 3.2).

Second, blast-hole crews drilled a pattern of holes ahead of the steam shovels in the pit. Each hole was sampled so locomotive conductors could be directed by mine engineers to haul the blasted rock to either the crushers or the waste dumps. Engineers tested the samples from each of these two types of drilling techniques at the assay office in Santa Rita to systematically map out the ore bodies to be developed and to determine shovel strategies (see figure 3.7). These calculations also changed Sully's earlier proposal to mine 60 percent of the disseminated ores with steam shovels to 90 percent, an astronomical cost-benefit considering the high expenses involved in underground methods.

Blast-hole drillers made shallow holes to lay charges for blowing up the rock to allow shovel engineers to carefully form benches in the descending pit. This "primary" blasting prepared the bulk of the material the steam shovels loaded onto haulage trains. Each blast-hole team consisted of a driller who operated the Cyclone drills

Map 3.2. *Chino Copper Company claim map showing ore bodies, stripped areas, ore and waste dumps, railroad tracks, and industrial and residential buildings, January 1, 1916.*

FIGURE 3.7. *Santa Rita Assay Office, ca. 1915. Assayers pulverized the ore samples churn drill crews brought to them and then added chemicals to create a solution that they could test for copper or other mineral content.*
COURTESY OF SILVER CITY MUSEUM.

acquired from the Sanderson Cyclone Drilling Company of Orrville, Ohio. There were also from one to three helpers who replaced drill bits and sharpened the dull ones, assisted in taking out the cuttings, and ensured that plenty of fuel, either coal or gasoline, was available to feed the machine's engine (see figure 3.8).

A separate crew of "hammermen" took care of "secondary" blasting duties. Using pneumatic drills, the hammermen drilled into boulders over six feet in diameter to place dynamite for pulverizing them so that the material could be handled by the 3.5-cubic-yard shovel buckets. They also implemented "dobing," from the term "adobe" (clay), placing a stick of dynamite in a dab of clay on top of the soon-to-be disintegrated rock. During

peak production periods in the 1910s and 1920s, Chino ran as many as thirty-two drill crews and twenty-nine drills over three separate shifts; hammermen no doubt accompanied most of the shovel crews as well. Through 1925, Chino purchased twenty-nine Cyclone drills (model numbers 4, 6, 12, and 14), the first eight being steam and the remainder gasoline versions. During the 1930s most of the Cyclones were retired and replaced by Bucyrus-Armstrong Model 29T electric drills (see Chino Drills in the appendices).[34]

Shovel crews used the prospecting data to work in tandem with the blast-hole drillers to carve the initial trenches into each of the ore bodies. After blast crews exploded thousands of tons of rock with dynamite

FIGURE 3.8. *Churn driller and his rig, 1912. Probably the gas-powered Cyclone No. 12, this drill pounded bits into the ore body to make holes for sampling and blasting. When the drill bits became dull, crew members sharpened them after heating them with a coke forge housed next to the rig. Drillers worked twelve hours straight, earning about $2.75 per shift in 1912, a figure that rose to $6.55 by 1916.*

directly in front of the digging machines, engineers swung their shovels' thirty-foot booms (or fifty-foot on the No. 7, Marion 100 steam shovel) over the materials, then dropped the buckets into the ore or overburden. They then lowered the boom and bucket into the rock. Using a series of levers, the shovel engineer next directed the full dipper over the ore cars. The craneman then tripped the latch to open the dipper, dropping the precious cargo (see figures 3.9 and 3.10). During the first year of pit mining the locomotive engineers drove their ore trains to stockpiles and waste dumps to the tune of 400,000 tons of ore divided into sulfide and oxide dumps to be milled later at the Hurley concentrator. Beginning in 1911, Atchison, Topeka, & Santa Fe locomotive engi-

neers took crushed ores in their trains to the Hurley mill (see figures 3.11 and 3.12 and lists of Chino Shovels and Chino Locomotives in appendices).

On his visit in 1922, T. A. Rickard described a typical mining scene: "We waited until the steam-shovel had finished loading the train of seven cars, meanwhile watching it digging manfully into the broken ore, puffing, rattling, and straining with the nominal energy of a hundred horses."[35] The noise of chains clanking, engines roaring, and ore crashing with each new loading or unloading action produced a dirty industrial scene. Officials, workers, and townspeople alike welcomed this scenario despite the incessant around-the-clock noise of pounding churn drills, revving shovel and locomotive engines, and the piercing screeches of steam cylinders and railroad whistles. Santa Ritans became inured to the industrial sounds and smells that characterized the once-serene mountain landscape. Children who grew up there knew nothing else during their lifetime. Quiet times and clear skies rarely existed over the next half century in this company town that sat below the protection of the Kneeling Nun. When the "normal" cacophony of industrial sounds, spewing of sulfurous odors, and puffing of black emissions ceased, in fact, company officials and miners and their families knew something was amiss in their world. Silence was not golden but a sign of gloomy shutdowns that periodically visited the mining town, such as those in 1921–1922 and 1934–1937.[36]

During the first decade of operations, Chino ran its modern machinery with hardly a hitch. As a result, Santa Rita's landscape dramatically changed in a few short years. The nonstop, around-the-clock blasting, digging, and hauling of raw material created giant gashes in the earth. The dumping of the waste rock formed new human-made mountains of crushed rock. Chino's efforts, in essence, began a rapid reconfiguration of the Santa Rita landscape, a kind of landscape that mining engineers throughout the American West would re-create at different locations again and again. From Bingham Canyon to Chino and Ray as well as many other future open pits, the new twentieth-century technologies meant that mining corporations would alter the terrain in the vicinity of the operations on a gargantuan scale. Such dramatic alterations in the landscape also required the reconfiguration of the watershed at Santa Rita. Chino engineers had to design water diversions, storage tanks, transport pipes, and other strategies in conjunction with the evolv-

FIGURE 3.9. *Steam shovel No. 5 with nine-man crew, ca. 1915. Purchased in July 1913, the No. 5 shovel sported a 3.5-cubic-yard dipper that could dig five tons of material in a single scoop. In 1927, electricians, machinists, and welders modified the machine, transforming it from a coal-powered rail unit to an electric shovel with caterpillar tracks. Prior to this labor-saving adaptation, the shovel crews consisted of nine workers: a shovel engineer, a craneman, a fireman, an oiler, and five pitmen. The electric machine required only four men. At the time this photograph was taken, the company paid shovel engineers $200 a month, the highest wage among non-management employees; cranemen earned $145 a month; firemen $4.00 a day; and pitmen about $2.50 a day.*

COURTESY OF TERRY HUMBLE.

ing landscape to route the streams around the pit, to co-ordinate the need for water, and to protect the area from potential floods.

At the Santa Rita pit the copper corporation's mining crews initially carved two narrow, V-shaped "valleys" into the foothills below Ben Moore Mountain. One trench—first known as the CYB for the Chino, Yosemite,

and Bliss claims—was soon branded the Northwest Ore Body. The other—the RTW for Romero, Texas Flat, and Whim Hill—was named the North Ore Body or Romero Pit. In 1913 they grew into each other to create one immense crescent-shaped pit that cut deep into the earth to uncover colorful bedrock strata. "Eight tiers of benches stood warm in the sunlight in many colors—white, red,

FIGURE 3.10. *Shovel No. 5 loading ore wagons, March 1911. Known as the "dimey" (because it was small and later assigned the number 10), this steam shovel scooped ore and waste with a smaller 1.5-cubic-yard dipper. Note the teamster not paying attention to the boom in an era when safety took a backseat to production. Minor miscalculations by the shovel runner or craneman threatened the rest of the crew and, in this case, the buckboards made of wood. In this scene, the shovel engineer is loading rich ore that the teamsters took to a dump for later transport to the Hurley mill.*
COURTESY OF TERRY HUMBLE.

green—with a reddish brown rock capping the orebody and a brown sump at the foot of the cliff of ore—600 ft. from top to bottom."[37] Beginning at the 6,340-foot bench (or elevation), the trenches stairstepped every 30 feet or so into the bedrock. Shovel engineers dug above and next to their machines on 100-foot-wide benches (see figure 3.13).

A dramatic blast filmed on June 25, 1913, characterized the massive operation's ability to transform the landscape in a single moment. Packing over $7,000 worth of explosives in numerous holes in the Hearst section, blast crews detonated twenty-five tons of dynamite for an audience of hundreds who had driven automobiles and

buggies to the site. The Lubin Motion Picture Company captured on film the massive explosion, whose rising dust nearly completely obscured the Kneeling Nun several hundred feet above the scene. Soon afterward shovel and locomotive crews got down to the business of hauling off the 200,000 cubic yards of rock. This new form of "mass" mining, as the open-pit engineers called it, also involved the calculated plan to eradicate the town of Santa Rita. This blasting and digging scene would be repeated on nearly a daily basis for generations to come, resulting in the ironic removal of the bedrock foundation of the copper town that by 1970 would disappear into "space."[38] The consequences would traumatize many of

FIGURE 3.11. *Chino steam Locomotive No. 10 under water spout, ca. 1916. The No. 10 "loco" was a 15 × 24 (15-inch diameter of steam cylinder and 24-inch movement of cylinder) H. K. Porter with a 0-4-0 designation (four drive wheels with no wheels on the front or rear truck). Chino purchased the No. 10 in 1912 and then sold it eleven years later. This 42-ton behemoth could pull eight 6-yard ore cars or four 12-yarders. In 1916, loco engineers earned $4.95 and fireman $4.00 per shift.*

FIGURE 3.12. *Chino steam Locomotive No. 13, ca. 1917. Chino purchased this 16 × 24 steam H. K. Porter 0-4-0 engine in September 1912 and then retired it in 1923. Note the safety shield (in front of the number 13) that protected workers from the dangers of exploding boilers.*

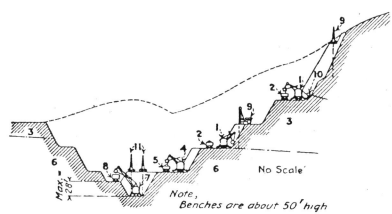

Typical section through the Hearst orebody showing the method of mining

1. Steam shovel loading waste.
2. Dump train hauling waste to dumps.
3. Waste to be removed.
4. Steam shovel loading ore.
5. Dump train hauling ore to crusher.
6. Orebodies.
7. Special S.S. making the initial cut (loading ore).
8. A., T. & S. F. 50-ton ore cars.
9. Gasoline churn drills drilling bank blast holes in waste.
10. Air drills drilling toe-holes and lifters.
11. Gasoline churn drills drilling bank blast holes in ore.

Dotted line: Original ground line.
Dot-and-dash line: Approximate line between ore and waste.

Note—Ore containing large rock is loaded in dump cars and hauled to the crusher. Ore containing rocks not larger than 12 in. is loaded direct in A., T. & S. F. ore cars and hauled to the mill at Hurley.

FIGURE 3.13. *Famed mining engineer/historian Thomas A. Rickard drew this diagram of a typical section through the Hearst ore body showing the method of mining. This drawing shows the coordination of churn-drill crews, steam-shovel operators, and locomotive engineers in the pit. From T. A. Rickard, "The Chino Enterprise—IV, Methods of Prospecting and Mining,"* Engineering and Mining Journal-Press *116, no. 26 (December 29, 1923): 1117.*

the longtime residents and leave an open pit the size of craters left tens of thousands of years ago by meteorites. This industrial endeavor also highlights the extraordinary ability of humans to alter the earth and its natural processes.

Chino utilized different kinds of explosives to prepare the rock for digging. For "soft ground," blast crews packed black pellet or Trojan bag powder into their targeted ore. They used a white nitro-starch powder for "all kinds of hard ground" unless it was wet; then the DuPont Corporation dynamite, Red Cross, was implemented (see figure 3.14). They filled churn-drill blast holes with black and Trojan powder. For piston and hammer drill holes workers employed the "Chino special," an ammonia powder manufactured in Louviers, Colorado.[39]

Accompanying the engineers and cranemen on the shovels were the fireman, an oiler, and the pit crew. The firemen, of course, shoveled coal into the boilers of the rectangular-shaped behemoths, carefully maintaining steam pressures that allowed for a high rate of efficiency and a certain measure of safety. The oiler lubricated the many moving parts of the engine, booms, axles, and wheels. Five additional men made up the rest of the crew. They mainly managed the heavy timbers used to support the outrigger arms and emplaced other ties with iron rail in front of the shovels to allow their advances forward. These laborers "lay the track, keep it clear, extend the water pipe to the shovel, and transfer coal from a car that stands on the track behind the shovel."[40] The collective effort of the shovel crews was a microcosmic example of the entire Chino operations and coordinated to apply technological and scientific efficiency in the name of profits. "In other words, the benches and tracks were planned to conform, as nearly as practicable, to the shape of the orebody, so that ore could be separated from waste as economically as possible."[41]

During blasting in these early years, all of the members of the crew hid behind their steam shovels in hope that none of the clastic materials would fall back to earth where they took cover. Shovel engineers often turned their cabs in the opposite direction of the blast, praying that no large boulders would roll down from the slopes above to threaten their safety. This danger at blasting times sometimes injured and even killed unprepared or unlucky crew members.

Maintenance and repair of the steam shovels, which sometimes failed in part because of the volatility of the boilers and the unwieldy nature of the moving parts, were done by repair crews known as "bull gangs." These skilled crews always worked in the pits to ensure that immediate attention could be given to broken-down ma-

FIGURE 3.14. *Powder crew, ca. 1911. Here a crew fills a blast hole with explosives to prepare the ground for mining. Composed exclusively of Mexican and Mexican American laborers, the powder crews conducted some of the most dangerous work in the pit. Note the lack of hard hats, steel-toe boots, and other safety measures in this early-day photograph. Powdermen did place their hats on top of loaded holes when locomotives or shovels approached with their ember-belching fireboxes.*
COURTESY OF PALACE OF THE GOVERNORS PHOTO ARCHIVES (NMHM/DCA).

chines (see figure 3.15). Usually composed of machinists, mechanics, and welders, these expert teams kept busy, working to keep the operations going at full capacity during good economic times. If all went well a single steam-shovel crew dug up about 2,000 tons of material per day. By the early 1920s the company operated fifteen shovels during three shifts of eight hours.[42]

Chino also owned a fleet of locomotives. Through 1914 the company purchased fourteen 15 × 24 and seven 16 × 24 Porter steam locomotives with seventy-two- and eighty-cubic-yard hauling capacities, respectively. From 1919 to 1929 it acquired twelve American 21 × 26 and two 24 × 26 "locos" and eight Baldwin 21 × 24 versions. Each machine had freighting capabilities of from seventy-five to ninety cubic yards (see figures 3.16 and 3.17). Initially, locomotive engineers piloted trains of six-yard "crab cars" that pulled a maximum of eight at

a time (see figures 3.18 and 3.19). This ore car, however, easily tipped over, or "turned turtle," requiring the company by 1916 to retire the volatile equipment in favor of twelve- and twenty-yard cars. Furthermore, because the crab cars and twelve-yarders also had to be dumped manually, the company soon began replacing most of them with twenty-yard automatic versions (see figure 3.20). These new cars offered greater efficiency and cost-effectiveness and proved safer, probably saving many fingers and hands. At maximum capacity during World War I, Chino ran twenty-one locomotives. Each loco towed five twelve-yarders (100 tons) or three twenty-yarders and one twelve-yarder (120 tons).[43]

Brakemen assisted the locomotive engineers. Among their varied duties, these skilled workers directed loco pilots with hand signals to warn of oncoming traffic or when they had to jump from their spot on the last

FIGURE 3.15. *Bull gang, ca. 1910. This repair crew worked in the mine nearby the unwieldy steam shovels that often needed mending on-site. Only three of the crew are identified: Henry Stanley (sitting on left); Harold G. "Tuff" Moses (lower right with hand on hip); and Sam Houghton (middle with pry bar). Moses, later appointed mine general foreman, was the younger brother of Horace, who served as general manager of Chino in the 1930s and 1940s. His longtime service to the Boy Scouts led to the establishment of Camp Tuff Moses in the Gila National Forest (see chapter 4).*
COURTESY OF THE NEW MEXICO STATE UNIVERSITY LIBRARY, ARCHIVES
AND SPECIAL COLLECTIONS (IMAGE #032003201383).

ore car to throw a rail switch. At night they did the same duties using red and green lanterns to communicate their planned actions. Many brakemen coveted the cushier switch-tender job. Stationed in small heated shanties at often-used rail switches, the "tenders" avoided many of the dangers inherent in working on moving trains, earned a slightly higher wage, and enjoyed refuge from inclement weather.

"Track gangs" laid the rail in the pits. Composed exclusively of Mexican and Mexican American laborers, these hardworking crews played a pivotal role in ensuring that locomotive engineers received their materials and then delivered them to the crushers (see figure 3.21). In the first two years of open-pit mining, track gangs laid

more than twenty miles of haulage lines. During peak wartime production Chino employed 132 trackmen. In addition to laying rail, pit labor included track cleaners, pumpmen, teamsters, track walkers (or "gaugers"), and engine coalers who refueled and rewatered locomotives at the chutes. Track laborers ensured that trains could haul waste to the dumps and ores to either stockpiles or the primary crushers. Teamsters transported coal, tools, and other necessities via wagons, which could get to hard-to-reach places inaccessible to trains. Concerted team efforts, from skilled locomotive and steam-shovel engineers, mechanics, and oilers to the less skilled laborers, kept the operations running quite smoothly during the initial phase of the open-pit era.[44]

FIGURE 3.16. *Locomotive with crew posing with children and men's lunch pails, ca. 1914. This loco is a 16 × 24 Porter 0-4-0 with its number faded. Victor Hugo Waggoner, who moved to Santa Rita in 1911, is standing on the far left and next to him is a man identified as "Fritz." Note that welders had not yet put a safety shield on this machine.*

COURTESY OF SILVER CITY MUSEUM.

The next step in the quest to process the ores involved crushers. Initially Chino crushed small tonnages using the Santa Rita Mining Company's dry and wet rolls at a small mill, probably the experimental one Sully used in 1908 for test runs on samples. The company soon realized, perhaps as early as 1909, that the machinery barely met the emerging demands of the burgeoning operations. Consequently, Sully worked with mill engineers William and F. G. Janney, W. T. MacDonald, and G. O. Bradley, all trained at the Utah Copper Company's mill in Garfield, to supervise the development of crushing and milling. Under their leadership, construction crews installed crushers over the next five years both at Santa Rita and at the milling complex in Hurley (see figure 3.22). The primary crusher at Santa Rita by mid-1912 had a capacity of 1,450 tons a day. Chino officials soon realized,

however, that this volume fell far short of their goals, especially since they had already planned to produce as much as 5,000 tons of concentrates a day (the milling of the 2 percent ores to produce as rich as 80 percent copper concentrate to be sent to the American Smelting & Refining Company [ASARCO] smelter in El Paso). Even though the Hurley mill housed coarse and fine crushers by 1911 to pulverize the ores in two and at times three steps, the primary crusher at Santa Rita hardly began to meet the concentrator's demands for ore, making haulage costs unacceptable. As a result, Chino constructed a larger primary crusher in 1914 at Santa Rita just south of the rail yard that the company soon realized still failed to provide the volume of crushed materials needed to run the concentrator at full capacity. Therefore, they built a second crusher section in 1917 adjacent to the Number 1

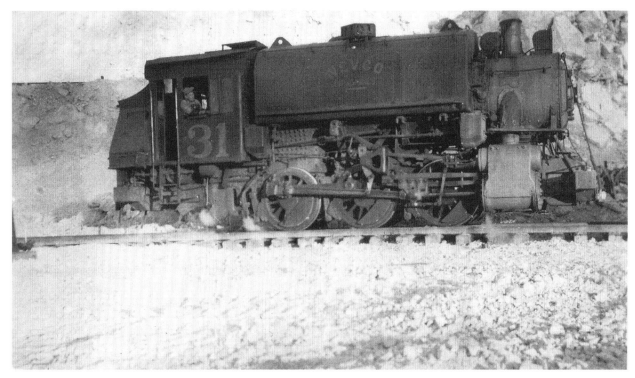

FIGURE 3.17. *Locomotive No. 31, ca. 1927. The Nevada Consolidated Copper Company purchased this 24 × 26 American 0-6-0 steam loco in 1926 as part of the mid-1920s rejuvenation program at Chino. Weighing 90 tons, the No. 31 could pull 300 tons of material in either ten 20-yard or six 30-yard ore cars. This engine worked until the 1940s, when the company retired steam in favor of electric locomotives.*
COURTESY OF DON TURNER.

crusher near the pit at Santa Rita (see figure 3.23). The ore cars dumped the materials above the crusher onto a metal grate called a grizzly that allowed the smaller rock to fall into ore bins. These small ore chunks were loaded directly into Atchison, Topeka, & Santa Fe cars for transport to the Hurley concentrator. Pieces of rock too large to pass through the grizzly fell into the Blake jaw crushers, which reduced the ores to a maximum of eight inches in diameter, also for haulage to Hurley by the Atchison, Topeka, & Santa Fe trains.[45]

From August 1911 to November 1912, Chino constructed five sections in its concentrator at Hurley. Though the company rarely ran all five 1,000-ton-a-day units, the mill produced 2,000 to 3,000 tons of concentrates a day during the first five years of production. Trains delivered the ore on spur tracks into storage bins on the west side of the crusher building at Hurley. The ore passed through various feeders and different-size grizzlies before conveyors took it to coarse and then fine crushers or Garfield rolls and then stored it in cylindrical steel ore bins placed on the west side of the concentration building.[46] "From the bins the ore is distributed to the [mill or concentrating] plant in a regulated [gravity] flow. The separation [from waste rock] and concentration [of the copper ore] are effected with jigs, Chilian [*sic*] mills, classifying and settling cones and vanners . . . The mill from here on is a lower structure with a slight descending grade, with an upper floor of classifying and settling cones and a lower floor of vanners [shaking tables]."[47] Some of the latter equipment, though, was also placed on the upper level (see figure 3.24).

Placed on an incline, the tables moved laterally to create dual separation actions: one a mechanically induced back-and-forth movement, and the other the natural force of gravity that pushed the materials downward. The combined actions separated most of the copper minerals from the waste rock. "The flakes of native copper can be seen moving amid the grains of black chal-

FIGURE 3.18. *Track gang repairs rail in November 1910. Note the track gang to the left of the 6-yard "crab car," the Romero mill in the background, the workers' homes below the dumps, and the steam shovel in operation in the mine cut.*

COURTESY OF TERRY HUMBLE.

FIGURE 3.19. *No. 500 railroad motorcar, ca. 1929. Powered by gasoline, this car carried the tools and machinery necessary for repairing rail in the pit. Note the blade in the front that was used for clearing track after blasting or accidental dumping.*

COURTESY OF PALACE OF THE GOVERNORS PHOTO
ARCHIVES (NMHM/DCA), NEG. NO. 162257.

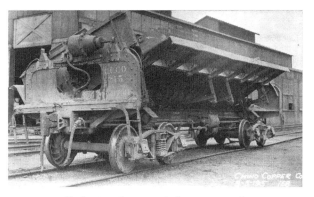

FIGURE 3.20. *Clark 20-yard automatic-dump ore car, August 5, 1915. The Chino Copper Company purchased twenty of these ore cars in 1914 to begin replacing the 6- and 12-yarders that had to be tripped and dumped manually. Many a dumpman's fingers were saved with this innovation. By 1917, Chino owned thirty-six of these safer and more efficient cars.*

COURTESY OF TERRY HUMBLE.

cocite . . . [and it] produces successive bands of minerals; first, one of dark-red cuprite, then bright flakes of native copper then the black chalcocite . . . and . . . bands of green-blue carbonates," all visible on the Wilfley ta-

bles. Some of this "middling" material required "further crushing [in Chilean mills] to liberate the mineral mixed with gangue." These reduced ores then went through classifying cones. These "classified products" were then treated on vanners. After passing through the vanners the concentrates went to settling tanks for dewatering and thickening. Tailings were either re-treated or sent to the dumps. Garfield tables were also used at certain stages in this process. Overflow water went to a large rectangular concrete reservoir north of the plant for reuse as part of the conservation program and to reduce costs and deal with water scarcity.[48]

Throughout the 1910s and 1920s, Chino officials struggled to reach concentration levels equal to those of the other porphyry operations. The first major effort to narrow this gap occurred in 1915–1916 when the company implemented "froth-flotation" milling.[49] Although gravity separation would continue to be used, mill engineer William H. Janney introduced newly designed flotation cells especially in treating the formerly stockpiled oxide ores. In the 1880s, mineralogist Carrie J. Everson of Colorado had accidently discovered the froth-flotation process. While washing some oil-stained concentrate sacks, she realized that metallic particles were attaching

FIGURE 3.21. *Track gang, ca. 1917. In this rare snapshot of an early track crew, workers take a break to pose for the photographer on the east side of the Hearst Pit. Exclusively Mexicans and Mexican Americans, these hardworking laborers laid track (twenty-four miles by 1917) for the shovels and locomotives, repaired the sometimes damaged 75-pound rail, replaced ties and spikes, and did other track-related duties. At the time, the track crews were in the process of replacing the 1909–1910 ties to the tune of 1,000 new ones a month. The track foreman, Gabriel Ontiveros (white shirt), is standing in front of the gang. Note the man (left center) wearing the pointed conical hat, popular headgear in Mexico at the time. The man to his right (with less pointed hat) is Conrado Padron, maternal grandfather of Willie Martinez, who furnished this photograph.*

COURTESY OF GUILLERMO MARTINEZ.

FIGURE 3.22. *The Chino Copper Company general office staff, March 31, 1915. These men worked at the mill operations in Hurley. Back row on porch,* left to right: *O. F. "Fritz" Riser, assistant mill superintendent; William H. Janney, mill superintendent; John M. Sully, general manager; Joe Bluett (in front of left column), mill office clerk; L. B. Wright (derby hat), stenographer clerk; F. M. Brown, purchasing agent; Ira Brown, general manager's secretary; Robert Bryden (in front of right column), mill office clerk; and Ruby Redding, stenographer clerk, mill office. Mr. Shedd, purchasing department clerk, is seated (on left banister). Seated on top row of steps,* left to right: *Mr. Phillips, purchasing department clerk; and Sam Bousman, engineer. On the next step,* left to right: *Douglas Dennis, messenger; and W. C. Maser, mill office clerk. Seated on right banister behind Sully's dog Sue: John Allen, cashier office clerk. On second step from bottom,* right to left: *Arnold Harris, engineer; David Boise, mill clerk; and L. Walz, cashier office bookkeeper. Seated on first step: Al Muse, mill superintendent secretary. Leaning against left post,* left to right: *George L. Webster, comptroller; and Bliss Moore, chief engineer.*

themselves to the bubbles that lathered in the soapy water. Lathering the bags literally separated the metal that she noticed glistening on the froth in the sun. "Using soap and water plentifully and being of an observant nature, she noticed that the sulphides floated to the surface of the water while the siliceous gangue sank to the bottom."[50] Applying this principle to separate metal from gangue in ore, mill engineers experimented with numerous oily additives to create froths from agitating slurry (a water-ore mixture). In what one metallurgist called "concentration upside down" (as opposed to gravity separation), froth flotation revolutionized milling in the early twentieth century.[51]

During the early 1910s, Chino mill engineer William Janney experimented with slurry mixtures and oils and then created a specific "agitating" cell to separate copper from gangue in the complex oxide as well as sulfide ores. This experimentation led in 1915 or 1916 to the implementation of froth flotation at the Hurley mill. Employing by 1923 eight "Janney mechanical air-type flotation machines" (see figure 3.25), Chino millmen sent slurry into the agitating machines spiked with various oily additives, from coal tar to creosote to pine oil and sulfur. Copper specks literally attached to the bubbly solution, to be skimmed off for further processing in Dorr thickeners similar to the final step in gravity separation.

FIGURE 3.23. *Primary crusher at Santa Rita, July 21, 1916. To meet the expanding ore needs of the Hurley concentrator, Chino constructed a second primary crusher, the No. 2 (to the right and below the grizzly). Note the locomotive above the No. 1 crusher and the automatic railcar dumping mined ore into the grizzly. The Atchison, Topeka, & Santa Fe Railroad contracted with Chino to haul the ores to Hurley. Note their loaded railcars ready for transport.*
COURTESY OF TERRY HUMBLE.

Both the gravity- and flotation-separated ores were then transported to settling tanks. In the flotation process, the waste materials settled to the bottom of the agitating machines for relatively easy piping to the tailings dumps. Recovery rates rose from about 70 to 79 percent when Chino combined flotation with the traditional gravity method. Chino's milling capacity also dramatically rose from an average of 3,000 to 6,000 tons a day of processed concentrates, facilitating peak production levels at the height of World War I.[52]

The dramatic expansion of the mining and milling operations required massive amounts of water. Water, in fact, was the key component in froth flotation. By 1920 the Chino enterprise used 600 million gallons a month. Because of the importance of water the company devised a conservation program that reached a recovery rate (re-used water) of 83 percent, or 500 million gallons, of the water circulated throughout the operations. Workers used water throughout the works. It was a key element in processing and transporting slurry (a water-ore mix). It allowed the piping of tailings to settling ponds. It was fed to the boilers of the steam-driven locomotives, shovels, and churn drills. It was essential at home for the workers and their families. The entire Chino community was dependent on water. This dependency on the precious liquid became increasingly apparent with the expansion of the works. The company regularly introduced more locomotives and steam shovels to mine still more ore; more mill equipment, like flotation cells and thickening tanks, to mill greater amounts of ore; and new water-hungry leaching and precipitating processes to produce still more concentrates.

As a result of the growing consumption of water, Chino had to vigilantly seek new sources as well as devise new conservation methods to decrease water losses. Company engineers clearly understood their dependency on water. Like high-profile mining engineer John Hays Hammond and others, who warned in the first two decades of the twentieth century of an impending resource scarcity crisis, they were aware of the paucity of natural resources. A shortage of the coveted resource water was their greatest concern.[53] Thus, conserving water became a key stratagem for the company. This realization hit home in the 1920s when the need for the essential resource became even more acute during a drought. A series of dry years, combined with water losses from evaporation at the 3-million-gallon reservoir and flooding caused by the destruction of Santa Rita Creek with the expansion of the pits, forced Chino's engineers to de-

vise plans to acquire more water and to conserve it at the same time. The company's water conservation program eventually addressed these concerns, resulting in careful management of the water. The result was the 83 percent recovery rate mentioned above.

The initial effort centered on gaining approval from the state for more cubic feet of water and then finding it underground. In 1915 water engineer James A. French of the state of New Mexico approved 27,366 cubic feet of water per second for use at the Hurley concentrator. Because of the drop in the water table at the Apache Tejo station the company decided to drill new wells and even tapped drain waters in dormant mines like the United States Smelting, Refining & Mining Company's Ground Hog Mine; by the mid-1920s, Chino pumped millions of gallons from the Combination and Black Hawk Mines as well. The new Bolton wells provided an additional 800

FIGURE 3.24. *Hurley concentrator, 1915. From 1910 to 1915, Chino constructed five 1,000-ton mill sections in the concentrator. At the time this photograph was taken, the mill was processing above capacity at more than 7,000 tons a day of copper ore. This figure gradually increased throughout the twentieth century with upgrades in milling technology. In the 1940s the concentrator produced about 20,000 tons a day, a tonnage that eventually reached 30,000 by 1980.*
COURTESY OF NEW MEXICO STATE UNIVERSITY LIBRARY, ARCHIVES
AND SPECIAL COLLECTIONS (IMAGE #03200550).

FIGURE 3.25. *Flotation cells in Hurley concentrator, 1915. Here workers tend to the flotation cells where most of the copper was separated from the waste material (or gangue). This innovation greatly enhanced the company's ability to process complex low-grade ores.*

gallons per minute (gpm); the K, I, 593, 700, Pinder, and Owen wells each offered about 500 gpm; various other sources offered anywhere from 100 to 1,200 gpm. The company even toyed later with the idea of transporting water all the way from the Gila River but compromised and used new sources at the Stark Ranch and Warm Springs sites.[54]

Engineers devised all kinds of strategies to use and reuse this steady supply of water. They designed elaborate pumping, storing, and transporting technologies. To pump the water, the company put different types of pumps in the water wells, dormant underground mines, and the Romero and Hearst Pits. Then the precious resource was transported through twenty-five miles of ten-inch redwood pipelines to the different processing locations. Workers also constructed storage facilities, such as a concrete-lined clear-water reservoir of 4.5 million gallons and two settling ponds of 15 million and

500,000 gallons. Another 375 million gallons were stored in the tailings ponds located south of the Hurley mill. During the wet months of late summer the company impounded rainwater in tanks and these new reservoirs. As early as 1912, mill engineers designed special wooden settling tanks used for allowing the ores to settle out of the gangue after milling to conserve water. As much as 67 percent of the muddy water could be reused, being recirculated in the milling process. By the end of the 1920s the company invested in more efficient Dorr thickeners that saved an even higher 70 percent of the water. In the 1950s and later, Chino implemented more efficient 250-foot-diameter thickening tanks, which "raked" the sandy tailings very slowly to thicken the waste material for transport to the tailings ponds. This water conservation technology also separated out the clearer water, recovering 80 percent of the precious liquid, which was then pumped back into the mill for reuse. Regardless, milling uses of water and evaporation at the storage "lakes" caused much of the loss; steam-powered machinery also took a substantial toll on water sources.[55]

By the mid-1920s, Chino engineers had to formulate additional strategies to supplement the new well and mine drainage sources. So they targeted Santa Rita Creek. Already dramatically reconfigured by the expansion of the pits, the stream regularly flooded during monsoonal rain showers. These massive inundations whisked away or damaged expensive equipment, ore dumps, and rail lines. They also caused flooding in the open pits, increasing expenses because of the costs of pumping out the water so mining could resume. As early as 1916, Chino's officials knew there was a major problem with flooding caused by the destruction of the natural watercourse. Yet the company procrastinated, not taking remedial measures until the 1925 flood forced its hand. In that year a massive deluge filled the bottom of the Hearst Pit so thoroughly that "the tops of smoke stacks and the tips of the booms of the steam shovels [were] peaking [sic] out above the water," according to longtime resident Paul Jones. "Some of the machinery was . . . totally submerged . . . Pools of copper water in the bottom . . . were a beautiful blue green and occasionally migrating ducks would understandably mistake the water for a normal lake and would alight there. It was tragic indeed to see many dead ducks floating on the face poisoned by copper water!"[56]

Chino officials decided it was time to take action. Consequently, engineers designed a water diversion to repair the dirt canal that Santa Rita Creek had become with digging and dumping. In December 1924, engineers produced a report that included six separate plans for dealing with the flood problem. They varied from a proposal to put in an earthen dam to the construction of a wooden flume and the drilling of an underground tunnel. In the end, the company narrowed it down to either a cement-covered flume or an open cement flume. The latter won out with an initial $675,000 price tag.[57]

Chino dedicated its new cement water flume in late July 1927 during the annual meeting of the New Mexico chapter of the American Mining Congress. Company engineer C. F. Ames presented a paper on the construction of the water diversion project that began back in March 1925 when shovel runners began digging the cut around the north rim of the Romero Pit (see figure 3.26). The undertaking required the company to move several homes and company buildings, initiating the first preliminary engineering plans to remove the houses and shops on the emerging island between the Romero and Hearst Pits; in the end, ironically, the island would be the last area of Santa Rita to be mined despite this early signal calling for the demise of the town in favor of mass mining. The water diversion channel, known locally as the flume or canal, was 7,500 feet in length, more than twenty feet wide, and eight feet deep with reinforced concrete walls and a floor twelve inches thick. On July 24, a day before the American Mining Congress meeting began, the flume passed its first real test when more than seven inches of rain soaked the area, nearly doubling the year's precipitation with a single shower. Ames, no doubt, proudly announced this success in his presentation. An engineering marvel designed to handle "a hundred-year flood," the canal never overflowed during its existence into the 1970s. Although the project ended up costing $140,000 more than predicted, the investment immediately proved to be cost-effective, nearly eliminating the need to pump water out of the pits and providing ample supplies for the new leaching processes. At the same time, the project buoyed the water conservation program, permitting the copper corporation to capture rainwater in various constructed basins, ponds, and tanks.[58]

Despite the unexpected expense of building the water diversion, the Chino operations proved exceedingly profitable as a result of the implementation of economies-of-scale technology. This success did not go unnoticed among the corporate giants of the copper industry, especially after the company invested $1.5 million in 1922–1923 to upgrade its concentrator and to purchase new machinery. Under mill superintendent J. T. Shimmin's guidance, engineers replaced worn-out hydraulic classifiers, Chilean mills, and Janney flotation machines with more efficient Dorr mechanical classifiers, ball mills, and Chino-type pneumatic flotation cells. The company also added McCully gyratory crushers to the Hurley mill complex by this time, further facilitating ore reduction efficiency. With these improvements, the concentrator's daily ore production capacity jumped from 5,000 to 10,000 tons in the mid-1920s, with an actual average of 9,200. In 1929 the mill's daily average reached nearly 11,000 tons per day. Leaching experiments also revealed that the best strategy for recovering the obstinate oxide ores was through implementation of the much cheaper precipitation of copper. Though not

FIGURE 3.26. *Cement water diversion on the north rim of the Romero Pit, 1927. Chino built this flume from 1925 to 1927 to divert floodwaters of Santa Rita Creek and to funnel the much-needed natural resource into catchment basins and storage tanks. A key component of the copper company's water conservation program, the diversion controlled floods and, during its life into the 1970s, never overflowed. A mile and a half in length, eight feet deep, and more than twenty feet wide, the channel also protected steam shovels and locomotives in the pits from summer washes. Children in Santa Rita often rode their bicycles in the cement aqueduct, putting in at the El Cobre Theater and riding the length of the "track" to its lower end, where they navigated around old tires, dead animals, and other debris.*
COURTESY OF TERRY HUMBLE

to be a large-scale part of production for another decade, leaching and precipitation made the Chino property still more attractive to industry giants like the Kennecott Copper Corporation. These water-dependent processes also highlighted the growing significance of Chino's water conservation program.

In addition, Chino's officials realized that as the pit grew, most of the oxide ore lay near the surface. The remainder of the deposit, which also contained substantial quantities of recoverable gold and silver and another soon-to-be important metal, molybdenum, was in sul-

fides. Consequently, Chino became a target for mergers with Ray Consolidated Copper Company (RCCC) in 1924 and then with Nevada Consolidated Copper Company (NCCC) in 1926. Even though the Chino and Ray units were each valued at about 30 percent of the NCCC's assets, the New Mexico property contained the largest ore reserves, at 125 million tons to 85 million for Ray and 68 million for Ely. [59]

Finalizing the first merger in February 1924 in New York City, Ray officials offered Chino $15 million at the meeting and then promised to update equipment and increase production. Investors received twelve Ray stocks for every Chino one. Soon the RCCC purchased new drills, locomotives, and the first electric shovel at Santa Rita. The company's view toward efficiency led to the remodeling of the steam shovels with new caterpillar crawler tracks to reduce the need for pit track crews and to lower operational costs. But it was the acquisition of a new Marion Model 350 electric shovel that highlighted the merger and exemplified an efficiency-minded business philosophy.

Purchased in June 1924, the gargantuan No. 14 shovel was the "talk of the town" because of its sheer size. Taking a full three months to assemble, it stood three stories high and required two sets of tracks to move into the ore body. The massive machine had an eight-cubic-yard bucket (more than double the capacity of the steam shovels) attached to an eighty-foot boom and it weighed 490 tons (see figure 3.27). Shovel engineers George Anderson, Gus Gerault, and Benny Bristow piloted the digging beast, putting it to work during three separate shifts from the beginning of its tenure. The No. 14 ran twenty-four hours a day for many years to come and in 1937 was renamed Electric Shovel No. 1 (ES-1) to distinguish it as the first of its kind in the open-pit mine. To accommodate the needs of the novel piece of machinery, new crews of men began to erect towers to string power lines in the benches of the pit. The Marion 350's enlistment signaled the start of a sixteen-year (1924–1940) electrification program at Chino. To keep up with the giant shovel's capacity, RCCC also invested in thirty-five new thirty-yard Clark railroad cars after testing a used version on loan to Chino from the Utah Copper Company.[60]

In May 1926 the Nevada Consolidated Copper Company purchased Ray for $46 million. The acquisition marked the eventuality of Chino becoming part of the Kennecott Copper Corporation (KCC), which held a 95 percent ownership in the Utah Copper Company. Utah, in turn, owned a 52 percent interest in the newly incorporated NCCC. Daniel Jackling believed that the three properties—at Santa Rita, Ray, and Ely—would fare better in the international market and give KCC more leverage in the industry if they were merged under one corporate umbrella. Annual profits at Chino also reflected the stability of the New Mexico investment. Though they did not reach the $10 million figure of 1917, profits averaged several million dollars a year in the 1920s, peak-

ing at $6 million in 1929 just prior to the economic downturn of the early 1930s.[61]

In the era of mergers, Chino Mines, as the New Mexico division became known, was part of a larger national consolidation phenomenon that permeated virtually all the major industries of the day. The post–World War I recession had led to a one-year shutdown at Chino from April 1921 to April 1922, a common fate for all the major porphyry properties in the western United States. Consolidation, corporate leaders believed, could protect industries during economic crises and facilitate control of production and, therefore, metal prices, especially in light of new competitive copper enterprises in South America, Canada, and Africa. To combat the perception that international competition threatened American domination of the industry, the copper giants, led by Anaconda, Phelps Dodge, and Kennecott, decided to form an international copper cartel. Though of limited effectiveness, the cartel facilitated America's ability to control international output and to secure a long-term cooperative relationship between the industry and the federal government. Increasing demands for the red metal to manufacture appliances, automobiles, and military hardware also solidified the probability for long-term profits at Chino and NCCC's other properties despite the impending crisis of the Great Depression.[62]

The infusion of funds into Chino's operations likewise ensured the continuation of the electrification program and offered financial stability despite about $2 million in losses in the early 1930s. By 1931 nine of Chino's thirteen shovels were electric, resulting in a savings of sixty-seven dollars per shovel shift, a cost-benefit that gradually improved over the next two decades. The NCCC had also purchased eight Baldwin steam locomotives in 1929 to keep up with the increased haulage demands of the more efficient shovels. The 1926 merger, which infused Chino with new capital and machinery, also pointed to the inevitability of the three properties of the NCCC becoming part of Kennecott's corporate fold.[63]

On June 12, 1933, the NCCC stockholders voted to sell all their assets to the Kennecott Copper Corporation. A Guggenheim family enterprise, KCC controlled the bulk of open-pit operations in the American West (the family also owned smelting giant ASARCO). This transaction, combined with the death of John Sully on July 15, marked a clear transition in the history of the Chino enterprise. A modest, approachable man, the fatherly

FIGURE 3.27. *Marion Model 350 Electric Shovel No. 14 with Bucyrus Diesel Model 30-B, No. 16, ca. 1927. In the mid-1920s, the new owners of the Chino operation, the Ray Consolidated Copper Company, invested in the gargantuan No. 14 to initiate the electrification of the pit. The smaller No. 16, with a one-cubic-yard bucket, caterpillar tracks, and diesel power, could get to places inaccessible to the steam shovels and the larger No. 14.*

COURTESY OF TERRY HUMBLE.

founding general manager had rapidly declined in health after an early June heart attack. Chino's workers genuinely mourned the loss of their mild-mannered boss. In their grief they may not have anticipated the dark days ahead on July 19 as they rode in their cars or on company buses in their longtime leader's funeral procession. Kennecott was the greatest copper corporation in the world at the time and Chino officials and the workers anticipated good times. The several-mile line of mourners that followed the man in the copper casket[64] to his final resting place in the Masonic cemetery in Silver City was probably not overly concerned with what was to come at Chino. Still, they had been hit hard by more than 1,000 layoffs during the previous two years. And they were dis-

tressed when the company announced it was reinstating a fifteen-day work month and more job cuts only days after Sully's burial.

The people of Grant County were not prepared a year later when Kennecott decided to suspend operations indefinitely. When the corporation's president, Daniel Jackling, made the lethal decision in October 1934 to "shut 'er down" at Chino, the news shocked the workers, many of whom had never known another employer. A kind of community hysteria traumatized the Depression-era Americans, who soon learned that KCC had other motives for the shutdown.[65]

The National Labor Relations Board (NLRB) would later rule that Jackling had closed Chino because company employees voted in September 1934 to unionize. A strident anti-union man, the copper magnate continued Chino's policy of blocking worker organization yet without the skillful sympathy of the late Mr. Sully.[66] At least seventy-one pro-union employees, in fact, would be blacklisted by Sully's replacement, Rone B. Tempest, a longtime protégé of Jackling. Many of them were personally escorted out of Santa Rita by the town's law-enforcement officers. Others were forced literally to move their homes out of town.

Despite the company's traditional stand on labor, a large majority of the employees favored union certification (outside of company officials, engineers, and office workers). The skilled workers voted to join American Federation of Labor (AFL)–affiliated craft unions. Simultaneously, the industrial laborers sought membership in the Congress of Industrial Organizations (CIO)–associated International Union of Mine, Mill, and Smelter Workers (Mine Mill). So they founded Local 63 at Santa Rita and Local 69 at Hurley.[67] New Deal legislation had legalized collective bargaining rights for unions the previous year for the first time in US history with congressional passage of the National Industrial Recovery Act (NIRA). Although the US Supreme Court in 1935 ruled the NIRA unconstitutional, Congress immediately responded that same year with the Wagner Act, which reinstated the workers' legal rights to seek collective-bargaining agreements with the soon-to-be-famous section 7a of the new law.

The workers of Chino had been used to company domination since the beginning of operations in 1909. Prior to the 1930s and 1940s they quietly accepted unjust features of employment: a sliding pay scale based on copper prices and profits, a dual-wage system that relegated Hispanic workers to lower-paid jobs with little chance for promotion except on the track gangs and powder crews, limited accident and death benefits, company-controlled health plans, no retirement packages, and intolerance of unions. Organized labor could have bargained to eliminate these unfair practices that were common traits in all the mining districts of the Southwest and elsewhere in the American West.[68]

Sully and his officials had put into place a paternalistic system of welfare capitalism. This ideologically formulated structure countered labor organization, which was inaccurately perceived and skillfully presented by company men as an anti-American foreign conspiracy

to undermine free enterprise. The general manager worked to maintain corporate hegemony over the workers even if it meant misleading the public about the effects of unionization. The company's adamant opposition to labor activism of any kind yet tacit approval of the Hispanics' *mutualista* (mutual aid society) Alianza Hispano Americano (Hispano American Alliance, or AHA) reveals the paternalistic sentiments of the company's power brokers. To honor this socioeconomic philosophy, the company provided homes, schools, hospitals, theaters, and baseball leagues to "take care of" the workers and their families as a traditional grandfather would his extended kin. By offering these benefits, the company hoped to deter pro-union men, who were seen as rebellious black sheep in the corporate "family."

Prior to the employees' vote in 1934 for union certification, only once did workers threaten to strike. That event occurred in 1912 when several steam-shovel engineers organized a one-day walkout. "Labor problems," as the company called worker grievances, surfaced on the morning of May 28, when shovel crews submitted a petition demanding an increase in wages commensurate with those at Bingham, Utah, and Copper Flat, Nevada. Every shovel runner, all but one of the cranemen (Harvey Forsyth), and two of the firemen, Luis Horcasitas and a Mr. Smith, had signed the appeal (see figure 3.28). The company suspended all operations and began an investigation. Soon Chino officials learned that Clarence Young, a steam-shovel engineer, and G. C. Hegler, his craneman, were the "instigators." "On leaving their shovel," the *Silver City Independent* reported five days later, "these two men persuaded the balance of the shovel men to quit work." Later that night they enlisted some of the trainmen. Sully fired all of them the next day.[69]

Mine superintendent Horace Moses decided to meet with the workers, agreeing to address their concerns by June 15 if they immediately returned to their stations. The following morning the discontented employees rejected the company's compromise, "insisting that they would not go to work until they received higher wages." Chino's officials acted swiftly by firing everyone involved, made them leave town altogether, and put them on a blacklist circulated to each of the western copper properties. Although production suffered over the next few weeks, Sully and his executive staff showed no tolerance for labor organization and, in this case, the threat of a strike. This paternalistic power structure punished labor organizers regardless of skill level while rewarding most loyal employees who obeyed company rule.[70] On December 23, in fact, Sully sent a personal note to eleven men. Among the recipients were T. J. Broadwell, a locomotive engineer, and Harvey Forsyth, a shovel hand (who would "earn" promotions to shovel engineer and then shovel department foreman soon thereafter). They had refused to sign the grievance. In his holiday note, Sully announced that they would be receiving a Christmas bonus for their loyalty to the company. He awarded them an extra month's wage, "as a token of appreciation for sticking to your post during the labor problems last May." Unfortunately, loyal fireman Luis Horcasitas never received a bonus letter, revealing that Anglos who abided by company law received preferential treatment over Hispanic ones.[71] Maintaining corporate hegemony over workers resulted in unfair practices by the company that targeted activist workers for punishment and passive ones for rewards unless, of course, they had a Spanish surname.

During the next few years Chino officials clarified for their workforce acceptable behaviors for employees, solidified their stand against unions, and supported mutual aid societies. At the same time, they subjected Latinos to discriminatory policies during layoffs. Even when some of the copper companies in Arizona recognized craft unions, Chino would not. Perhaps the isolation of southwestern New Mexico facilitated their ability to block any kind of unionization. Because of borderlands politics in the 1910s during the Mexican Revolution and World War I, especially after Pancho Villa's violent exchange with the community of Columbus, New Mexico, in 1916

FIGURE 3.28. *Chino steam-shovel and train crews on "strike," May 29, 1912. The men posing for this picture walked off the job the day before this photograph was taken. They hoped to pressure the company into raising their wages to the level of similar positions in Utah and Nevada. Chino responded swiftly, firing all of them, and then sent a list of their names to copper properties all over the American West.* COURTESY OF SILVER CITY MUSEUM.

Steam Shovel and Locomotive Men Strike, Santa Rita, N. M.

and the Bisbee Deportation a year later, the company justified using strong-arm intimidation tactics against Mexican *and* Mexican American workers. Regardless, the company imposed its hegemonic will over all the employees, Hispanic *and* Anglo. In February 1916, for example, when Sully and Moses learned of an initiation ceremony of the Order of Locomotive Firemen scheduled for several Chino train crewmen, most of whom were Anglo, they fired everyone on the list. It included two brakemen, six firemen, and one loco engineer. The adamant anti-labor officials then sent these men's names to copper companies in Arizona, Nevada, Utah, and Montana. Corporate hegemony over workers extended to all the major copper properties in the American West. "Malcontents" who asked for higher wages and threatened a walkout in 1923 received the same treatment at Santa Rita. The company soon replaced these discharged workers with thirty Mexican men recruited in El Paso. In another incident soon after the 1916 complaints, Chino executives sent a warning to its security officers to be on the lookout for Hispanic labor recruiters from Clifton, Arizona. According to rumors, they had targeted the Santa Rita–Hurley complex for recruitment into the Western Federation of Miners.[72] Law enforcement officials in the company towns vigilantly sustained corporate power over the employees.

The company addressed the mutualista Alianza Hispano Americano quite differently. Viewed by officials as devoid of union and political leanings, the AHA had founded in 1906 Victory Lodge No. 21 at Santa Rita (see figure 3.29). Present since the 1870s throughout the American Southwest, mutual aid societies had bolstered morale among Mexican workers seeking opportunities in the United States. By the early twentieth century, AHA chapters surfaced to offer members some insurance and burial benefits as well as places for social activities like shooting pool, playing cards, and drinking beer. Similar to Masonic lodges and ladies' auxiliaries among Anglos, the AHA served as a kind of ethnic social club that took care of its own in times of crisis, saving the company costs for assisting accident victims and their families while also providing a generally positive atmosphere for workers during off-duty hours and layoffs. Rarely, however, did AHA chapters get involved in controversial on-the-job disputes, and there exists no record of Lodge No. 21's involvement in labor issues at Chino. Arizona's Hispanic union organizers, in fact, criticized mutualistas

because they seemed to substantiate the stereotype of "docile" Mexicans. They also provided financial benefits that labor leaders believed were the responsibility of the copper companies. The society did nothing, for instance, when Chino officials unfairly laid off Latino laborers during slowdowns and then reassigned Anglo employees to some of those positions. The AHA clearly served two company purposes: as a benevolent association to counter unionization, and as a Hispanic-identified social club that tacitly approved of occupational and residential segregation practices at the mine operations and in the company towns of Santa Rita and Hurley.[73]

By the mid-1930s, Chino's Hispanic workers believed that unionization was imperative to provide a vehicle for eliminating the paternalistic injustices imposed on them from the beginning of operations. For them labor rights would soon come to mean civil rights.[74] Because the company identified their ethnicity with the type of work they were hired for and then allowed to perform, they resented the limits to their opportunities. As late as 1950, lower-paid "laborers" were almost exclusively Mexican and Mexican American. The laborer title became synonymous with "Mexican job." José M. Martínez, for example, began his tenure at Chino in 1918 as a track gang member only to retire in that same position in 1953. Wage compression also penalized some of the workers, as was the case in 1934 when newly hired laborer Julián Horcasitas got $3.65 a day compared with Francisco Costales's $2.60. Costales, however, had been a track laborer with the company since 1923. Countless instances of this kind of unfairness must have infuriated many of the Hispanic workers, who without a collective voice felt powerless to do anything. Mine Mill, they believed, could negotiate to do away with these discriminatory practices.[75]

Another concern for the Latino workers was safety, especially concerning the dangers inherent to the laborers' type of work. Each year several men, most with Spanish surnames, lost their lives at the New Mexico operations (see Chino Fatalities in the appendices). Though there were various reasons for the deaths, the fatalities commonly involved falling boulders, flying blasted rock, railroad accidents, and mishandling of machinery and explosives. In July 1911, for example, when Eugenio Molina opened a can of black powder, it exploded in his face, burning him so badly that he died thirty-six hours later in the Santa Rita Hospital. Others died from negligence, as

FIGURE 3.29. *The Alianza Hispano Americano Lodge No. 21, ca. 1921. Mexican and Mexican American mine workers gathered at Lodge No. 21 (founded in 1906 four years before the Masonic lodge) to socialize and play table games. Before modern-day benefit packages, the AHA offered Chino's laborers life and burial insurance. The building burned to the ground on December 13, 1933.*

COURTESY OF SILVER CITY MUSEUM.

was the case of Emilio Escarsiga in April 1913. Instructed by his foreman not to move a flatcar loaded with ties and rail, he chose to remove the wheel chock, thinking he could control the car's speed. Unfortunately, he miscalculated and the car jumped forward. One man leapt off but Escarsiga stayed aboard. When the heavily laden flatbed slammed into another loaded waste train, he was instantly crushed to death. Anglos were not immune either to these kinds of dangers, as was the case in August of that same year. Head machinist M. H. Longnickle, shovel engineer F. Martin Whitman, and fireman William Byrd all died instantly when the percussion from an ore blast violently threw the shovel they were hiding behind into the mountainside, crushing them (see figure 3.30).

In 1913 alone, thirteen men lost their lives. Three years later seven fatalities visited Santa Rita. Although the company instructed the men to be careful, unsophisticated safety training, employee negligence, and the realities of working under dangerous conditions could not eliminate severe accidents and fatalities during the life of the mine. Cruz Sereseres was electrocuted to death in February 1917 at the new crusher. Pitman Ausencio Herrera was killed near steam shovel Number 4 in August when a large boulder seemed to appear out of nowhere from the blasted bank above, rolling over the unsuspecting victim. Twenty-seven-year-old locomotive engineer Joseph Worrel died in February 1918 in a freak accident when an ore slide struck a steam shovel, causing

FIGURE 3.30. *Steam-shovel accident, 1917. The shovel runner took on too much weight when he scooped the large boulder still in the dipper in this photograph. Though no one was killed in this accident, some workers were not so fortunate on other occasions.*
COURTESY OF PALACE OF THE GOVERNORS PHOTO ARCHIVES (NMHM/DCA).

the crane section to swing around and strike the cab of a passing locomotive, breaking its steam pipe and scalding the unfortunate young man. He perished two days later. Clyde Lucas died in August 1925 when a swing chain from one of the shovels came loose and viciously struck him in the head. In March and June 1927, first Manuel Reyes and then Ysario Padilla lost their lives in rockslides.[76]

Accidents often left workers maimed. In March 1912 trackman Bentura Minjares lost a leg, and then in August track cleaner Ysiquare Lovato lost his left arm at the shoulder after each was hit by a train. When steam-shovel engineer Marion Portwood was inspecting his digging machine in August 1913, a gust of wind blew his jumper into the cog wheels, pulling him and his arm into the moving teeth of the machinery. Knowing his life was in danger from loss of blood, he calmly walked to the nearby Santa Rita Hospital, where Dr. Carrier amputated the mangled arm. In July 1919 mechanic W. C. Rock lost an eye in the machine shop after cutting a bolt with a hammer and chisel. The violent force caused a metal chunk to fly upward and deep into the soon-to-be-

removed eye. In April 1923, Victor Crittendon and D. C. Reynolds both suffered severe injuries when a hoist chain on the Number 8 shovel broke loose and pummeled each man. Crittendon received a laceration on his head and Reynolds a broken femur. Numerous such accidents visited the mine works annually before the 1950s.

Perhaps even more common were near misses. On September 25, 1925, for example, five powdermen nearly suffered grave injuries, with one barely escaping with his life. Just as the crew lit twenty-five packs of dynamite called "dobies" during secondary blasting, a boulder suddenly rolled down the embankment and pinned Alfonso Sias against a rock near the blast area. Yelling for help, Sias was relieved when Matilde Robledo and Teodoro Ríos came to his aid. Meanwhile, Simon Sias jerked the burning fuses from their targets. In his heroic effort, Simon disarmed thirteen of the dobies before the other twelve began to explode in sequence. Fortunately, the first blast forced the boulder that had trapped Alfonso to move just enough for his two rescuers to free him. They all then scurried for cover behind the shovel while the remaining dobies exploded. Miraculously none of them was injured. Three years later Robledo and Ríos won bronze medals for their actions from the Carnegie Hero Fund Commission of Pittsburgh, Pennsylvania.[77]

These dangerous conditions, combined with the workers' desires for better safety, higher wages, clearer lines of promotion, and a seat at an employee-company bargaining table, inspired them in 1934 to unionize. Laborers as well as skilled employees initially sought representation collectively in a single union, the International Union of Mine, Mill and Smelter Workers, even though many of the Anglos in the early 1940s would leave Mine Mill to join their respective craft unions. Sully's successor, Rone B. Tempest, a reserved and pensive man who was distrustful of unions, did understand the emerging hopes of Chino's employees. Yet he was unwilling to support union certification. So almost immediately on taking over as general manager, he established in July 1933 the Employee Representation Plan.[78] His expectation in appointing compressorman Joseph I. Kemp as the first employee representative was to short-circuit labor-organizing efforts through the creation of a company union.

Kemp, however, turned out to be a pro-union man despite his own good wage and believed management treated the workers unjustly. Within a few months of his appointment, he wrote to Mine Mill's executive secretary requesting union cards and successfully gained a charter on March 24 for Local 63. Tempest and Jackling were livid. Only eight months earlier, coal miners at Chino's subsidiary, the Gallup American Coal Company, had rejected general manager Horace Moses's invitation to start a company union. The ensuing ugly labor struggle forced the hand of the employees, most of whom were Hispanic, and they soon joined the communist-connected National Miners Union. Kennecott's officials would have none of this at Chino. When Chino employees of all stripes—from skilled machinists and locomotive engineers to track cleaners and other common laborers—began joining Locals 63 and 69 to the tune of 300 members by the summer of 1934, Jackling immediately planned for a lengthy shutdown. He then used the economic distresses of the Depression to justify his actions. Paternal traditions and corporate hegemony came under threat with the workers' action. Historic segregation of workers based on ethnicity and occupation was coming under scrutiny, as were the company's regular paternalistic practices of a sliding pay scale and a dual-wage

system. Consequently, the officials' unwillingness to negotiate served initially to break down longtime barriers between Anglo and Hispanic employees, who united to organize collectively. In September the workers voted 302 to 221 in favor of certification in Mine Mill.[79]

By February 1935 the officers of Local 63 boldly began efforts to alter company-labor dynamics at Chino. That month the new union officials filed a complaint with the regional labor board of the NLRB. They claimed that the corporation had shut down the previous year to thwart the formation of unions. Already rebelling against occupational segregation practices, the nine elected representatives of the local—six Anglo and three Mexican Americans—"charged" Kennecott with "interference, restraint and coercion."[80] The company, they claimed, closed the operations because of officials' visceral resistance to the recent certification process and the workers' "concerted activism for the purpose of collective bargaining." The brash new labor leaders argued that the company violated section 7a of the NIRA. Initially, the federal labor board ruled in the company's favor after hearing testimony from witnesses for both sides (ironically, Pablo Arellano, a track foreman, and Plutarco Pedraza, a powder foreman, supported the company's position, showing how management allied across ethnic lines in working to suppress labor activism). By May, though, the NLRB agreed to review the case on Mine Mill's appeal.[81]

The NLRB did not rehear the case until May 1938, sixteen months after the Chino Mines Division restarted operations. As many as 104 witnesses testified before a federal examiner during early May and mid-June hearings to the packed Grant County Courthouse in Silver City. Mine Mill reiterated its charges against Kennecott for violation of section 7a (now part of the Wagner Act), marching ninety pro-labor witnesses before examiner Joseph Kiernan of the regional NLRB headquartered in Los Angeles, California. Fourteen company supporters gave their views too.

The hearings revealed that the company had intimidated workers prior to the vote to unionize. Assistant mine superintendent Roy Grissom, for example, had asked Felipe Huerta in 1934, "Why are you such a strong union man?" Huerta said he hoped to have a process for improving workplace conditions. The manager retorted, "It is better for you to think it over or not vote for the union because this camp has been working and operating for twenty-five years and you have been protected. If the Union is successful the camp will stop for a year and you will lose your work, possibly you will not get it again . . . leave those outsiders out." The company official also told José Portos that employees needed "to be careful and vote the proper way, especially the Mexicans." Several of Chino's officials—mine superintendent Harry "Cap" Thorne and mill foreman Claude Dannelley among them—said similar threatening things to other Hispanic workers. Testimony revealed that Antonio Cruz apparently "got under Thorne's skin by 'constantly exciting and constantly agitating the Mexicans.'" He was threatened with termination. Such treatment compelled union organizers to hold meetings in secret at various locations during their efforts to seek an employee vote on certification.[82]

Pro-company testimony revealed that workers had also made threats of their own. Dutch Stewart allegedly informed the minister of the Hurley Community Church that Kennecott's machinery, its shops, and even company houses would be destroyed if obstructions to labor organizing continued. They would also "get" general manager Tempest and mine superintendent Thorne. Tempest purportedly feared that the union men would kidnap his children, so he began carrying a pistol. The general manager's suspicions must have been palpable because he decided to take his own life on May 7 in the midst of the initial period of testimony. No doubt conflicted over his loyalties to the company and knowing of his own illegal behavior in personally putting together a blacklist of union workers, Tempest was at a crossroads. He was not sure what action to take on the eve of being ordered by NLRB's officials to turn over the records that would have revealed his guilt. So he killed himself in his garage instead of showing up to testify. He never had to face the humiliation of his actions. In the end, the workers' testimony, their willingness to vote for unionization, and now evidence that the company closed because of the formation of the union led the NLRB to rule in November 1938 in favor of Local 63. The board ordered Chino to rehire seventy-one of the blacklisted workers, give them back pay, and allow them to return to their jobs as members of the union.[83] Mine Mill had won the case at least momentarily. Corporate hegemony took a big hit with Tempest's tragic death symbolizing the emergence of a new order at Chino.

Kennecott, of course, appealed the NLRB decision and the case was reopened in December 1939. Again, the federal labor authority ruled against Chino, directing the company to rehire the blacklisted workers and pay them lost wages. Another company appeal followed, this time before the Tenth US Circuit Court, which in September 1941 reversed the NLRB's ruling. An NLRB counter appeal to the US Supreme Court on behalf of Local 63, however, resulted in finality for the case. In April 1942, in fact, the nation's highest court upheld the earlier NLRB ruling. The court declared that Kennecott had indeed violated the Wagner Act. Soon thereafter sixty-seven of the aggrieved employees were rehired and given compensation. The copper workers' rights to unionize were legitimized and management-labor relations could go forward. Ironically, the traditional occupational segregation practices of the company were soon reinstated but not by management. Skilled workers, many of whom left Mine Mill in 1941, decided instead to join craft unions representative of their jobs and their Anglo ethnicities (see figure 3.31).[84]

The Supreme Court decision boosted Mine Mill's efforts to entice more Chino workers into the labor collective. Union officials, emboldened by the defense mobilization effort, scarcity of labor, and patriotic contributions of Hispanic Americans to the impending war, already were targeting Mexican Americans for membership in Locals 63 and 69. In a May 1941 meeting of Local 63, in fact, vice president Julián Horcasitas reassured skeptical attendees that "the union is here to stay come hell or high water." Over the next few months several hundred mine and mill workers at Santa Rita and Hurley joined the union under principally Mexican American leadership. Though Mine Mill membership would continue to include Anglos (especially those who had been blacklisted), the emergence of the Chino Metal Trades Council, a collective of eleven AFL craft unions, altered the cross-ethnic makeup of the CIO-affiliated union. Anglos who associated their ethnicity with their jobs chose the craft unions over the industrial one. Yet because of the traditional segregation policies of these skilled collectives, the Latino workers, even if they held craft jobs, which began to be available to them because of new federal laws and wartime labor shortages, chose to join Mine Mill.

This reality during the 1940s gave the industrial union a near monopoly on bargaining at Chino. The historic exclusion of Mexican Americans from craft union membership, Chino's future labor relations director Barney Himes would later contend, worked to empower

FIGURE 3.31. *Skilled workers gathered for this snapshot outside the maintenance shops in downtown Santa Rita, November 10, 1937. Standing, left to right: William A. Smith, Bill Potten, Francis "Merle" Hickman, Gumesindo Ortiz, Fred Spanable, Bill Emerson, Marshall Early, D. L. Beilue, Mr. Reed, Manuel Gutierrez, Jimmy Bourne, and Riley Mabben. Sitting, left to right: Jim Blair, Fred Adams Sr., Jack Craig, Tom Creswell, Russell Holmes, Ed Peters, Elmer Votow, Henry Perrault, Johnny Hickman, Henry Murphy, Allen Peters, Mr. Smith, and G. A. Stewart.*

the Hispanic workers. They trusted Mine Mill. In the end, Himes observed, the union held "the largest, most crucial of all [bargaining] units . . . transferring great bargaining power [from the craft unions] to . . . Locals 63 and 69." Ironically, Mine Mill had used exclusion of Mexican Americans as a recruiting tool only a generation earlier. Now the labor organization championed the Chicanos' cause, which was reflected in the dramatic difference in which Kennecott treated its workers for decades to come in terms of bargaining for wages, promotions, seniority, and other expectations.[85]

With this success Mine Mill soon won new rights for Chino's workers. Foremost among the achievements was an "Agreement Covering Wages and Working Conditions," effective July 7, 1942, between Kennecott and Mine Mill Locals 63 and 69. Company-labor negotiators agreed to numerous provisions "in order to promote harmonious relations between the Company and its employees and to facilitate the peaceful adjustment of differences." In the eight-article contract the company agreed to various worker demands. Chino officials contracted in Article I to recognize Mine Mill "as the sole collec-

tive bargaining agent" of the employees and to in no way interfere with new hires' rights to join the union. They also claimed, they "will not discriminate against any employee . . . because of union membership . . . or . . . race, creed, color or national origin." The company representatives in Article II accepted the eight-hour workday and agreed to pay any overtime at one-and-a-half times the normal wage. Article III clarified the role of seniority at Chino, ensuring that employees earned credit from the day they gained employment regardless of transfers and missed time from illness, accidents, vacations, leaves of absence, and military service.

Article IV addressed policies concerning promotions. Advances, the agreement read, "shall be made available to all and shall depend upon an employee's qualifications, experience and suitability for the position." The next article defined the new grievance policy. It focused on seeking solutions with supervisors and company officials first before taking unresolved conflicts to a Joint Arbitration Committee. The committee was to be made up of two company and two union representatives and one agreed-upon "independent," non-prejudiced member, "preferably

FIGURE 3.32. *Pinder Shops, August 5, 1943. With the inevitable growth of the open pit, Kennecott officials decided in the late 1930s to move the maintenance repair shops from the downtown area to the Pinder claim, as seen here in this photograph. At these shops, machinists, mechanics, electricians, welders, and other maintenance crew members repaired shovel parts, locomotives, and other machinery. The workers mined this entire area by the mid-1950s, requiring the company to move the shops again to their present location less than a mile to the west.*
COURTESY OF TERRY HUMBLE.

a citizen of good repute, [and] resident of Grant County." Workers were instructed to file their grievances within five days of the "discipline, demotion, or discharge." Those fired because of "intoxication [or] willful disobedience of safety laws" surrendered their rights to a hearing. Time off was covered in Article VI, which awarded employees regular pay during vacations and time-and-a-half for Independence and Labor Days and on Christmas. The parties agreed in Articles VII and VIII to the "prevailing wage" (beginning at $4.60 a day for unskilled and $7.50 for skilled employees) during the course of the contract, to run through June 12, 1943.[86] A new era in labor-management relations had dawned at Chino.

Labor's gains at Chino by the early 1940s mark a transition in worker-management relations at the New Mexico copper enterprise. The timing was fortuitous as well. Kennecott had just completed a five-year rejuvenation program (1937–1942) that ensured long-term production, unprecedented profits, and employment opportunities for generations to come. Investing nearly $9 million, Kennecott upgraded Chino's mining equipment with fleets of larger electric shovels and locomotives and also completed the electrification of the pit. The company built the new Pinder shops to maintain the new machinery (see figure 3.32). It revamped the concentrator at Hurley with new Symons cone crushers, Marcy ball

FIGURE 3.33. *Kennecott smelter at Hurley, January 22, 1939. From 1937 to 1942, Chino completed a five-year rejuvenation program to upgrade mining and milling technology as well as construct a new copper smelter, seen just after its completion in this photograph. Note the Santa Rita Store, the largest building at the end of Cortez Street on the left, and El Tejo movie theater on the right side of the street. The 501-foot stack symbolized Chino's coming of age in the copper industry. Later Kennecott added another taller stack. When the current owner of the operations, Freeport-McMoRan Copper & Gold Inc., demolished the two stacks on June 5, 2007, locals lamented the end of an era at Chino.*

FIGURE 3.34. *The "island," downtown Santa Rita, 1942. This scene of downtown Santa Rita reveals the encroaching pit as it eats away at the landscape. Always the "elite" part of town, this section, known as Quality Hill, was the location of the general manager's home, the hospital, the teacherage, the Chino Club, the company store, the post office, and the fancier homes of the copper town. The copper water pond in the foreground became a favorite "forbidden" playground for the more adventuresome children. During the 1940s and 1950s, Kennecott steadily removed all the buildings to make way for the expansion of the pit.*
COURTESY OF TERRY HUMBLE.

mills, and two-stage regrinding classifiers; constructed a new molybdenum milling section (profits from "moly" through 1954 equaled the entire cost of the rejuvenation program); and increased leaching and precipitation production. Most importantly, the corporation invested $5 million to build a new smelter whose manly 501-foot smokestack rose above the Hurley mill complex to symbolize the modernization of the operations (see figure 3.33). As a result of the rejuvenation program, Chino

Mines began annually producing more than 100 million pounds of copper, a figure that would remain constant through 1980 with exceptions in 1946, 1959, and 1967 (the last two were strike years). Profits and wages also reflected this success, ensuring the gradual increase in overall benefits to the company's workers in the decades after World War II.[87]

Chino had come a long way since 1910 when the No. 2 shovel dug the first bucketful of ore. The pit ranked fifth in production internationally and third domestically at the peak of World War II.[88] By this time workers had also won legal protections under federal law and certified representation in Mine Mill and the craft unions. Chino was no longer dependent on ASARCO for smelting. With other advances in technology, employment and profits were secured for decades to come. Although the expansion of the pit had reduced Santa Rita's main section of town to an island (see figure 3.34), the company town was still accentuated with company-built homes, stores, a hospital, clubs, and the mine works themselves. Under Sully's paternalistic tutelage these remnants of the past began to seem like relics in the face of the new monuments to the future—organized labor and modern technology. No longer did the workers want to be treated like dependent children, and they would fight unfettered corporate hegemony as a result. The carefully planned and constructed industrial and social landscapes were in transition as a result. Still, the workers understood the inevitability of the demise of Santa Rita at the hand of massive economies-of-scale technology and they would even support the company's fight against environmental regulations. The progression of the gigantic open pit's benches translated into prosperity for the future. Smelter smoke signified good-paying jobs. The rise of Santa Rita as a company town and then its eventual disappearance into "space" under the assault of newer technology are the focus of the next chapter.

NOTES

1. George H. Hildebrand and Garth L. Mangum, *Capital and Labor in American Copper, 1845–1990: Linkages between Product and Labor Markets* (Cambridge, MA: Harvard University Press, 1992), 96–103; E. B. Alderfer and H. E. Michl, *Economics of American Industry* (New York: McGraw-Hill Book Co., 1950), 92–94.

2. Chino Copper Company (hereafter CCC), *First Annual Report*, June 1, 1910; CCC, *Report*, March 25, 1911; CCC, Monthly Report, #14, January 12, 1911; *Silver City Enterprise*, July 1, September 30, December 9, 23, 1910; *Silver City Independent*, May 24, September 27, December 20, 1910, February 28, 1911; T. A. Rickard, "The Chino Enterprise—II: Later History and Formation of Present Company," *Engineering and Mining Journal-Press* 116:19 (November 10, 1923): 810; Arthur B. Parsons, *The Porphyry Coppers* (New York: Arthur B. Parsons, 1933), 3–5. By the early 1930s, Chino would rank third among all copper mines in the world.

3. T. A. Rickard, "The Chino Enterprise—IV: Methods of Prospecting and Mining," *Engineering and Mining Journal-Press*

116:26 (December 29, 1923): 1116; Rickard, "The Chino Enterprise—II," 804; John Sully II, "Copper King," www.desert exposure.com/200710/200710_santa_rita_copper.php, accessed October 2007; *Silver City Independent*, January 17, February 7, June 6, 1905.

4. Rickard, "The Chino Enterprise—II," 804.

5. Rickard, "The Chino Enterprise—IV," 1113; Rickard, "The Chino Enterprise—II," 807. See Parsons, *The Porphyry Coppers*, 12, who claimed that "this technical progress [in developing the porphyry mines in the American West and elsewhere] was attained by engineers engaged directly and indirectly in the exploitation of the Porphyry Coppers . . . Hundreds, perhaps thousands, of individual engineers contributed. Scientists, researchers, mechanical and electrical engineers of whom many were in the employ of manufacturers of equipment, mine operators, mill-men, ore-dressers, metallurgists, all played . . . an important part. New methods—even processes—were devised; old ones were improved. New devices and machines were invented." Also see Clark C. Spence, *Mining Engineers and the American West: The Lace-Boot Brigade, 1849–1933* (Moscow: University of Idaho Press, reprint edition, 1993), chapter 11, "A Most Important Missionary of Civilization," 361–370, which shows the mining engineers' belief in their role in bringing the benefits of modern technology to places like Santa Rita in the American West. Parsons, *The Porphyry Coppers*, 209, complimented Sully's effort by comparing his modest $10,000 expenditure to engineer Krumb's (spelled Kraumb elsewhere) $150,000 price tag for examining the Utah Copper (Bingham Canyon) at the same time.

6. *Silver City Enterprise*, January 29, 1909; Rickard, "The Chino Enterprise II," 804, 807–809; John M. Sully, "Chino Copper Company: The Story of the Santa Rita del Cobre Grant and Its Development from Discovery to the Present Date" (reprint from Official State Book for Distribution at Panama-California Exposition at San Diego, 1915), 12, 17; Parsons, *The Porphyry Coppers*, 208–209; Sully, "Copper King."

7. Rickard, "The Chino Enterprise—II," 804–806; CCC, *Ten-Day Report*, May 24, 1922. Rickard interviewed Sully concerning the incorporation of the Chino Copper Company and then stayed at Santa Rita for about a week to examine the properties firsthand; see Melissa D. Burrage, "Albert Cameron Burrage: An Allegiance to Boston's Elite, 1859–1931" (master's thesis, Harvard University, 2004). This outstanding thesis reveals the many and varied financial ventures of A. C. Burrage and how he represents the Boston elites' and the Rockefeller Syndicate's pervasive roles in copper mining in the United States and elsewhere in the world.

8. Burrage, "Albert Cameron Burrage," 221–249, 276–304.

9. Ibid., 805.

10. Burrage, "Albert Cameron Burrage," 221–249.

11. Quoted from Rickard, "The Chino Enterprise—II," 805.

12. Quoted from Sully, "The Chino Copper Company," 12.

13. Rickard, "The Chino Enterprise—II," 806, 810; Sully, "The Chino Copper Company," 12; Burrage, "Albert Cameron Burrage," 223–225.

14. Rickard, "The Chino Enterprise—II," 806; Burrage, "Albert Cameron Burrage," 240, 244.

15. Rickard, "The Chino Enterprise—II," 805–810.

16. *New York Times*, May 20, 1909; also see http://en.wiki pedia.org/wiki/Henry_Huttleston_Rogers, accessed July 2010.

17. Sully, "The Chino Copper Company," 12.

18. Rickard, "The Chino Enterprise—II," 806; Parsons, *The Porphyry Coppers*, 211–212; Hildebrand and Mangum, *Capital and Labor*, 77–78.

19. Rickard, "The Chino Enterprise—II," 806. Sully reported in 1915 that Chino "has an authorized capitalization of 900,000 shares of a par value of $5 each. Of this authorized capital 870,000 shares are issued, leaving 30,000 shares in the treasure"; see Sully, "The Chino Copper Company," 14. Also see Burrage, "Albert Cameron Burrage," 244, 248.

20. Rickard, "The Chino Enterprise—II," 806–807; Sully, "The Chino Copper Company," 14; Burrage, "Albert Cameron Burrage," 227–230, 234–235. A. C. Burrage also experimented with steam-shovel technology, advocating with Sully and Jackling this technique at the Chino open pit.

21. Sully, "The Chino Copper Company," 14; Parsons, *The Porphyry Coppers*, 3.

22. James O. Clifford, "Interesting Review of Chino's Mines and Methods," *Mines and Methods* (August 1912): 549; CCC, *Second Annual Report*, February 27, 1912, Monthly Reports #29 and #30, May 3, 1912, June 3, 1912.

23. "The Chino Copper Company," *The Mogollon Times* (1913): 36; *Silver City Daily Press*, November 29, 1945; Dale Giese, *Forts of New Mexico* (Silver City, NM: Dale Giese, 1991), 18; Claude A. Dannelley, "Memories of Hurley," 3, Special Collections, Miller Library, Western New Mexico University, Silver City.

24. *Mattie N. Dowell vs. Chino Copper Company*, Civil Action No. 4524, Sixth Judicial District Court, Grant County, New Mexico, October 1, 1912, 1–2.

25. Ibid., 5. Dowell eventually sold the land to the Chino Copper Company; see "Warranty Deed," *Miscellaneous Deed Record*, Grant County, Book 69, March 3, 1924, 188.

26. See Duane A. Smith, *Mining America: The Industry and the Environment, 1800–1980* (Lawrence: University Press of Kansas, 1987), 115–116; and Donald J. Pisani, *From Family Farm to Agribusiness: The Irrigation Crusade in California and the West, 1850–1931* (Berkeley: University of California Press, 1984), 162–165. Mining companies in California, Colorado, Montana, and other states where settlers lived closer and in larger numbers to the mine works faced complaints like Dowell's and from towns and cities far more often. Like Chino in this case, the companies either bought the lands or paid the landowners for the right to pollute. Many companies used the same strategy involving air pollution cases.

27. See John D. Wirth, *Smelter Smoke in North America: The Politics of Transborder Pollution* (Lawrence: University of Kansas Press, 2000), 1–79.

28. Sully, "The Chino Copper Company," 13.

29. Rickard, "The Chino Enterprise—II," 810.

30. *Fourteenth Census of the United States, 1920*, Grant County, New Mexico, Precinct 13, Santa Rita (Washington, DC: GPO, 1921), 77–111; *Fifteenth Census of the United States, 1930*, Grant County, New Mexico, Precinct 13, Santa Rita (Washington, DC: GPO, 1931), 119–132, 135–160. The census figures for Santa Rita are as follows: 1860: 214; 1890: 133; 1900: 1,874; 1910: 1,951; 1916: 3,500; 1920: 3,565; 1926–1927: 5,000; 1930: 3,889; 1940: 2,565; 1950: 1,862. Also see Zaragosa Vargas, *Labor Rights and Civil Rights: Mexican American Workers in Twentieth-Century America* (Princeton, NJ: Princeton University Press, 2005), 46–61, 265–270, and David J. Weber, "'Scarce More Than Apes': Historical Roots of Anglo-American Stereotypes of Mexicans," in David J. Weber, ed., *Myth and History of the American Southwest* (Albuquerque: University of New Mexico Press, 1987), 153–167, for two discussions of racist treatment of Mexicans and Mexican Americans in the Southwest. Also see James W. Byrkit, *Forging the Copper Collar: Arizona's Labor-Management War of 1901–1921* (Tucson: University of Arizona Press, 1982), for a discussion of the Bisbee Deportation.

31. Copper precipitation was an ongoing process. When a shovel crew uncovered timbers in 1915 from one of the Spanish mine shafts, they discovered that the timbers contained an average of 7 percent copper that had precipitated into them over the previous eighty years or so, revealing the power of geologic forces in redistributing the copper even during the life of the mining operations; see Rickard, "The Chino Enterprise—III," 985.

32. Rickard ("The Chino Enterprise—III," 985) claims that the Spanish-speaking laborers called the porphyry ores by a slang term, *mosceado*, rooted in *mosca*, or fly, perhaps to denote "fly-specked," a metaphor for the greenish black specks in the ore. The correct spelling is *mosqueado*.

33. Parsons, *The Porphyry Coppers*, 344–345, and for elaboration on porphyry geology, see 338–354; Clifford, "Interesting Review," 547–548; T. A. Rickard, "The Chino Enterprise—III, Geology of Santa Rita," *Engineering and Mining Journal-Press* 116:23 (December 8, 1923): 981–982.

34. Sully, "The Chino Copper Company," 13; Sully to Jackling correspondence, January 1, 1910; CCC, *First Annual Report*, June 1, 1910, *Annual Report* (1912), 8; CCC, Weekly Reports, March 6, May 8, June 12, July 16, 23, 30, August 17, 1909; *Silver City Enterprise*, January 7, March 25, May 14, August 6, 20, September 30, 1910; *Silver City Independent*, June 8, 1909, May 3, 1910; n.a., "The Chino Copper Co.," *Mines and Minerals* 32:11

(June 1912): 668; Rickard, "The Chino Enterprise—IV," 1118; also see Parsons, *The Porphyry Coppers*, 355–368, for elaboration on churn-drill prospecting and calculating of ore bodies.

35. Quote from Rickard, "The Chino Enterprise—III," 985.

36. Ibid.; CCC, *First Annual Report*, June 1, 1910; CCC, Monthly Report, March 25, 1911, 1119.

37. Ironically, the Bingham Mine's first shovel started on the 6,341-foot bench and the Nevada Consolidated Mine at the 6,300-foot bench; see Rickard, "The Chino Enterprise—IV," 1117.

38. CCC, Monthly Report, March 5, 1913; CCC, Eighth Quarterly Report, October 25, 1913; *Silver City Independent*, July 1, 8, 15, October 21, 1913; Rickard, "The Chino Enterprise—III," 985; *Silver City Enterprise*, June 27, 1913. Eventually the Society for People Born in Space would be established in the 1970s to memorialize the onetime town of Santa Rita.

39. Rickard, "The Chino Enterprise—IV," 1118–1119.

40. Quoted from Rickard, "The Chino Enterprise—IV," 1119.

41. Quoted from ibid., 1117.

42. CCC, *First Annual Report*, June 1, 1910; CCC, Monthly Report, March 25, 1911; see Parson, *The Porphyry Coppers*, 369–392, for elaboration on steam shoveling at open-pit operations.

43. Rickard, "The Chino Enterprise—IV," 1119–1120; CCC, First Quarterly Report, May 7, 1912, Third Quarterly Report, October 26, 1912, Fourth Quarterly Report, February 8, 1913; Sully to Jackling correspondence, August 1912; *Silver City Enterprise*, March 29, October 18, November 8, 1912.

44. Rickard, "The Chino Enterprise—IV," 1118–1119; CCC, *Annual Report*, 1913; Sully, "The Chino Copper Company," 13.

45. Charles A. Dinsmore, "Mining in the Santa Rita District, New Mexico," *The Mining World* 31:21 (November 20, 1909): 1028; CCC, Second Quarterly Report for 1912, July 31, 1912; Sully to Jackling correspondence, August, 1912; John M. Sully, "Milling the Ore of the Chino Mine," *Mining and Scientific Press* (March 30, 1912): 465; T. A. Rickard, "The Chino Enterprise—V: Concentrating the Ore at Hurley and Smelting at El Paso," *Engineering and Mining Journal-Press* 117:1 (January 5, 1924): 13; also see Parsons, *The Porphyry Coppers*, 212–213 (and page 436 for a drawing of a Chilean mill).

46. Crushing and milling of ores were a very complex process that required delivery of the materials through various pieces of machinery and phases of reduction. This description is simplified and abbreviated. To see the flow sheet of the early crushing and milling processes, see John M. Sully, "Milling the Ore of the Chino Mine," 464–466; also see T. A. Rickard, "The Chino Enterprise—V," 13–15, for upgrades through 1922 in milling, especially the introduction of the flotation process.

47. N.a., "The Chino Copper Co.," *Mines and Minerals*, 669–670; Hildebrand and Mangum, *Capital and Labor*, 84.

48. Ibid.; Sully, "Milling the Ore," 466; Parsons, *The Porphyry Coppers*, 431.

49. For elaboration on the flotation process, see Parsons, *The Porphyry Coppers*, 441–445.

50. *Silver City Enterprise*, March 31, 1914.

51. Hildebrand and Mangum, *Capital and Labor*, 53, 64; see Dawn Bunyak, "The Inventor, the Patent, and Carrie Everson: Defining Success," *Mining History Journal* 12 (2005): 9–24.

52. Rickard, "The Chino Enterprise—V," 13–14.

53. See Smith, *Mining America*, 82–86. Hammonds, Waldemar Lindgren, and other mining engineers warned of the need to conserve key natural resources like coal, water, and timber to ensure the continuation of "efficient" mining in the twentieth century. Large-scale corporate efficiency appeared most obviously in the strip and open-pit mining strategies of the iron and copper industries. Although profitability and the perpetuation of mining were their main concerns, they participated in various conservation discussions, including attendance at the National Conservation Commission meetings, to promote efficiency and conservation of natural resources in the mining industry. Smaller operators, ironically, had a much harder time effectively implementing conservation practices because of the costs of using large-scale technologies to achieve resource savings, as exemplified by the costs Chino incurred to acquire and conserve water.

54. John M. Sully, "Milling the Ore of the Chino Mine," *Mining and Scientific Press* (March 30, 1912): 464; J. T. Shimmin, "The Hurley Mill, Water Supply and Power Plant," *American Mining Congress* (July 25–26, 1927), 4; *Silver City Enterprise*, July 9, 1915; Fred Hodges, *Milling Methods at the Hurley Plant of the Nevada Consolidated Copper Company, Hurley, New Mexico*, Information Circular 6394, US Bureau of Mines (Washington, DC: Department of Commerce, 1931), 15; Paul M. Jones, *Memories of Santa Rita* (Silver City, NM: Silver City Enterprise, 1985), 11; William H. Goodrich, general superintendent of mines, *Annual Report, The Chino Mines Division, Kennecott Copper Corporation* (1944), 16; F. C. Green, general superintendent of mills, *Annual Report, The Chino Mines Division, Kennecott Copper Corporation* (1945), 17; "The Chino Copper Company," *Mines and Minerals* 32 (June 1912): 669–670; Nevada Consolidated Copper Company, *Annual Report* (1927), 8; Thorne to Sully, *Reports*, October 14, 26, November 4, 14, 25, December 14, 1925, January 5, June 15, October 18, November 18, 1926; *Silver City Enterprise*, February 13, April 24, October 26, November 14, December 14, 1925.

55. J. O. Clifford, "Interesting Review of Chino's Mines and Methods," *Mines and Methods* (August 1912): 551; Hodges, "Milling Methods," 11; Shimmin, "The Hurley Mill," 4; Kennecott Copper Corporation, *Annual Report* (1942), 4; Horace Moses, *Annual Report* (1946), 11–12, 17; US Bureau of Mines,

Minerals Yearbook (Washington, DC: GPO, 1946), 1506; *Chinorama* 12:4 (December 1958): inside cover; *Chinorama* 7:4 (July–August 1961): 13.

56. Jones, *Memories*, 115–117. At this early date, the company did not address water pollution aside from the strategy of allowing it to drain into Santa Rita and Whitewater Creeks and allowing it to filter into the groundwater system of company-owned lands. The Dowell case reveals the corporate strategy until congressional passage of the Clean Water Act of 1967 and later clean water acts when EPA and New Mexico Environmental Improvement Division officials began to enforce cleanup through fines and threats of fines (see epilogue).

57. Chino Copper Company, *Water Diversion Plans*, December 1, 1924.

58. C. F. Ames, "Diversion of Santa Rita Creek at Chino Mines," presented at Santa Rita, New Mexico Chapter of American Mining Congress, July 25 and 26, 1927; *Silver City Independent*, April 14, 1925, July 19, 26, 1927; *Silver City Enterprise*, March 13, 1925, September 2, 1927.

59. Parsons, *The Porphyry Coppers*, 218–225; Nevada Consolidated Copper Company (hereafter NCCC), *Twentieth Annual Report* (1926), March 18, 1927; NCCC, *Twenty-Second Annual Report* (1928), March 22, 1929; NCCC, *Twenty-Third Annual Report* (1929), March 21, 1930 (on mill tonnage per day in mid-1920s). The mill treated about 3,000 tons a day prior to the new 10,000-ton increase even though the capacity was at 5,000 tons a day.

60. *Silver City Enterprise*, January 11, February 22, July 4, 1924; *Silver City Independent*, January 25, February 19, March 25, July 29, September 2, 1924; H. A. Thorne, superintendent of mines, to Sully, *Reports*, January 21, February 11, March 1, June 14, 25, July 5, September 4, 1924; Sully to Jackling, *Reports*, April 30, September 23, 1924.

61. *Silver City Enterprise*, July 16, 1926; NCCC, *Annual Reports*, March 28, 1927, March 21, 1930.

62. See Hildebrand and Mangum, *Capital and Labor*, 112–115, for a discussion of the impact of the cartel, which seems to have been limited but reveals US producers' efforts to try to control the copper market.

63. The Chino Mines Division lost $500,000 in 1931, $1 million in 1932, $323,000 in 1933, and $52,000 in 1934; see NCCC, *Twenty-Fifth Annual Report*, March 15, 1932, *Twenty-Sixth Annual Report*, March 10, 1933, bimonthly and monthly reports, Chino Mines, 1933; H. A. Thorne to Rone B. Tempest, Monthly Reports, Chino Mines, 1934; H. A. Thorne to Sully, Monthly Reports, June 20, 1929, February 5, May 20, 1930, January 5, 1931; Sully to Jackling, Reports, September 3, October 20, 1931.

64. It took eight pallbearers—R. B. Tempest, Horace Moses, H. A. Thorne, Fred Hodges, D. W. Boise, Plutarco Pedraza, E. J. O'Brien, and Percy Wilson—to carry the copper casket. There were also twenty-six honorary pallbearers listed, all An-

glo. Pedraza was the single Hispanic among the pallbearers, revealing Sully's (or at least the succeeding company officials') loyalties to occupational and residential segregation.

65. Kennecott Copper Corporation (hereafter KCC), *Eighteenth Annual Report*, April 4, 1933, *Nineteenth Annual Report*, April 7, 1934; *Silver City Independent*, June 6, 1932, June 6, 13, July 4, 11, 18, 25, 1933; *Silver City Enterprise*, July 21, 1933.

66. Jackling had been strongly opposed to any union organization since the turn of the century; see Philip J. Mellinger, *Race and Labor in Western Copper: The Fight for Equality, 1896–1918* (Tucson: University of Arizona Press, 1995), 92–93, 116, 119, 125.

67. *Silver City Enterprise*, September 18, 1934; *Silver City Daily Press*, July 2, 1940; initially the Chino workers voted to join the International Union of Mine, Mill, and Smelter Workers as an associate of the American Federation of Labor until Mine Mill was expelled in 1934 from the AFL soon after the KCC workers' vote to join. In 1935 the Congress of Industrial Organizations was organized and Mine Mill joined that consolidation until its expulsion from that organization in 1950 because of its association with communists; see Vernon H. Jensen, *Nonferrous Metals Industry Unionism, 1932–1954* (Ithaca, NY: Cornell University, 1954), 26–29, 264–269.

68. See James W. Byrkit, *Forging the Copper Collar*; A. Yvette Huginnie, "A New Hero Comes to Town: The Anglo Mining Engineer and 'Mexican Labor' as Contested Terrain in Southeastern Arizona, 1880–1920," *New Mexico Historical Review* 69 (October 1994): 323–344; Mellinger, *Race and Labor*; and Vargas, *Labor Rights*, for extensive discussions on Mexican, Mexican American, and Anglo labor issues in the copper industry in the Southwest in the first half of the twentieth century.

69. *Silver City Independent*, June 4, 1912.

70. Rickard, "The Chino Enterprise—IV," 1120, claimed that the petition was not an attempt to unionize: "This [one-day walkout] they did as individuals; it was not a unionized gesture of protest. It was not a strike."

71. CCC, Monthly Reports #30 and #31, June 4, 30, 1912; *Silver City Enterprise*, May 31, 1912; *Silver City Independent*, June 4, 1912; Sully correspondence, December 23, 1912; Sully to Moses, correspondence, October 1, 1912.

72. Sully to Moses, correspondence, October 26, 1916; Thorne to Sully, *Report*, June 2, 1923; Sully to Jackling, *Report*, January 3, 1924; also see Mellinger, *Race and Labor*, 35–58, 66–72, 154–173, for an extensive discussion of Latino labor activism at Clifton, Arizona, in the first two decades of the twentieth century. Clifton's proximity to Santa Rita, only seventy miles to the west, would seem to have had an influence on Latino labor activism at Chino, but there seems to be little evidence of it.

73. *Silver City Enterprise*, May 11, 1906, April 26, 1907; Mellinger, *Race and Labor*, 9, 19, 41, 45, 127, 134, 140, 197–198; Sully to Jackling telegram, August 6, 1914; Sully to Babbitt correspondence, November 3, 1914; also see James McBride, "La

Liga Protectora Latina: A Mexican American Benevolent Society in Arizona," *Journal of the West* 14:4 (October 1975): 82–90.

74. Vargas (in *Labor Rights and Civil Rights*) contends throughout his important study of Latino workers in the United States in the 1930s and 1940s that they believed unions, legitimized through federal law, were the key to their gaining fuller rights of citizenship along with work-related gains. Mine Mill served that purpose at Chino from the reopening of operations in 1937 until its absorption into the United Steel Workers of America in 1967 (see chapter 5).

75. Ellen Baker, *On Strike and on Film: Mexican American Families and Blacklisted Filmmakers in Cold War America* (Chapel Hill: University of North Carolina Press, 2007), 32–33, offers these examples. She also provides a chart that shows that of the 484 "Mexican" employees at Chino in 1930, 420 worked as "laborers," which means that most held the lowest-paid jobs and with very little opportunity for promotion. Of the 335 "Anglo" employees only 5 were "laborers."

76. *Silver City Enterprise*, July 7, 1911, August 22, 1913, February 15, 1918, August 21, 1925, March 18, 1927; *Silver City Independent*, July 4, 1911, September 2, 1913, February 27, 1917, February 19, 1918; CCC, *Accident Report*, April 16, 1913; Horace Moses to Sully, March 2, 1917; Anderson to Sully, September 2, 1917; Thorne to Sully, *Reports*, March 21, June 21, 1927.

77. CCC, *Accident Report* (1912); *Silver City Enterprise*, August 9, 1912, August 8, 1913, August 1, 1919, April 20, 1923; *Silver City Independent*, August 5, 1913, May 1, 1928; Thorne to Sully, *Report*, September 28, 1925.

78. Employee representation plans originated in the 1910s with the Rockefeller-owned Colorado Fuel & Iron Company of Pueblo, Colorado, and they were initiated to thwart unionization: see H. Lee Scamehorn, *Mill & Mine: The CF&I in the Twentieth Century* (Lincoln: University of Nebraska Press, 1992), 6–81.

79. By the early 1940s, stewards' instructions in Mine Mill included various duties shop stewards were to carry out, and among them was to "report all unsafe working conditions to immediate supervisor in charge"; see "Stewards Instructions,"

March 6, 1941, Box 864, International Union of Mine, Mill, and Smelter Workers Collection (hereafter IUMMSWC), Western History Collections (WHC), Archives, University of Colorado, Boulder. There were a few Anglo employees who joined Mine Mill but the craft unions still excluded "Mexican" workers; see Baker, *On Strike*, 23, 32–33, 48–51; Huginnie, "A New Hero"; *Silver City Enterprise*, September 19, 1934.

80. Ironically, the Anglo and Hispanic workers with very few exceptions would be segregated during the early 1940s by union membership that would be divided between Mine Mill and the respective craft unions, thus reinstating the tradition of occupational segregation. The battle to change this discriminatory practice would be ongoing at Chino well into the post–World War II period; see Baker, *On Strike*, 50.

81. *Silver City Enterprise*, February 15, April 5, May 17, 1935.

82. Quotes in this paragraph come from Baker, *On Strike*, 52–53, who quoted from the records of the NLRB hearings; also see *Silver City Daily Press*, April 25, May 4, 5, 6, 9, June 14, 17, 1938.

83. Baker, *On Strike*, 53, 55; *Silver City Daily Press*, November 22, 23, 1938; *Silver City Enterprise*, November 25, 1938.

84. *Silver City Daily Press*, December 18, 1939, July 2, August 26, 1940, February 7, April 17, July 25, August 11, 1941; *Nev. Consol. Copper Corp. v. NLRB*, 122 F.2d 587, 593 (10th Cir. 1941); *NLRB v. Nev. Consol. Copper Corp.*, 316 U.S. 105, 107 (1942).

85. Baker, *On Strike*, 55–62.

86. "Agreement Covering Wages and Working Conditions between Nevada Consolidated Copper Corporation, Chino Mines Division, and International Union of Mine, Mill and Smelter Workers and Locals Nos. 63 and 69," July 7, 1942, Box 864, IUMMSWC, WHC, Archives, University of Colorado, Boulder.

87. See A. B. Parsons, *The Porphyry Coppers in 1956* (New York: American Institute of Mining, Metallurgical, and Petroleum Engineers, 1957), 125–135, for a discussion of the Chino rejuvenation program of 1937–1942.

88. Parsons, *The Porphyry Coppers in 1956*, 10–11.

IV

Santa Rita

THE COMPANY TOWN AND THE COMMUNITY

By 1970 Santa Rita no longer existed. The steady march of the mining operations ate away at the hilly terrain. The open pit grew to an enormous size and the copper town actually disappeared. The shovel crews so thoroughly reconfigured the landscape that there was no earthen foundation left for the workers' homes and the company's store, hospital, and workshops. Although Santa Ritans knew of this impending destruction, which materialized over

three decades, they felt an acute sense of loss. Now they had only their memories of the community. It dawned on many former residents that where Santa Rita once lay, there was nothing but space. This quirk of fate was not lost on three well-known Santa Ritans who decided in the mid-1970s to memorialize their birthplace with the founding of the Society for People Born in Space.[1] Harrison Schmitt, Gilbert Moore, and Ted Arellano, the founders of the society, worked for the National Aeronautics and Space Administration. Their vocations starkly reminded them of the irony of having been born in what was now "space" above the bottom of the open pit; interestingly, astronaut Schmitt himself had traveled into space in 1972 as the lone geologist on Apollo 17. These men and other members of this exclusive club met

nearly every year from 1981 until 2000 to remember their lives in the now defunct town. They wanted the "community" to live on.

In the year of the last of these reunions, Sheila L. Steinberg conducted a sociological study of Santa Rita. She discovered in nearly 300 responses to her survey and in interviews with "informants" that there still remained a feeling of community.[2] One respondent remembered that the "community was just like a family." The reminiscences contained repeated mentions of the positive "tight-knit" and "friendly" atmosphere in the mining town. Others recorded their feelings of strong bonds. One former resident wrote that "Santa Rita is HOME . . . My death plans are written and paid for and my family is in agreement and it gives me comfort to know my ashes will be scattered at Santa Rita." Another penned: "We will always be a part of the Santa Rita Community. It is part of us and we are part of them." "I feel home sick," another recalled, "even after having lived in California for over forty-six years." "Santa Rita is like the glue to our bond," one person declared. In a separate interview, Jesus "Tuti" Chacón reminisced that he "never met any better people than the people from Santa Rita in the respect that they were very united . . . they were very much like brothers and sisters." The copper town may have disappeared but a strong sense of community still survived fully thirty years after its demise.[3]

They also reminisced about the Kneeling Nun, an obvious symbol of their attachment to the former town and to the place. "It represents a wonderful community that exists only in our collective memory!" "It's a beautiful landmark." "And [it's] the only thing besides the pit left of Santa Rita." Others claimed the rock monument "reminds me of home"; it "is spiritual and I love it." It "represents the family and community of Santa Rita." In response to many locals' fear that the company was going to tear down the landmark, one person assigned a mystical quality to the Nun. "She's special. She's always been there and is a part of me and she should always be there. They can dig somewhere else." They willingly accepted that mining destroyed their town. But they also tenaciously opposed the obliteration of Santa Rita's historic icon.[4] She brought comfort to a community of people who could no longer go to the place where they grew up and raised their families to remind themselves of their past. For many of them she was their divine guardian.[5]

The former Santa Ritans also recognized the key role of the mining operations in their lives. Nearly 90 percent of the respondents' fathers had worked for Kennecott, leading them to reveal a mix of both "positive" and "negative" memories of life in the copper town. "Santa Rita's destroyer [Kennecott]," one person acknowledged in the survey, "is from the source that supported the families, educated the children, built our churches, supplied our recreation, took care of our health and birthed our new citizens and buried our dead." Santa Ritans realized that their hometown was going to disappear and, though they were saddened by the inevitable loss, many knew its demise brought economic benefits to the community. "Everybody worked at the mine, they made the community."[6]

For the informants interviewed for the study, life in Santa Rita was not always fair nor was it entirely free. "The company had a say-so in everything," one interviewee candidly revealed. "Kennecott built the school and owned it." Although a fervent sense of community persisted, it was often limited to an individual's experiences within his or her own ethnic group. The pervasiveness of segregation, which separated Hispanics and Anglos at the mine works and in the residential areas, caused tensions that elicited painfully truthful comments. The "white community had better homes," one person lamented, and "these homes were unreachable" for Mexican Americans.[7]

A clear ethnic and class hierarchy defined the residential patterns in Santa Rita. From the founding of the company town in 1910 until the 1950s, Hispanics lived on the east side of town in neighborhoods from west to east named Espinaso del Diablo (Devil's Backbone), El Barrio de la Iglesia, El Cañon, and El Barrio de Los Osos ("Bears," named for the large-in-stature Gonzalez family). Later, in the 1940s, the company erected homes in a new barrio, Las Casas Nuevas, located north of El Cañon. Company officials and Anglos inhabited Booth Hill, Ball Park, Santa Rita Hill, and Downtown in the west half of the mining town (see figures 4.1, 4.2, and 4.3). Anglo residents called their section of town "Santa Rita" and the Hispanic part "East Santa Rita" or "Mexican Town," reflecting a nomenclature of segregation. Hispanics viewed all neighborhoods collectively as "Santa Rita," usually not making a distinction between "Santa Rita" and "East Santa Rita." The town's cemeteries as well as those nearby in Hanover and on the Georgetown Road were also segregated.[8]

From the inception of operations in 1909 the Chino Copper Company literally built the town from the ground up. Consistent with welfare capitalist philosophy, the paternalistic company officials believed they knew what was best for the workers and their families and set about to construct the "ideal" mining community with amenities not often available in similar remote western mining towns. From the start, general manager John Sully, with direction from corporate board member Daniel Jackling, reordered the residential landscape. Because the town's predecessors had built haphazardly, Chino virtually scoured the surface of the earth of buildings and vegetation to make way for a "comfortable" planned industrial community. The horizontal slate was

FIGURE 4.1. *"Downtown" Santa Rita, 1910. This photograph shows the center of the emerging downtown of Santa Rita. Directly below the Romero works* (center background) *is the newly completed teacherage (two-story, bicolored structure). Note the general manager's home (peaked roof to left of teacherage), the Santa Rita Store (four buildings right of teacherage with four-panel window above white door), and the Downtown School (L-shaped building). This scene shows the copper camp before it became a substantial, planned company town.*

wiped clean for an entirely new vertical creation, a company town. A reporter for *The Mogollon Times* described the company's initial actions in reformulating the community. "During the formative period of this great corporation the ground was denuded of every vestige of undergrowth and timber. Numberless houses, shanties, shacks and Jacals [*sic*] were razed to the ground, rendering the site of Santa Rita unrecognizable, excepting the Kneeling Nun and the old . . . fort. Each day saw trainloads of lumber, machinery and supplies and building material unloaded."[9]

Although the reporter exaggerated the extent of the company's razing of old Santa Rita, especially concerning businesses and more substantial homes in the downtown, Chino officials took complete control of the reordering of the residential landscape. Sully and his family moved into the already built general manager's residence of the former Santa Rita Mining Company. Officials retained any buildings that met the standards for the emerging controlled community that Sully hoped would entice family-oriented workers to create the perception that this new town would not be like the fleeting "Wild West" mining camps of lore. Company "law" saw to it that this Old West phenomenon would not be repeated either. In essence, corporate authorities planned for a "model" company town similar to others constructed in the American West. Sully could, he believed, watch over the community and play the role of a caring but firm father figure, offer reasonable amenities, and work to ensure that employees looked to him and the company for their needs rather than seek such benefits from unions.[10]

FIGURE 4.2. *Santa Rita "Main Street" on payday, ca. 1914.* Note the workers lined up at the Santa Rita Company Store on payday, the automobile in front of the post office, and the arroyo with foot and auto bridges in the foreground. Two workers pose with their lunch pails (right center) as well. On the left side of the main street are, front to back: *company garage (with sign in front) where the company housed a fleet of Packard vehicles; H. M. Derr's barbershop; and the ten-room general office building, location of the general manager's and other officials' offices. Also note the electric lines, a sure marker of the emerging company town.*
COURTESY OF TERRY HUMBLE.

FIGURE 4.3 (below and pp. 114–115). *Panoramic vista of Santa Rita, ca. 1916. This extraordinary photograph shows the entire town of Santa Rita as well as the mine works. Directly left of the pit is the downtown, where the emerging "island" is visible. The first homes of the Booth Hill neighborhood are barely evident behind the smokestacks of the old Romero on the edge of the pit. Also note the boardinghouse (large two-story white building at the far right on p. 113 and far left on p. 114), the Elite Café (large T-shaped building, bottom right).*
COURTESY OF SILVER CITY MUSEUM.

Chino financed the construction of hundreds of homes for its Anglo workers, who reaped the rewards of a segregated society. Hispanics built their own homes, usually the board-and-batten style, because the company discouraged the construction of *jacales*, mud-covered single-room dwellings common in Mexico. Contractors began building three-, four-, and five-room residences by December 1909 for Anglo families on the lower slope of Santa Rita Hill. The company built homes on a need basis until 1941. Like company housing throughout the American West, the cottages were box-type, made of lumber lined with building paper, and painted white, green, or brown. Company reports recorded that contractors erected these modest homes "for accommodation of the white employees." Prior to the late 1930s the company equipped nearly half of the dwellings with plumbing and electricity. Officials then assigned these first new homes numbers in the 500s. Instead of naming the dirt streets, Chino's officials decided to number

all the houses (and mine buildings) and rented post office boxes to those who could afford them. The Anglo neighborhoods eventually reached house numbers in the 1200s and those in the barrios in the 1400s. Single Anglo workers often resided in the boardinghouse after its construction in 1912 (see figure 4.4), and then the Santa Rita Hotel after its erection in 1929 (see figure 4.5). The boardinghouse sat near the top of Santa Rita Hill and the hotel near its foot.[11]

From 1916 to 1921 and then in the 1940s, the company built employees' cottages on Booth Hill. Anglos considered these homes numbered in the 700s a step up in the pecking order in comparison to the first structures built at the foot of Santa Rita Hill. Upper Santa Rita Hill[12] homes built from 1917 to 1927 and numbered in the 900s were considered equal to those on Booth Hill. Commonly, when a worker moved out of a house the company took applications from those interested in "moving up." Officials then often assigned the vacated residence to a

loyal worker whose occupation reflected that of the other inhabitants of the neighborhood. Mainly foremen and craftsmen, for example, inhabited Ball Park, which was constructed from 1926 to 1940 with house numbers in the 1100s. This assignment of houses reveals that the company often distributed homes by rank and seniority similar to the town's segregation based on ethnicity. The most elite neighborhood was Downtown. The homes there were numbered from 1 to 200 and were the town's largest dwellings, most having been built before 1909 by the Santa Rita Mining Company. They were reserved for the general manager, superintendents, and, if there were vacancies, some of the foremen and craftsmen.[13] In the

1950s the company moved the Downtown structures, which were on the "island" surrounded by the pits like a moat, to the new neighborhood of South Hills to the west. Revealingly, this reformulation of "elite hill" that moved the highest-ranking officials out of the noisier and dirtier section created a socioeconomic residential pattern common to other industrial landscapes in the United States. The prevailing westerly winds also ensured cleaner air for this new neighborhood. Ironically, the lower-paid Hispanic employees lived farthest from the industrial nuisances, contrary to conditions experienced by most workers in smokestack enterprises in the nation. Still, wind did sometimes carry emissions into the barrios.

FIGURE 4.3 (continued). *The largest white frame building in the foreground is the Orpheum Theater. Directly above and slightly left of the theater is the neighborhood of Santa Rita Hill. Above and to the right of the Orpheum is the district of Espinaso del Diablo, the dividing line between the Anglo and Hispanic sections of town. Right of this Hispanic neighborhood is the Barrio de la Iglesia (in the vicinity of the Catholic church near the top of the hill, p. 115). To the extreme right in the panorama is the Barrio del Cañon (note the* bonque *or bunkhouse, the long one-story building below and right of the church). Also note the Hotel Owen (large two-story structure, bottom left). The Alianza Hispano Americano Lodge No. 21 is the white building just right of the theater in the Y of the two dirt streets (p. 114).*

Mexican American housing conditions differed from those of the Anglos. Rather than build new quarters for the Hispanic workers and their families, the company leased lots to these laborers and then encouraged them to build their own homes. Lots rented for 50 cents a month, a much smaller fee than the $15 to $26 charged to Anglos for renting company housing. The company did construct bunkhouses, or *bonques*, for single Mexican and Mexican American workers. Chino also rented lots to some of the inhabitants of the barrios to erect tents over wooden platforms. Regardless, Hispanic leaseholders rarely had access to indoor plumbing, electricity, and heating, utilities common to many Anglo households. Their plumbing consisted of a water pipe that was either sticking through a wall in the homes or a water pump in the yards. Women washed clothes with washboards and heated water and cooked on woodstoves outdoors. They "relieved" themselves in outhouses or outdoors. They had "poor" barrios, Guadalupe Fletcher lamented.

"Everybody had the same. Toilets outside. Wash your clothes outside, boil them in a big tub, on wood fires."[14]

These distinctions in amenities based on ethnicity sent a clear message to the Hispanic workers that they were considered "inferior" to the Anglos from the company's point of view. As was true in most mining camps in the Southwest, corporations designed their company towns to segregate Mexican and Mexican American workers from Anglos. Lower wages for Hispanics ensured that they could not afford to live in the "Anglo" section of town.[15] Kennecott did eventually construct company housing for Mexican Americans in the late 1930s in El Cañon and in the 1940s in Casas Nuevas in conjunction with new federal anti-discrimination laws and the rise of the labor movement. When her family moved into the Casas Nuevas barrio, María Martinez remembered that the "house was a luxury compared to the one they had before because it had a bathroom inside which was a big thing to her."[16]

FIGURE 4.4. *Dining room inside the boardinghouse, ca. 1912. The table is set for fourteen hungry workers. Note the kerosene lamp hanging above the table, the brass tokens left of the inside doorway, and the unusual wallpaper. The inside doorway led to the men's rooms where they could pick up their brass tokens, which probably denoted who had eaten. The dormitory burned in 1928, requiring the company to scramble to house the 100 men who boarded there.*

Not surprisingly, Chino built Hurley in the 1910s with a similar plan in mind. Officials designed the mill town to be segregated along ethnic and class lines, but in this case with named streets in a traditional grid pattern suitable to the flat terrain. Mining corporations throughout the Southwest often used grid patterns in laying out roads and residences in their company towns. Engineers did this to either replace the traditional Hispanic plazas or not plan for them at all, as was the case in Hurley and Santa Rita.[17] Mexican Americans lived north of the Atchison, Topeka, & Santa Fe rail line in a residential enclave eventually known as "North Hurley" and Anglos to the south and west of the millworks. Similar to Santa Rita, the housing schemata placed the Hispanic workers and their families farther from the noisier, dirtier section of town (like Santa Rita, this pattern was abnormal

in comparison to other industrial towns in the United States). The superintendents and other Anglo employees resided nearest to the operations so that any "problems" could be addressed immediately.

The close proximity of the superintendent's home to the plants allowed the man in charge literally to hear any abnormalities in the crushing and concentrating processes. Mill superintendent Frank Thayer, in fact, was known in the late 1930s and early 1940s to leave the family supper table or to get up in the middle of the night when the mill shut down or showed signs that threatened the company's expectations for full-throttle production. He and future mill managers lived just east of the mill complex on First Street, known as "bosses row," and just one of many named streets. The farther the mill workers resided from the mill, the lower their rank in the

FIGURE 4.5. *Santa Rita Hotel, July 1929. Chino constructed the Santa Rita Hotel in 1929 for $60,000 to replace a boardinghouse that had burned and to provide more rooms for the rapidly expanding workforce of single Anglo workers. Completed in August, the two-story hotel sported fifty-three rooms and one suite and was located just east of the Hotel Owen. The Santa Rita Hotel closed in the mid-1960s with the expansion of the pit and was buried under one of the mine dumps.*

COURTESY OF TERRY HUMBLE.

company. Consequently, Hispanic employees, who were paid less and had fewer opportunities for promotion, lived farther from the mill town's cacophonous clanking crushers, dusty tailings dumps, and irritating smoke emissions, especially after the addition of the smelter in 1939. Few would have recognized this irony in light of the unfairness of being assigned poorer-quality housing and having fewer opportunities for advancement in the company.[18]

In addition to housing, the company decided to construct various buildings in Santa Rita to provide essential services for the creation of a full-service community. During this formative construction period from 1909 to 1918, General Manager Sully's chauffeur drove the taciturn boss around town in his new Studebaker E-M-F. From the backseat the top company boss could care-

fully oversee like a proud papa the raising of the company hospital, a post office and Masonic Lodge, the staff houses, new additions to the company store, and other buildings (see figures 2.12, 4.6, 4.7, and 4.8). By fall 1910, contractors had completed a men's dormitory and mess hall (see figure 4.9). They also affixed a twenty-room addition to the boardinghouse that with fifty new rooms could accommodate 100 men. A year later the company financed the materials so that parishioners, mostly Hispanic, could erect a new Catholic church (see figure 4.10). Accommodating the Hispanics' religious needs, the company hoped to attract family-oriented workers.

Placing a high value on elementary and junior high education, Chino also constructed the Hill School complex. The company put up three buildings—erected in 1914, 1917, and 1918, respectively—to complete the ed-

FIGURE 4.6. *Construction of company hospital, January 1, 1911. The Chino Copper Company constructed the Santa Rita Hospital to provide medical services for its workers and their families as part of the establishment of the company town. The wood-frame building contained two wings, one for beds and maternity patients and another for surgery. With twenty-six rooms and a full medical staff, patrons considered the hospital one of the best healthcare facilities in the Southwest prior to World War II. The company periodically expanded the hospital to meet the growing demands of an increasing workforce. Chino closed it in October 1954 during plans to enlarge the pit. Rose Mary Marquez was the last baby to be born there, her mother later remembering that workers were moving out beds and equipment as she convalesced after childbirth.*

COURTESY OF TERRY HUMBLE.

ucational compound (see figure 4.11). In the 1920s the copper corporation built two additional schools: the Sully School at Santa Rita and the Hurley High School. Officials also closed the Downtown School, built by the Santa Rita Mining Company at the turn of the century, and then razed it to expand the pit soon after completion of the Sully School.

Entertainment facilities proved important too. Al Owen, a trusted friend of Chino who had assisted the company in acquiring the Pinder/Slip claim, moved into the "saloon" building put up in 1912 at the site of the old concentrator. He was also awarded a monopoly on serving alcoholic beverages for his loyalty. In 1918 the company built the Chino Club exclusively for Anglos (see

figure 4.12). A year later it added outdoor tennis courts. Officials jettisoned plans for a swimming pool because of the instability of the ground as a result of daily blasting in the pits. The company erected a similar club in Hurley that year with bowling alleys in the basement, and then sometime in the 1930s, added a pool. Anglos engaged in various extracurricular activities at the club, including bowling, playing pool, swimming, reading, dancing, and chatting with friends. A decade later the company constructed the Casino, a club reserved for Hispanic employees. This hangout, along with the Alianza Hispano Americano lodge, served as the site for various Mexican American festivities, from weekend musical and dancing performances to Cinco de Mayo and Mexican

FIGURE 4.7. *Santa Rita Post Office and Lodge Hall, December 1, 1910. This newly built structure housed the US Post Office and confectionary with newsstand downstairs and the Masonic lodge upstairs until 1953, when it was moved and then dismantled. John W. Turner served as the first postmaster in this building and C. H. Fox as the manager of the soda fountain. Santa Ritans were more familiar with Dave Cliver, who held the position of postmistress from 1941 until the post office closed in 1971 (after 1953 she served at a new post office building). The Masonic Lodge offered its members as well as the Order of the Eastern Star (wives of the Masons) and the Rainbow Girls (the daughters' organization) a location for their civic and recreational activities. The company store (left) and the Kneeling Nun can be seen in the background.*
COURTESY OF TERRY HUMBLE.

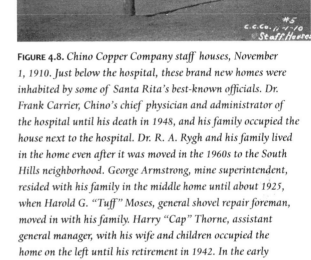

FIGURE 4.8. *Chino Copper Company staff houses, November 1, 1910. Just below the hospital, these brand new homes were inhabited by some of Santa Rita's best-known officials. Dr. Frank Carrier, Chino's chief physician and administrator of the hospital until his death in 1948, and his family occupied the house next to the hospital. Dr. R. A. Rygh and his family lived in the home even after it was moved in the 1960s to the South Hills neighborhood. George Armstrong, mine superintendent, resided with his family in the middle home until about 1925, when Harold G. "Tuff" Moses, general shovel repair foreman, moved in with his family. Harry "Cap" Thorne, assistant general manager, with his wife and children occupied the home on the left until his retirement in 1942. In the early 1950s, Kennecott moved all three structures to South Hills. Eventually each of them was moved in the 1960s to Silver City.*
COURTESY OF TERRY HUMBLE.

Independence Day celebrations. Daniel Marrujo, who grew up in Santa Rita in the 1920s and 1930s before becoming a trackman at Chino, remembered such community gatherings. He fondly reminisced about a "band, a regular band, and they [the celebrants] were all made up of Hispanic people, a lot of them from Mexico." They "felt strongly about their national origins" while they were making a new home in the United States (see figure 4.13).

Occasionally the corporation hosted special guests in a makeshift saloon called the Chino Bar. The American Institute of Mining Engineers, in fact, hosted their annual conference in September 1916 in this impromptu watering hole in downtown Santa Rita (see figure 4.14). The company also had constructed steam heating and electric light plants by 1911 to provide those utilities to each of these buildings and to Anglo households in the downtown that could afford the extra two- to three-dollar monthly charge. Likewise, workers put in an extensive

system of water pipes and fire hydrants.[19] Additional service-related buildings went up in 1916, which included a mortuary, an icehouse (the former school that was actually renovated), and a residence for the Catholic priest, Father Gerey. Officials also had to plan for cemetery space for future use and had to relocate graves with the growing of the Hearst Pit.[20] Sully believed that the stability of the community depended on offering the employees services that any "good" parent would provide for the health, education, and welfare of their children from cradle to grave.

Several private businesses also emerged in these early years. Investors Pedro "Pete" Gómez, Luis Horcasitas, and Miguel Barela joined in 1909 to finance and erect one of the most popular enterprises, the Orpheum Theater (see figure 4.3). Soon abandoned by his partners, lone proprietor Gómez welcomed his new patrons on opening night in March 1910. An active miner, the theater owner had leased portions of the Pescado, Carrasco, and

FIGURE 4.9. *Santa Rita dormitory and mess hall, November 1, 1910. This photograph shows the fourteen-room men's dormitory just before its completion. The two-story building to the right is the ten-room mess hall. Behind the dorm is the teacherage, also a ten-room, two-story structure, which housed the company town's teachers as well as some of its nurses. Note the Romero works puffing smoke in the background.*
COURTESY OF TERRY HUMBLE.

Chino claims from 1901 to 1906 before investing in the trendy cinema. He featured films on "one of Edison's best machines" in the large frame building. One of the few Hispanic-owned businesses in Santa Rita, the "moving picture house" allowed Chicanos and Anglos to view features together although in segregated seating. Purchased by Chino in 1919, the Orpheum eventually showed "talkies" beginning in 1931 after the Radio Corporation of America installed a speaker system. Santa Ritans celebrated the new "rage" since they no longer had to go to Silver City to experience films in sound. The theater also hosted numerous community events, including local plays, traveling troubadours, basketball games, boxing matches, and eighth-grade graduations. The mining company tore down the deteriorating building in the

FIGURE 4.10. *Santa Rita's new Catholic church, March 1, 1911. In December 1910, Father Morin deeded the first church lands to the Chino Copper Company in return for a new lot on the Cobre and Mimbres mining claims on the hill in East Santa Rita. The mining company then funded and constructed this second Santa Rita Catholic church that could seat 200 parishioners and became the symbol of the Hispanic neighborhood of Barrio de la Iglesia. A few years later, Chino built a parsonage inhabited by the priests, the longest termed being Father Marius Gerey (1916–1937) and Father Miguel Estivill (1937–1944). When this stone structure burned in January 1942, Kennecott financed the raising of a third Catholic church.*
COURTESY OF TERRY HUMBLE.

FIGURE 4.11. *Hill School, ca. 1940. This rare photograph of the Hill School complex shows two of the three structures of the educational facility (center and left). Known as "La Ratonera" among the Hispanic students, the Hill School symbolized the transition of young Spanish-speaking students to bilingual English speakers (see discussion of schools later in this chapter). Terry Humble fondly remembers this view of the school with the Kneeling Nun in the background because the photographer took this picture in front of his home.*
COURTESY OF SILVER CITY MUSEUM.

early 1940s and then replaced it with the new El Cobre Theater (see figure 4.15). Opening to a full house with Abbott and Costello's *In the Navy*, El Cobre offered a full menu of films that featured Spanish programming Monday and Tuesday nights and children's shows, usually "shoot 'em ups," on Friday evenings. Like Jim Blair, who enforced residential segregation, Pete Cody, manager and "bouncer" of El Cobre, enforced separation of Hispanics and Anglos at the features. As Elvira Moreno Chacón recalled, "if any dared to cross over to either side" of the theater, "Pete Cody, the bouncer, would throw them out rudely or make them go to their side." In the early 1960s the company buried the popular theater under one of the giant waste dumps (see figure 4.16).[21]

Company favorite Al Owen and his brother John constructed a new hotel in 1913 just east and south of the Orpheum. Sporting thirty-five rooms, the two-story hostelry offered its patrons rooms with steam heat, electric lighting, and bathrooms with plumbing until it closed in 1925 (see figure 4.3). Next to the hotel, P. B. Samaniego opened the Zapatería, or shoe shop (see figure 4.17). Within a few years private restaurateur George Ohnsman opened the Santa Rita Café, later acquired by the mining company and known as the Elite Café, an exclusively Anglo eating destination (see figure 4.18). In the late 1910s this café featured the cooking skills of Wong Lee, one of the few Chinese inhabitants of the copper town. Hispanic restaurant-goers traded with the San

Nicolas Café. Many other temporary businesses sprang up in Santa Rita from time to time, among them the Manhattan Confectionary and Candy Shop, Carrillo's Saloon, tailor shops, stage lines, the Santa Rita Garage and other automobile repair shops, and barbershops.[22]

Law enforcement "services" also played a pivotal role in the formation and maintenance of the company town. From the onset of operations Sully counted on Santa Rita's chief of police James Blair and a team of security officers to enforce his authority. Under Sully's direction, Blair wielded near-absolute power over the community in establishing company rule. A college-educated

Texan and former schoolteacher, Blair had migrated to Grant County in the 1890s because he suffered from asthma. After a short stay in Deming, he moved to Pinos Altos, where he cultivated his talents as a "song and dance man" and bartender in the town's main saloon. Soon he won the election for county sheriff, a position he held for four years (1898–1900 and 1902–1904). In the interim he had served as head of security for the Santa Rita Mining Company. These experiences and his gregarious personality caught Sully's eye in 1909 and the general manager hired the Texan as special investigating officer for Chino. In 1912 Blair earned the title of chief of police in Santa

FIGURE 4.12. *Chino Club in Santa Rita, ca. 1918. A popular spot for Anglo employees of Chino, this club housed a library, confectionary, dance hall, pool tables, and slot machines. Officials jettisoned a plan to construct a swimming pool (because of earth-shaking daily blasting) in favor of a two-lane bowling alley in the basement. This structure also served as a makeshift hospital during the influenza epidemic of 1918–1919. The company built a nearly identical Anglo club in Hurley. Chino moved the Santa Rita club in 1953 to the area just west of the Ball Park neighborhood and then refinished the dance floor and bowling alleys. In about 1965, the club was moved (without the bowling alleys) to the corner of Thirteenth and Santa Rita Streets in Silver City.*
COURTESY OF TERRY HUMBLE.

FIGURE 4.13. *Brass band of Santa Rita, June 17, 1917. Sitting ready for a concert in front of the Orpheum Theater, this all-Hispanic band played at numerous celebrations and fiestas in Santa Rita and vicinity. When the* Chinorama *published this picture in its July 1956 issue, asking readers to identify the members, Gumesindo Ortiz wrote that many of them worked for Chino's track gangs (see figure 3.21, which was published on page 19 of the July 1956 issue of* Chinorama *with this photograph) and he provided many of their names but without pointing them out in the picture. They were Professor José Alonso Pajares, J. Jesus Sustaita, Pasqual Luna (with X on right leg and grandfather of Richard Luna, who loaned this photograph), Gabriel Ontiveros (track gang foreman in figure 3.21), Pedro Aguilar, Pedro Gómez, Miguel Bustos, Juan Flores, Juan Antonio Lopez, Julio Jaques, Rosendo Carrillo, Luis Cordova, José Muela, and Fernando Rodriguez.*
COURTESY OF RICHARD G. LUNA.

Rita (see figure 4.19). His loyalty to the company and no-nonsense style earned him a reputation as a strong-arm enforcer of company rule.

The peace officer, however, did tolerate the mixing of Anglos and Chicanos. He himself had married Inez Diaz and lived with her and their five children in "Mexican Town," even though Blair Canyon, named for him, was near the dividing line between the segregated neighborhoods. Despite Blair's tolerance, however, tensions still arose between Chicanos and Anglos. Paul M. Jones, author of *Memories of Santa Rita*, remembered that an adult-imposed segregation by each ethnic group

FIGURE 4.14. *Chino Bar, September 18, 1916. When the American Institute of Mining Engineers held their annual conference in Santa Rita, the company transformed one of the garages into this bar. Note that Chino offered free drinks to the visiting engineers.*
COURTESY OF SILVER CITY MUSEUM.

limited the mobility of children in the copper town. "Woe to the young boy who was foolish enough to cross this barrier alone! He generally got clobbered!"[23] "On the east side was the Mexican town, and over from it across La Canyon," Robert Gardner recalled, "that's where you wasn't supposed to go or you'd get beat up." Ramón Arzola and Terry Humble told of a daily routine of rock throwing and fisticuffs. The conflicts "always started out as rock fights," Arzola claimed, "but then ended up in fist fights . . . [and] the kids used to end up with black eyes and bloody noses." The Santa Rita boys were so notorious locally for their rock-throwing escapades, Humble mused, that girls in the mill town of Hurley would not date the boys of the mine town of Santa Rita because at some point they had been pelted by the contentious rowdies.[24]

Girls were discouraged from mixing too. Josephina Enriquez experienced an adult-imposed segregation from the mother of an Anglo friend as well as her own

mother. One day she "walked with her friend to her . . . house. The little girl introduced her to her mother and the mother just shut the door on her." On her arrival home, the perplexed Chicana told her mother what had happened and "her mother scolded her for going over there."[25] The company's policy of discrimination tacitly made its way into the minds of the adult population, which imposed segregation on the children very similar to the way white and black grown-ups kept their children apart in the South and urban centers.

Under Sully's direction Blair imposed a specific kind of company law. With the assistance over the years of various deputies—among them Lon Portwood and Lyman Garrett, brother of the famous sheriff Pat Garrett—the chief of police periodically ran off "strikers" (such as the shovel engineers in 1912), union organizers, petty criminals, and other "bad men" or "agitators," as the company called them. "Dad fought to keep the unions out," his son Frank remembered. The definition of those who were considered undesirable citizens transcended the traditional criminal or troublemaker model because the company also used law enforcement to monitor morals. Longtime resident Paul Jones, for example, claimed that "the company had absolute and ironclad control over who lived in town and if one exceeded the limits of company-defined decorum, he was politely, but firmly, invited to leave town. Some of the acts that precipated [sic] such an invitation were excessive family fights, flagrant immorality, especially if it involved someone else's spouse, stealing, and other greater or lesser sins as defined by the company, whose word was law, and there was no appeal."[26]

Ironically, Blair skirted the law when it came to alcohol. A well-known drinker, he commonly warned the local bootleggers when the state liquor inspector planned his visits. "They'd pick the one [producer] with the best color [of shine]," Blair's son recalled, "and that's the one they'd pass around to the group [of Blair's friends]. They didn't destroy the still [just] confiscated all the whiskey . . . There was no animosity in the raiding." Rafael Kirker was amused that Blair had regularly alerted his "Tío Juan" (Uncle John), who would hide his still and elixir just prior to the raids. "Moon-shiners must have been one of the least of his worries, or he must have been one of Juan's customers."[27]

FIGURE 4.15. *El Cobre Theater, December 3, 1941. Like the Orpheum before it, El Cobre Theater was one of the most popular entertainment spots for both Hispanics and Anglos, even though patrons sat in segregated sections. The picture house featured films in Santa Rita from 1941 until it closed in the early 1960s and was buried under the dumps.*
COURTESY OF TERRY HUMBLE.

FIGURE 4.16. *El Cobre Theater movie calendar, October 1952.*
COURTESY OF TERRY HUMBLE.

FIGURE 4.17 (ABOVE). *Zapatería (shoe shop) and jeweler's shop, ca. 1920. This photograph shows P. B. Samaniego's zapatería (and a jeweler's shop) that stood just east of the Hotel Owen. A prominent Hispanic businessman, Samaniego also managed the bunkhouses in East Santa Rita. In 1929, the company razed the zapatería to make room for the Santa Rita Hotel. José Cosme Acosta opened his own shoe shop in that same year just north of the old zapatería, mending shoes there until he moved his business to Bayard in the early 1960s. Don Cosme repaired shoes there until the early 1990s.*
COURTESY OF SILVER CITY MUSEUM.

FIGURE 4.18 (BELOW). *Santa Rita Café, ca. 1929. Built in October 1925, this eatery served Anglos only and was known as the Elite Café to its patrons as well as Hispanics (who ate at a separate restaurant, the San Nicolas Café). Owner George Ohnsman closed the establishment in mid-1927 and it was left vacant. After the boardinghouse burned in February 1928, however, the new owners, the Chino Copper Company, refurbished the building, known by most locals as the Santa Rita Café after that. Workers and their families could get a "decent" meal there in large part because the mining company allowed them to make payroll deductions at the café. The restaurant burned in May 1956, never to be rebuilt.*
COURTESY OF SILVER CITY MUSEUM.

FIGURE 4.19. *Jim Blair, chief of police and security, on horseback in Santa Rita, ca. 1920s. Santa Rita's most influential security officer from 1909 until his death in 1941, Blair was a tough lawman with a soft touch. He carried out the company's strong-arm enforcement of the law as well as morals, yet was tolerant of integration of the community. He married Inez Diaz about 1912, living with their five children on "Blair Hill" in the Hispanic section of Santa Rita. A former Grant County sheriff, Blair held a deputy sheriff's commission (1904–1941) throughout his law enforcement career in New Mexico.*
COURTESY OF TERRY HUMBLE.

Protection of company property greatly concerned Blair in his role as the company's top lawman. Security officers made their rounds at night, for example, to enforce a 10:00 PM curfew, to check the locks on company buildings, and to serve as the alarm system in case of mischief. During the shutdowns of 1921–1922 and 1934–1937, Chino hired a "skeleton crew" of security guards to watch over the mill yard, the tailings dams, the power plant, the Apache Tejo pumping station, and the town sites of Santa Rita and Hurley. One of the better-known duties of Blair and his deputies was to accompany Sully on the tenth of each month when he traveled from Santa Rita to Hurley to deliver the mill workers' payroll, which

sometimes exceeded $30,000. This well-known practice, in fact, led to one of the most celebrated outlaw stories in Santa Rita's history. At 10:00 AM on August 10, 1911, three armed robbers held up Sully and two passengers, Messrs. Bruff and Bradley, as they made their way toward Hurley in his Studebaker. Fortunately for the company, Wells Fargo agent C. W. Marriott, escorted by armed deputy H. E. Muse, had already taken the payroll to Hurley on a special train in a coincidental break from the normal routine. Distraught when their search for the booty turned up nothing, the criminals stole all three men's money and watches and Sully's valuable diamond. As soon as the flustered victims reached Hurley, the composed general manager telephoned officials in Santa Rita.

Assistant superintendent Horace Moses, Constable Lon Portwood, and Reese Jackson, probably a company deputy, immediately mounted their horses and began a pursuit of the bandits. About four miles from Santa Rita, deep in the boulder-strewn mountains south of the Kneeling Nun, the trio dismounted and began tracking the dangerous crooks. Aware of their pursuers, the robbers set up a trap for the unsuspecting trackers, who soon found themselves being instructed to put their hands in the air and drop their weapons. The offenders then smashed the officers' rifles against a boulder, holstered their pistols, and took flight on their horses, deciding after a brief debate not to take their shoes. The next day Grant County sheriff Herb McGrath formed a six-man posse that included Blair, Garrett, and four other well-known lawmen—Tom Moore, Elmo Murray, George Sargent, and Ed Head. Although each of them was renowned for "getting his man," a weeklong search failed to turn up the criminals. McGrath posted a $1,000 reward but with no result. In November, Luna County sheriff Dwight Stephens and a posse in pursuit of an escaped prisoner and two of his accomplices killed John Greer in a running gun battle that cost the lives of two of the deputies. Reese Jackson later identified the deceased outlaw as one of the three robbers in the Chino heist attempt. Despite multiple efforts and several harrowing incidents, the Santa Rita lawmen themselves never captured the other two Chino bandits. One was arrested by other New Mexico lawmen and then hanged in 1913 in Santa Fe, and the other allegedly died in 1920 in the influenza epidemic.[28]

Blair and his team also had to deal with serious crimes, such as murder. The chief of police, for example, investigated the murders of Catarina Muñoz in 1913, Perfecto Sanchez in 1914, and Eusebio Salas in 1915. Muñoz's murderer, Reymundo Gonzales, later pleaded guilty and was sentenced. Sanchez's killer, Apolinar Gonzales, escaped, purportedly being murdered himself in Los Angeles twenty years later. Eusebio Salas's alleged assassin, Cornelio Pedroza, was captured the day after the murder but avoided prosecution when the key witness failed to show for the trial. Another well-known crime involved robbery and murder. In April 1921 two young Mexican teenagers, Eleuterio Corral, age sixteen, and Rumaldo Lozano, age seventeen, killed Grant County jailer Ventura Bencomo in a jail break. Corral had

been convicted of attempted larceny and Lozano of the robbery of an elderly Hispanic Santa Ritan. Several days later Blair and McGrath detained the boys after a gunfight in separate abandoned mine tunnels near Fierro, a few miles northwest of the copper town. Found guilty of first-degree murder in May 1921, the two Mexican youths received the death sentence, which the New Mexico Supreme Court upheld in December. They were hanged in Silver City on January 20, 1922.[29]

The Augustine "Pat" Ruiz family in late 1930 welcomed Blair's services. Robbed of their life savings of more than $5,000, discovered when Mrs. Ruiz found a hole sawed in the floor under their bed where the thieves had entered the home, they appealed to Santa Rita's law officer for help. Acting on a tip, Blair and Deputy Juan Serna traveled to El Paso and arrested Rafael Rivera. Fortunately for the Ruizes, Rivera had spent fewer than $100 and the elderly Santa Ritans, to their surprise, got back more money than they thought they had lost. Rivera earned a sentence of nine to twelve years in the state penitentiary. A year later Blair's team solved a break-in at Gallardo's Confectionary, the culprit, Rafael Turrey, getting an eighteen-month to two-year term in prison for taking $30 worth of candy, gum, and cigarettes. Later that year Santa Rita's lawmen saw to it that two youthful criminals, Jimmy Reed and Milton Madden, would learn their lesson for an unreported crime with terms at the State Industrial School in Springer, New Mexico.[30]

By the mid-1920s, Chino had built jails in both Santa Rita and Hurley. Most occupants were incarcerated for minor infractions, such as public drunkenness or fighting. Such offenders were usually released the next morning in time to get to work. By the 1930s and 1940s most of the inmates were Mexican Americans in large part because the police patrolled their sections of the towns almost exclusively. Company security guard Juan Serna, in particular, targeted Hispanic men. Once in custody, they were severely beaten as part of a concerted effort to enforce a moral code tacitly prescribed by Chino officials and the citizens of the barrios. Serna's tactics led one Santa Ritan to claim that he "was hated by nearly all of the Mexican population in this small town." Others remembered the lawman more fondly in his role as a school-bus driver for Chicano Santa Ritans attending the Hurley High School.[31]

During the two world wars Chino took extraordinary measures to maximize security at the Santa Rita–

Hurley complex. When the New Mexico secretary of state wrote Sully in 1917 to warn him to be ready for "threats" from Mexico, especially after the Pancho Villa incident at Columbus, the Chino official assured him that Santa Rita was prepared for just such a possibility. The company had at its disposal two Colt machine guns and 19,000 rounds of ammunition that workers had carefully secured in one of the mine buildings. To further assure the state official, he wrote to him that many of the workers could handle rifles and other firearms in case of an attack.[32]

After Blair's death in 1941, Joseph "Cap" Mitchell was hired as head of security (see figure 4.20). The new head lawman had at his disposal a contingent of fourteen permanent guards who protected company property and kept order. A former machinist's helper, Mitchell had served as Santa Rita constable in the 1920s and as a member of Blair's security force in the 1930s before taking on the duties of security chief. Soon after the United States entered World War II, he and the watchmen were trained in military tactics by an executive officer of the US Army Headquarters at Fort Bliss, Texas. Although at times arbitrary, especially during the initial years of company law, the corporation's security efforts worked to maintain a low crime rate despite some examples of prejudicial treatment of Hispanic men in particular.[33]

The company store also played a central role in the lives of the workers. The isolation of the copper camp gave Chino a monopoly on trade. A popular gossip-gathering place, the Santa Rita Store sold "meat, ice, groceries, hardware, clothing, shoes, hay, coal, gasoline, furniture, kerosene for lamps, pharmaceuticals, guns, ran a mortuary, and often acted as a bank." Robert Gallardo remembered all these items but especially the butcher block and the glass refrigerator cases full of fresh meat. Generally high-quality goods, the products could be expensive because of the company monopoly and the isolation of the camp, especially prior to the proliferation of the automobile in the 1950s. Bill Wood remembered that in the 1930s, "every time they'd bring in some rare food, even bananas . . . everybody in town would say, 'hey, they got bananas,' and everybody rushed down there. Coconuts were really a delicacy. And strawberries. All they had was maybe an 8 foot long vegetable counter and everything was still in the box. You'd go down there and grab one before they were all gone." Accommodating deliveries also enticed employees to buy from the company mercantile. In the early years, Pancho Lucero delivered goods by horse and wagon, customers trusting him to enter their homes and leave items on their kitchen tables in their absence. Later, from the 1930s to 1950s, Coy Matthews and Hub Coker regularly took orders by phone so that Mel Rivera and Lupe Hernandez could make the deliveries by truck during store hours from 7:00 AM to 5:00 PM.[34]

With a cornered market, Chino established a scrip system common to most company towns to facilitate control of trade in household goods. Workers had to purchase books of coupons called *boletas* with values ranging from one to twenty dollars. This system offered workers access to a wide array of goods but at higher prices. The company gave employees the choice of being paid in cash or boletas, which often meant the difference between a squandered paycheck and food on the table. Wives and mothers preferred the coupons, making them very popular. The company used this preference as well to extend credit, a financial trap that often allowed families to build a debt similar to peonage experienced by tenant farmers in the South and other company-town employees in the West. The company created an entire department at the Santa Rita Store to handle this extensive credit scheme. Paul Jones, who worked in the late 1930s for the store, recalled: "All Chino employees were eligible for a charge account and when an employee drew his pay, his bill at the store would be deducted, leaving him only the balance. Therefore one's credit was good and at no risk to the store. There were very few cars, and not even many horses owned by the employees in those days, and most walked. Thus most employees were bound to the company store." The company also cracked down on "sharpies" who bought boletas at discount prices and then tried to horn in on the monopoly. Security officials "escorted [these profiteers] out of town with the suggestion that [t]he[y] not return," clarifying for everyone in Santa Rita who ran the town.[35]

During the shutdowns of 1921–1922 and 1934–1937, the company store served a dual role in the lives of the employees—one a benefit, the other a risk. Because the company extended credit even when operations halted, unemployed workers and their families had access to food and other supplies to maintain at least a marginal standard of living. This "advantage" could turn into a nightmare when debts mounted and perpetually obligated them to the company on renewal of employment. Consequently, the workers gave up some of their free-

FIGURE 4.20. *Joseph Clarence "Cap" Mitchell, ca. 1953. Born in Georgetown, New Mexico, in 1886 to John Vincent, a "Cousin Jack" (a skilled miner from Cornwall, England), and Marcella Vasquez, Mitchell grew up in the Mimbres Valley, helping to haul vegetables to market in Santa Rita. He earned his nickname, "Cap," when as a young boy he refused to wear a ten-gallon cowboy hat and instead donned a cap. In his mid-thirties he gained employment as a machinist with Chino. In the 1930s he began his law enforcement career with the mining company, serving as a watchman, the constable, and head of security while also serving as a Grant County deputy sheriff. He retired five years before his death in 1958.* COURTESY OF ELMO MITCHELL.

dom. On the other hand, most Americans, especially during the Great Depression, suffered a grimmer fate. As a result, the company offered some hope for those men and their families who could not sign on with New Deal agencies and who obeyed company rules. This kind of company paternalism created worker dependency and, for some, gratitude toward the copper corporation. It also may have discouraged employees from seeking assistance from unions.[36]

The company also placed great emphasis on schools in its desire to maintain a "stable" community. The Chino Copper Company established a typical educational system in Santa Rita. Children attended the Downtown School prior to the addition of the Hill School complex after 1914 and then the Sully School after its erection in 1927 (see figure 4.21). Beginning in 1924, the company bused the few high school students from Santa Rita to the new Hurley High School. Company-employed teachers taught a curriculum commensurate with school systems in other parts of the United States. In grades one through six, individual classroom teachers, such as long-timers Mary Lee Archer (1918–1962), Clara (ca. 1920–1936) and Maude (ca. 1926–1944) Trevarrow, and Inez Huff (ca. 1908–1910 and 1918–1950), taught a range of subjects from mathematics and English to science, art, music, and physical education. Once the children passed sixth grade they took a junior high curriculum. During the 1920s, 1930s, and 1940s, for example, Grace Ames offered reading and English, and Elizabeth Keough, home economics, geography, social studies, and mathematics to students at Sully. Other teachers taught additional courses in science, mechanics, and history.

Interestingly, the company placed education almost exclusively in the hands of women. Chino, in fact, hired fewer than five male teachers (there were more than fifty total) throughout the history of public education in Santa Rita. Those men who did teach, such as Robert Galloway in the 1910s and 1920s and Hiram Nunn in the 1940s and 1950s, also served as principals and superintendents. On the other hand, the company often filled the principal slots at the Hill and Sully Schools with successful female teachers. Louise Snyder, for example, served in the 1910s in that capacity. Elizabeth O'Mara did so in the 1920s and 1930s and Deborah Archer in the 1950s. Some of the teachers earned degrees at New Mexico Teacher's College in Silver City but most came from other parts of the United States, generally from the greater Southwest.

Enrollment in Santa Rita varied from about 450 students annually in the first ten years of the open-pit era to peak numbers around 650 in the 1920s (though approximately 1,100 children ages six to eighteen lived in Santa Rita). In the 1930s to 1950s period, 400 to 600 students matriculated in the first eight grades at Santa Rita. From 20 to 100 high school students from Santa Rita enrolled in the Hurley High School from the 1920s until the early 1960s.[37]

The company closely monitored the schools in Santa Rita throughout the town's history. Officials carefully hired qualified teachers to teach, paid their salaries, and provided them with housing in the "teacherage." Since Anglos were also known as "Americans" and all of the Hispanics as "Mexicans," the company chose Anglo educators with an eye to "Americanizing" the immigrant children. In the same light, school board members usually worked for the company as well and were always Anglo, such as highly placed officials Horace Moses, Harry "Cap" Thorne, and Harvey Forsyth. The company's intimate relationship with the schools sometimes resulted in interesting and talked-about marriages. Santa Rita principal Nina Light, for example, in 1918 married John W. Brock, a prominent mining engineer at Chino. Teachers Winifred Cummings and Dorcas Reeder in 1930 betrothed craneman John "Jack" Wesley Hardin and locomotive engineer Bob Mackey, respectively. Another educator, Carolyn Beals, two years later hooked up with drillman Clyde Osborne.[38]

FIGURE 4.21. *Sully School, 1927. Named for Chino's first general manager, the Sully School was open from 1927 until the 1970s. Company-employed teachers offered Santa Rita's children a curriculum for the first through eighth grades. Typical of public schools throughout the United States, this facility housed substantial classrooms, a library, a music room, a science laboratory, and, after 1951, a gymnasium. Forty-year teacher Mary Lee Archer put the "X" on the left side of the photograph to mark the location of the classroom where she taught from 1927 until her retirement in 1962 (she taught at the Downtown School for five years prior to arriving at Sully). Kennecott razed the building in 1981.*
COURTESY OF TERRY HUMBLE.

From the start of company-controlled schooling teachers taught students the importance of being "good" citizens of the United States. Classes with the highest attendance rate, for instance, earned the right to fly the national flag. Patriotic officials also insisted that Mexican American students learn English before being integrated with Anglos at school. As a result, the teachers segregated non-English-speaking Hispanic students from Anglo ones in the first four grades at the Hill School (see figure 4.11). Some of these students resented the segregation and as a result named the school La Ratonera,[39] or "Little School for Mice." Jesus "Tuti" Chacón recalled that "Chicano and Mexican children who could not speak English attended school there until they learned to speak English." He felt discriminated against and resented that the girls had segregated restrooms, "Anglos on one side [of the hall], Chicanos on the other side." Olga Viola Chavez even remembered her classmates being chastised for speaking Spanish. "I don't think they tried to be bilingual because then we were punished for knowing another language . . . They made us learn English fast [too]." Unfortunately for Chavez, school became a "frightening experience" on a day-to-day basis. The teachers' intimidation tactics no doubt were designed to Americanize the Chicanos so that they could function according to the company's prescribed community ethic as well as more efficiently once they began employment at the mine works.[40]

Still, this effort failed to eliminate Spanish from the community, as Paul Jones recalled. What "we spoke on the play ground had little resemblance to that taught by the teachers. It was a mixture of Spanish and English with perhaps a few Indian words thrown in for good measure. Maybe one could call it Spanglish. In any case, it was effective, and we had no trouble communicating. As we progressed through school, the Mexicans became more adept at English and we Gringos improved our Spanish. It was not unusual in later years to witness a conversation between two people with one speaking English and the other replying in Spanish."[41] Clearly, the efforts to Americanize Hispanic students did not eliminate their desire to use their native language. Eventually, in fact, bilingual students gained an advantage in the work world, ironically, though, in the era after Santa Rita disappeared.

A more complete look into life in the schools reveals that teachers integrated the Hispanics and Anglos in an era when segregation of the two ethnic groups was common in the Southwest (see figures 4.22 and 4.23). Photographs show Anglo and Hispanic schoolchildren posing together with their teachers. This ethnic mixing played a key role in the future of Santa Rita and Grant County, generally. Even before the US Supreme Court ruled in *Brown v. Board of Education* in 1954 against "separate but equal" facilities for white and black children in public schools, Santa Ritans were already beginning to break down ethnic barriers through integration in the schools. This gradual cross-cultural intermixing eventually reduced the workplace and educational limitations placed on Hispanics in the mining town despite tensions brought about by these changes, especially in the early 1950s when Mine Mill Local 890 struggled to ensure full citizenship rights for Chicanos.

The highlight of Santa Rita's school year centered on the eighth-grade graduations at the Orpheum Theater. These festive occasions symbolized both the students' success as scholars and the common belief among many Santa Ritans that their formal education had come to an end. The Reverend Harold Johnson, pastor of the Community Church from 1929 until his retirement in 1965, often gave "inspiring" speeches at the convocation exercises. Prior to the early 1930s nearly all of the graduates were Anglo. By the early 1930s, however, Hispanics made up close to 50 percent of diploma recipients. In the early 1940s, 85 percent of these graduates came from Chicano homes. The percentage leveled off at about 65 percent in the early 1950s, paralleling the approximate proportion of Hispanic employees at Chino, which stood at about 60 percent from the 1910s on.

Obviously, Mexican American families placed a high value on at least an eighth-grade education, especially with the rise of the labor movement in the post–World War II period. On the other hand, fewer Hispanic students graduated from Hurley High School than Anglos in this same period. Chicano parents expected their sons to start work at the mine as soon as the company would hire them. Lilly Durán Delgado, one of thirteen siblings, for instance, remembered that "as soon as you were old enough to do for yourself . . . you were expected to . . . work." Bernard "Barney" Himes, longtime labor relations manager at Chino, heard one company official say as they were driving by a school playground of Chicano boys that "there's my next generation of track laborers."[42] For generations, parents and company officials pressured

Figure 4.22. *Hill School kindergarten class, 1920. Teacher Jennie Chester's kindergarten class pensively poses for a group picture on the stairs of Hill School. Unfortunately most of the children have not been identified. According to one of the pupils, Myrtle Barber, the second girl directly left of the teacher with a white dress and tousled hair, Maxine Anderson and Ramon Rodriguez are in the front row; Clara Roybal is in the second row; Rose Esquibel and Esperanza Garcia are in the third row; and Lloyd Jennings is the first boy on the left in the back row. Note the dress of the children and the boys' hats in their hands.*

COURTESY OF TERRY HUMBLE.

young Chicano men (and many Anglos for that matter) to quit school at about the age of sixteen to begin their occupational careers.

Hispanic girls experienced paternalistic obstacles to their educations as well, with few of their fathers prior to the 1960s allowing them to pursue a high school diploma. This pervasive "machismo" limited Spanish-speaking women to traditional wifely and motherly roles. Even after Chicanas took the lead in picketing the Empire Zinc Company in the early 1950s in the famous Salt of the Earth strike, Hispanic men expected their wives to return to their traditional female roles after the dust had

settled from the labor dispute. Consequently, company and familial pressures limited the Hispanic students' educational opportunities, especially prior to the 1970s.

Still, there were exceptions. One inspirational case involved a woman who defied her father. Mrs. Cruz Galván Valenzuela pursued her educational interests and earned a bachelor's degree in education at the State Teacher's College. She taught at the Hanover school in the 1940s and later facilitated the establishment of bilingual education in Grant County. Another Chicana, Amelia Herrera, gained local fame in the early 1950s for her spelling prowess, which landed her in regional and

134 | S A N T A   R I T A

FIGURE 4.23. *Seventh- and eighth-grade classes on the east steps of Sully School, 1928. Unfortunately, many of the students in this photograph are unidentified. Myrtle Barber, pictured with the group, remembered some of her peers' names:* first row, *Halleen Beatty, Teresa Storer, Della Binson, Glenna Trevarrow, and Vera Mae Kennedy;* second row, *Rosemary Head, Clarabelle Henry, Clorinda Casdin, and Lucy Lucero;* third row, *Sara ?, Frieda Turner, and Sara Dunckhost;* fourth row, *Myrtle Barber (second from right) and Elena Chavez;* fifth row, *Rafael Saenz, Johnny Deemer, William Pegler, Wilson Crittenden, Everett Martin, Clarence Henry, and Jimmy Blair. Barber noted on the back of the photograph that these individuals also appear: Lloyd Jennings, Dennis Murphy, Freddy Clayton, and Jimmie Smith. The principal teachers for this group were Edith Hanes, history; Winifred Cummins, math; and a Miss Jones.*
COURTESY OF TERRY HUMBLE.

national bees. These and other cases suggest a gradual generational shift in valuing education beyond eighth grade for the emerging Chicano culture.[43]

A transition in attitudes among Hispanic Santa Ritans did take place eventually. Lilly Durán Delgado's experience reflects this shift. She remembered, for ex-

ample, that her parents—Louis and Marta (Guiterrez) Durán—in the 1940s and 1950s "didn't say too much . . . about our getting an education." However, when the local priest asked her father to serve as the president of the Catholic school, "he came to understand the importance of education," a realization that many Hispanic parents experienced in the 1950s, 1960s, and later. Her oldest brother may have influenced the family's attitudinal change because he attended high school. Although she did "know there was some discrimination in the beginning" at Hurley High School and there were separate buses for "Mexican" and "Anglo" students, "things were a lot better" by 1956 when she graduated. "Discrimination was felt by some maybe more than others, especially if you had little." Poverty could be an additional marker for discrimination. But in the end Durán believed that most of the teachers, who were "all Anglos . . . really wanted to help their pupils . . . They were with them and behind them . . . [And] they would stay after school hours" to ensure that Hispanic students learned and that their parents knew their progress and the importance of education to their success. Working-class Anglo families experienced a similar transition, especially when signs of the end of open-pit mining as early as the 1970s forced a rethinking of how best to provide for their families. Education could be the ticket to a prosperous future in the post-industrial world.[44]

The company also provided a substantial and affordable healthcare system in Santa Rita. Another feature of paternalism, the company hospital played a vital role in the care of the workers and their families. Under the supervision from 1910 to 1948 of Dr. Frank Carrier, the hospital staff offered medical services for any member of a worker's family. For full coverage a single employee paid $1.50 a month and a worker with family members $2.50, an affordable expense at a time when wages ranged from about $4 to $10 a day. The isolation of Santa Rita gave workers no other choice but to pay the health insurance premiums, which it turned out were far cheaper than anywhere else in the Southwest. Rafael and Nicanora Kirker, for example, believed that "one of the advantages of working" for Kennecott "was the availability of medical care . . . for a small fee deducted from the workers' wages." Although the "benefit" could be abused, "most people would seek care only when absolutely necessary," hinting that the company may have also tried to keep medical costs down by discouraging doctor's visits.[45]

In addition to treating and caring for injured workers, Dr. Carrier, other doctors, and a staff of nurses delivered about 4,000 babies at the Santa Rita Hospital. Known affectionately as "El Orejón" ("Big Ears") to "La Raza," Carrier gained the trust of the workers, both Chicano and Anglo. Despite this trust, the hospital, like the company store and to a lesser extent the Hill School, symbolized ethnic prejudice and "fatherly" care in the paternalistic mining town. During his nearly forty years of service, in fact, the popular physician segregated patients in separate treatment rooms. Still, Dr. Carrier and his staff of Chicana nurses' aides (see figure 4.24) by the 1940s commonly made house calls to Anglo and Hispanic households. They delivered babies of both ethnic groups at the hospital; set up quarantine tents during typhoid, cholera, and influenza epidemics to protect all the town's inhabitants (the doctor even had to treat three smallpox cases); and, of course, treated injured mine workers. The "Spanish flu" epidemic of late 1918 greatly concerned Carrier, who treated more than 180 patients in the hospital and makeshift triages at the clubhouses in Santa Rita and Hurley. Six years earlier Carrier did several appendectomies on a host of Santa Ritans, including John Sully and the unfortunate Timothy McNamara, who waited too long to seek treatment and died after surgery. This rash of appendicitis led Sully to write to Jackling that "Doc has been taking out so many appendixes here lately he has got afraid of himself," an ironic reference to Carrier's own recent appendectomy.

Mexican American parents, especially prior to World War II, often chose to stay at home for births. As was common in much of Mexico and rural America, mothers gave birth under the supervision of midwives. Teresa Holguin, for instance, brought more than 100 Hispanic Santa Ritans into the world from the 1910s to the 1940s. The longtime baby deliverer, distrustful of modern medicine, used herbal mixtures to facilitate the births as well as the mothers' recoveries. She also offered herbal remedies, preferred by many Hispanics in the mining town, for other ailments.[46]

Company officials also believed that religion promoted their ideal for a "stable" family-oriented community. Consequently, Chino provided the funds and sometimes the workers to construct the local churches. In 1911, for example, the company paid to raise the second Catholic church in Santa Rita. In this same light, the corporation deeded an additional tract of land in Hurley

to build a similar house of worship there. Later, when Santa Rita's holy structure burned in an electrical fire, Kennecott funded in 1943 much of the reconstruction for a third Catholic church in Santa Rita (see figure 4.25). Sully and later general managers realized that the vast majority of Hispanic workers and their families revered the Church. They looked to their priests for moral guidance as well as spiritual services, such as baptisms, confirmations, marriages, special masses, funerals, and other religious functions.

Under the tutelage of Father Marius Gerey, two generations of Hispanic Santa Ritans (and a few Anglo Catholics) received their religious instruction from the compassionate reverend. Known affectionately as Father "Jerry," the native Frenchman served as the copper town's principal priest from 1916 until his death in 1937. He trained the most serious youths as altar boys in the case of males (see figure 4.26) and as Hijas de María (Daughters of Mary) for girls. Among the more celebrated of the acolytes, Chano Merino, who later became president of Local 890, served the church in his youth under Father Gerey's instruction. Merino's experience cultivated a quiet confidence in him. As early as sixth grade the future labor leader stood up for other Chicano children at school and in the community during the initial phases of La Raza's fight for equality in Santa Rita. Young Chicanas tended to learn traditional female roles in their churchly activities. The Hijas, for example, wore white veils and white dresses to celebrate their messiah's mother. In their dedication to Mary's memory they performed many services for the Christian community. As Soccoro Herrera Lopez recalled, "the sisters at the church used to have lots of fiestas to celebrate the saint's [sic] days. We used to dress up like saints." The youngsters also raised funds during the month of May (or "month of Mary") to assist the poor. The priests also encouraged the children to join the Catholic Youth Organization, which often sponsored dances and parties hosted by Father Leony.

The most popular saints' celebration, of course, was Santa Rita Day, observed every May 22. For a day and a half in 1919, for instance, Father Gerey and many parishioners organized a big festival that featured fireworks, a band, a mass, and lots of tamales, enchiladas, and other tasty Mexican foods. This event also spurred the formation of the Benito Juarez and Estrella Clubs. Interestingly, Father Gerey also sponsored Boy Scout Troop 104, made up entirely of Chicano youngsters. In the late 1930s

FIGURE 4.24. *Santa Rita Hospital staff, 1947. Pictured standing,* left to right: *Frances Kimble, Josepha Guadiana, Betty Ortega, Kathleen Marshall, Patrocinia Carrillo, Mrs. O. K. Gibbs, Fortunata Arellano, Andrea Llamas, Etta Martin, Connie Valerio, Juanita Muñoz, Epifania Ontiveros, unknown, Maria Llamas, Lucita Martinez, Rufina Martinez, Nurse Jones, Celia Hurtado, Pepe Arellano, Olivia Costales, Ms. Trigg, Frances Castañeda, Irene Henry, Mrs. Meyers, Felipe Martinez, Lorraine Bonnell, Emily Himes, and Polly Quost; kneeling in front,* left to right: *Dr. B. A. Johnson, Dr. Frank N. Carrier, unknown, and Dr. John H. Rives.*

Father Miguel Estivill succeeded Gerey. The new priest carried on the religious instruction and guidance that Catholic Santa Ritans had learned to count on from the inception of the company town. As with their support for the Alianza Hispano Americano, Chino officials welcomed these Christian-sponsored activities because they instilled in locals devotion to tradition rather than other "radical" institutions like labor unions.[47]

Prior to the Reverend Johnson's arrival in 1929, Protestant Santa Ritans attended various churches. The Santa Rita Baptist Church was the most prominent house of worship when he first moved to the copper town. Episcopalians, Presbyterians, Methodists, and non-denominational parishioners, however, did not have a distinctive permanent church to serve them. This situation changed dramatically with the appearance in the mining camp of the soon-to-be-loved Johnson. His predecessor, C. M. Christiansen, had resigned in October after three years' service. His critics claimed he failed to put forward the effort to increase membership, offer more services to the community, and revive the Boy Rangers and Boy Scouts.

Born in South Dakota in 1898, Harold Eugene Johnson (see figure 4.27) arrived with his family in early December. The budding family immediately moved into the well-provisioned parsonage on Upper Iron Hill. He was already an ordained Congregationalist minister with some college course work at the University of New Mexico in Albuquerque, and his experience with the Boy Scouts of America over the previous twenty years greatly

FIGURE 4.25. *Santa Rita Catholic Church, 1943. On the morning of January 13, 1942, Manual Guitierrez, returning home from work at the mine, discovered that the Catholic church was on fire. Men quickly arrived with the fire truck but little could be done to save the spiritual sanctuary because of poor water pressure from the Iron Hill tanks. Unfortunately, the parish lost many of the church's priceless (in terms of sentimental value) ornaments and furnishings even though firefighters saved the rectory. Understanding the significance of the holy structure to its employees, Kennecott decided to fund the rebuilding of the church (the third Catholic church in Santa Rita's history) that is featured in this photograph. Note the small ledge level with the top stair. Local children made it a challenge to walk all the way around the church on the two-inch-wide shelf.*
COURTESY OF TERRY HUMBLE.

appealed to the Protestants in Santa Rita. On his arrival, the Reverend Johnson welcomed all Protestants to the Community Church (see figure 4.28) to listen to his sermons and participate in churchly activities. His dedication inspired a rapid increase in membership and activism in the Ladies Aid and Christian Endeavor Societies as well as other service-related groups in the mining town. The revered minister also initiated annual traditions, such as sunrise services on Easter and Christmas, Labor Day picnics, and charitable events.[48]

Santa Ritans' fondest memories of Johnson, however, centered on his role as the sponsor of the Boy Rangers and the Boy Scouts and as an advocate for the Girl Scouts. Working with scoutmaster Frank Thurber, who founded Troop 105 in 1918 with membership in the Yucca Council of El Paso, Johnson dedicated himself to

FIGURE 4.26. *Father Estivill and altar boys, May 3, 1943. Dedication day for the new Catholic church in Santa Rita. Pictured here are Father Estivill and the acolytes,* left to right: *front row, Gregorio "Gori" Chavez, Frankie Merino, Johnny Fernandez Cardenas, Arturo Herrera, and Reynaldo Gómez; back row, Mike Barrajas, Jesus Ojeda, Severiano "Chano" Merino (future labor leader), and Lorenzo "Lyncho" Gómez.*
COURTESY OF TERRY HUMBLE.

the education of Santa Rita's young Anglo men. From time to time he and Thurber took the youngsters on day excursions, usually near Santa Rita. The scouts learned armchair versions of local geology, zoology, and botany, hiking under the minister's supervision (as he rode Old Blue, a burro) to well-known natural sites. He regularly accompanied them to Sapillo Creek, the 'I' Box Ranch, the Three Circles Mimbres ruins, Cameron Creek, Big Foot and Map Cave in Rustler's Canyon, and, of course, the Kneeling Nun. At this latter revered town monument they periodically replaced the wooden crosses previous generations of boys had placed on the top of the rock monolith (see figure 4.29). During the summer Johnson and Thurber took the impressionable adventurers on ten-day pack trips on horseback or on foot into the Gila wilderness. There they taught the boys the skills of scouting—camping, fishing, woodcarving, tracking—and the importance of cooperative companionship. Johnson often recruited US Forest Service rangers to give the boys instruction in forestry, game management, and appreciation of the outdoors. Forest ranger C. D. Wang, for example, supervised a tree-planting project in 1937 when the Santa Rita boys transplanted 2,500 saplings on the game refuge east of the Mimbres Ranger Station in the Gila National Forest.

The scouts played an educational role from the 1920s to 1960s in the lives of many of Santa Rita's youth. The schedule of presentations in 1931 serves as an example of the instructional opportunities offered by membership in the scouts. Dr. B. A. Johnson tutored them in first aid and personal and public health. Shovel foreman Harvey Forsyth instructed them in agronomy, botany, and poultry and natural sciences. Chino tinsmith George McCormick taught them metal craftsmanship, and blacksmith H. A. Sarver gave a smithing demonstration. Harold "Tuff" Moses offered the young men tips on knot tying and rope splicing and rigging. Other skilled men introduced them to bugling, swimming, mapping, pioneering, and surveying. Even though Troops 104 (Hispanic) and 105 (Anglo) were segregated, Johnson and scoutmaster Manuel Sepulveda began in the 1940s to organize activities, from skills demonstrations and local hikes to camping trips that included both of the Boy Scout groups. On February 10, 1940, for example, Sepulveda shared his knowledge of mining, safety, and woodworking at a combined event where Johnson demonstrated blacksmithing and camping skills. The Protestant minister, in fact, had worked the year before with Santa Rita's Father Miguel Estivill, sponsor of the Chicano troop, to rejuvenate "Spanish American" Troop 104, whose membership and activities had fallen off in previous years. In recognition of the Reverend Johnson's contributions to the scouts, the national office awarded him the Silver Beaver award in February 1936 at a meeting where boys from both groups also received awards for their various achievements during the year. Noted southwestern artist Reuben Gonzales and Joe Luna earned similar accolades in the 1950s for their dedication to Troop 104 (see figure 4.30), which remained very active in the Chicano community in the post–World War II period.[49]

The Girl Scouts played an active civic role in Santa Rita as well. Under the tutelage of Elizabeth Scheele, Ethel McKnight, Winifred Cummins, and Rosemary

FIGURE 4.27. *The Reverend Harold Eugene Johnson, 1960. Pastor Johnson served his congregation at the Santa Rita Community Church from 1929 until his death in 1966. Revered and loved, the minister initiated numerous spiritual traditions and organizations, including the annual Easter sunrise services, Labor Day picnics, the Pilgrim Fellowship, and other religious programs. He was most remembered, however, for his dedication to the local Boy Scout troops. This portrait was taken to honor him as the 1960 Grant County Citizen of the Year.*
COURTESY OF TERRY HUMBLE.

Head, from the late 1920s on, these young female Santa Ritans engaged in numerous activities. They hiked to the Kneeling Nun, attended summer camp, and learned how to swim, play basketball, and survive in the Gila wilderness, similar to the Boy Scouts. On the other hand, they also sponsored events that validated the traditional female roles expected of women in Santa Rita. The young women, for example, held sewing and canning "parties" to assist the poor, especially during the Great Depression. During the gloomy thirties they stuffed more than 1,000 sacks with fruit, candy, nuts, and popcorn at Christmastime. After putting together these charitable packages, they handed over their gifts to the Boy Scouts, who made deliveries to hundreds of families, Anglo and Chicano. In 1934 alone, "Santa Claus" gave out 1,260 such "stockings" to local children.[50]

During the two shutdowns at Chino in 1921–1922 and 1934–1937, the mining company and civic organizations cooperated to ease the hardships of many Santa Ritans. General Manager Sully, for example, used company funds to pay for rail transportation for the single Mexican workers hoping to return to jobs in their home country during the first of these closings. Disembarking at Juárez, the multinational workers were received by Mexican government officials who took them to agricultural jobs in Chihuahua, similar to cross-border labor exchanges, legal and illegal, in place today. Chino officials allowed most of the Mexican workers with fami-

lies to stay on company property rent-free. As many as eighty-six impoverished families received direct assistance during the one-year cessation. Among the provisions were shoes for the poorest children, food baskets, and Christmas goodie bags for all of Santa Rita's youngsters. Sully took the workers' impoverished conditions personally, reflecting his strong belief that he was like a parent to them. He had clarified his duty from the onset of operations in 1909 "to take care of a great number of employees" in the company towns. In his fatherly way he also encouraged those who could afford it to give needier families a chicken to go with the rice and beans provided for the holidays by Father Gerey of the Catholic church. The company strongly approved of the efforts of the Girl and Boy Scouts in these benevolent endeavors.[51] Again, company-sponsored and -supported activities precluded

FIGURE 4.28. *Santa Rita Community Church, 1940. Located on Iron Hill, the Community Church served all of the Protestants in Santa Rita after its dedication in May 1940. The Reverend Johnson gleefully moved his parishioners from a building downtown (formerly a saloon, assay office, and high school) to this new structure. Note the two boys, Bill and Ben Bean (who lived nearby), on the steps of the church, which carpenters would soon adorn with a wooden cross. In 1949, the congregation added a parsonage next door where Pastor Johnson lived with his family. When the pit expanded in 1965 to this area of town, the parish moved the church and parsonage to the Arenas Valley.*
COURTESY OF TERRY HUMBLE.

FIGURE 4.29. *Boy Scout Troop 105 below the Kneeling Nun, 1955. One of the longest-standing traditions of the Santa Rita Boy Scouts (and Girl Scouts) was to hike to the Kneeling Nun 1,500 feet above the copper town. On this occasion, the boys went with their scoutmaster the Reverend Johnson and Troop 104 (not pictured) to replace the missing wooden cross at the top of the rhyolite megacolumn. Each scout carried tools or replacement pieces—such as two-inch pipe, pipe wrenches, hammers, rock drills, cement, sand, and water—to complete the task of putting up a metal crucifix. Pictured standing,* from left to right: *Robert Herkenhoff, Richard Reece, Jerry Porter, Tommy Martin, Jackie Wynn, the Reverend Johnson, and Eugene Herkenhoff. In the truck bed,* left to right: *Terry Humble, James Martin, James Craig, and Freddie Smith. The boys' handiwork lasted for thirty years until the 1980s, when, unfortunately, vandals tore down the cross and tossed it over the cliff.*
COURTESY OF TERRY HUMBLE.

the need, from the point of view of corporate officials, for a union.

Collectively, the company and civic organizations worked together through the Welfare Committee. This organization coordinated the company's and private organizations' charitable donations during the Great Depression. The committee, for instance, encouraged private citizens to give food, clothing, and wild game to the poorest of the unemployed. Male Santa Ritans with the means sponsored massive rabbit drives and wood-hauling bees. The Boy Scouts and Girl Scouts as well as the Benito Juarez and 4-H clubs raised money at dances and collected canned goods and milk for the least fortunate schoolchildren. Many other civic organizations, such as the Moose Lodge, the Mexican White Cross, the Alianza Hispano Americano, and the American Legion and its Women's Auxiliary, also put forward similar efforts to assist the underprivileged. Some men hunted deer and then sold the meat in town at affordable rates. For its part, the company provided hot meals to students at the Sully and Hill Schools, donated trucks and fuel to haul wood and game, and again offered laid-off Mexican

FIGURE 4.30. *Boy Scout Troop 104, 1943. Sponsored by Father Estivill, Troop 104 poses on the steps of the Santa Rita Catholic Church for a group photograph. Pictured,* left to right: *top row, Manuel Sepulveda, assistant scoutmaster; Armando Fletcher, assistant patrol leader; Robert Macias, assistant patrol leader; Rudy Herrera; Octaviano Llamas; Alfonso Valenzuela; Jesus Rodriguez; Frank Gusman, bugler; Manual Avalos, patrol leader; and Mike Sepulveda, scoutmaster.* Middle row, *left to right: Julian Jimenez, assistant patrol leader; Armando Lucero; Manuel Guadiana; Rafael Ambriz; Jesus Aranda; Ramón Ortiz; Conrado Sandoval; Raymundo Trujillo; and Lorenzo Gómez.* Bottom row, *left to right: Gonzalo Muñoz, Tony Rodriguez, David Vargas, Johnny Saenz, Sam Saenz, Gregorio Chavez, Armando Avalos, Henry Alvarado, and assistant scoutmaster Oscar Rojas.* Front left row, *left to right: Victor Torres, patrol leader; Ted Arellano, senior patrol leader; Agapito Ambriz, patrol leader; and Everisto Torres, patrol leader.*

COURTESY OF TED ARELLANO.

workers rail rides to El Paso. Others panned for gold near Vanadium and Pinos Altos to supplement their meager incomes. Ambitious workers took their families to other southwestern mining towns like Bisbee, Arizona, in search of jobs.

During the two-year shutdown, officials allowed many of the unemployed workers and their families to stay in company housing free of rent and utility costs. For much of this time many of the employees' families also received free medical benefits. The company, however, drew the line when it came to those who defied the corporate "family." Chino officials, for example, excluded pro-labor employees from these benefits. Numbering about seventy men, they and their families were excluded from these perks because of their efforts to certify Mine Mill Locals 63 and 69 in the late summer of 1934. Regardless, most of the jobless did seek public aid through the Federal Emergency Relief Agency, the Civilian Conservation Corps, and the Works Progress Administration. Many men worked on the Bear Canyon Dam project. Others maintained highways and roads. Some completed assignments at CCC camps at Redstone, Whitewater, Walnut Creek, the Upper Mimbres Valley (Camp Sully), and elsewhere.[52]

Even before Santa Ritans sought distraction from these economically hard times, they looked to sports for diversion from the mundane daily work routine at the mines. Baseball, boxing, and bowling enticed many of the locals. Baseball enthralled Santa Ritans more than any other form of entertainment. As early as 1901, miners started the tradition of playing baseball as part of the Fourth of July celebrations. Over the next decade the American-born sport became the most popular form of amusement in Santa Rita and elsewhere in the Southwest.

In 1912, southwestern New Mexicans formed the famed Copper League. The mine workers' league initially fielded teams from Santa Rita, Hurley, Tyrone, Silver City, and Fort Bayard (see figure 4.31). Eventually teams joined the league from Bisbee and Douglas, Arizona; El Paso, Texas; and Juárez, Mexico. Like other mining companies in the region, the Chino Copper Company took baseball seriously. Officials recruited many a "miner" to work at the pit for a good salary so that they could hit home runs and field pop flies after work. As with the control over the schools, the stores, and the hospital, the company played a key role in the formation of the Copper League and the maintenance of competitive teams. Chino's officials, for example, paid local store owner L. H. Bartlett $150 as early as 1910 to manage the Santa Rita Diggers. By 1913 W. H. Janney, an engineer at the Hurley mill, served as league president. Other company officials—Horace Moses, Harry Thorne, and Dr. Frank Carrier, among other Chino associates—met in early spring to determine the schedules for the league. Chino's superintendent John Sully often threw the first pitch on opening day as well.

The Santa Rita Diggers (later Miners) and Hurley Concentrators (later Millers) rarely lost a game in these early years. Annually, these dominant ball teams competed against each other to win the coveted MacNeill

Cup, named for Charles M. MacNeill, president of the Utah Copper Company and one of the financial backers of Chino. More than 1,200 fans turned out in September 1914 when the two teams played the "game of all games" in a playoff that ended with the Concentrators winning the cup. A dispute between the Diggers and the umpires in July of that year nearly cost Santa Rita a chance to play in the series. League president Janney resigned at the height of the squabble after the team withdrew from the Copper League in protest. Eventually cooler heads prevailed and the Santa Rita team was reinstated.[53]

Santa Rita's baseball players in 1918 made national news. In April of that year the Copper Leaguers defeated the Chicago Cubs in the first of two exhibition games. Already having visited Santa Rita the previous year, posting a 14 to 6 victory over the Diggers, the Cubs agreed to play one game on April 5 in Santa Rita and a second the next day in Deming. In the first game the Santa Ritans emerged victorious 6 to 5 in a hotly contested ten-inning thriller. Although the Cubs got revenge in Deming the following day (6 to 0), the previous day's contest brought wanted national attention to the southwestern league. Chino officials used the victory to entice ball players from across the United States to play for pay in the Copper League. At a time when many mine workers earned about $135 a month, baseball-playing employees raked in as much as $250 during the season.[54]

The southwestern mining companies' desire to field professional teams led in the mid-1920s to the recruitment of banned major league players. Kicked out of the majors after the infamous 1919 "Black Sox" scandal when the Chicago White Sox allegedly "threw" the World Series, these players found jobs and new teams in the Southwest. The nicknamed "Outlaw League" enticed several fallen professional ball players. Lefty Williams pitched for the Fort Bayard Soldiers. Buck Weaver took up his glove for the Douglas squad. Eddie Ciotte and Chick Gandil joined the Chino Twins (a combined Santa Rita–Hurley lineup). Several other ex–major leaguers ended their playing and coaching days in the Copper League. They included Roy Johnson, who starred in Kansas City; Jimmie O'Connell, out of San Francisco and New York; Ben Diamond, from Tulsa; and others. Johnson served as the Fort Bayard manager, soon earning a reputation for "rowdyism." Living up to the outlaw reputation, he landed several punches to the face and body of one of the umpires during a game against Santa

FIGURE 4.31. *Santa Rita Ball Club, 1912. The copper town's favorite pastime was baseball. Pictured standing,* left to right: *Ben Bristow, center field; Jack Dempsey, left field; Mickie O'Boyle, third base; H. E. "Red" Moseley, manager; and an unidentified scorer.* Seated, *left to right: H. Young, catcher; Brown, first base; J. D. "Danny" Nolan, pitcher; unknown batboy; Hall Forsyth, shortstop; Horace Moses, right field; unidentified team member. Note the famed "CCC" logo on the uniforms in recognition of the team's sponsor, the Chino Copper Company.*
COURTESY OF SILVER CITY MUSEUM.

Rita. The "rude" comments of the Santa Rita fans enraged Counts, the second baseman for the Soldiers, so much that he whipped a ball into the grandstands, nearly hitting an onlooker in the noggin. The manager and his player later earned undisclosed fines, punitive measures that did little to erase the memory of the incident.

Despite this confrontation and others like it, Santa Ritans revered the game and the players. The reaction to the death of fifteen-year league veteran Herman B. "Rube" Weeks at forty after a two-month illness in 1927 reveals the depth of love the fans felt for their players. Hundreds showed up for the funeral. The list of six pallbearers read like a who's who of Chino officials. In the late 1930s, Kennecott constructed a new housing addi-

tion just north of the old baseball field, inspiring officials to name the neighborhood Ball Park. This Anglo quarter sported some of the finest homes in Santa Rita, again reflecting the value placed on the game in the copper town (see figure 4.32).[55]

The "Ball Park" became a popular spot for fans as well as Santa Rita's young men (see figure 4.33). Mine workers joined the Twilight League, playing sixteen games during a season. Four of the five teams' names during the 1927 season reflect the players' positions at Chino: the Carpenters, the Nut Crackers (machine shop), the Pencil Pushers (office workers), and the Drillers. The last of the teams was the all-Hispanic Indians. Rarely were Hispanics allowed to play in the Copper League.

Although company officials excluded most Chicano players from playing pro ball, they did encourage them to sport recreational league teams. In the end, Hispanics experienced a form of discrimination on the diamond identical to that at work. They could rise only so far in baseball as well as in the company. As a result, they endured discrimination on and away from the job. Despite these disadvantages, the 1930 Santa Rita Miners, made up mostly of Chicano players—Sepulveda, Vargas, Cano, Garcia, Esquibel, Cordero, Gomez, Lucero, Walters, Varela, Saenz, and Hernandez—with Pete Lucero at the helm, rarely lost against the all-Anglo teams. When other teams could not play them (or would not), they divided up into two teams, the Indians and the Tigers. Despite their prowess as ball players, Chicanos were not welcomed into the Grant County league probably until the 1940s.[56]

Ralph Kiner proved to be Santa Rita's most famous baseball player. Called up to the majors in 1940 at the age of eighteen, Kiner, a strapping six-foot-two-inch slugger, went on to play for the Pittsburgh Pirates. He led the major leagues in home runs in each of the first seven seasons of his career, a record that still stands today. The outfielder's batting prowess earned him the reputation as the Pirates' greatest right-handed home-run hitter. Only Willie Stargell, a lefty, blasted more home runs in the legendary team's history. Elected to the National Baseball Hall of Fame, Kiner went on to become a well-known baseball announcer. Locals remembered his uncle Warren's bakery in Santa Rita.[57]

Baseball and softball played a lasting role in the daily lives of Santa Ritans as well. After the "big league" games, the children lined up to play. Even though teams were segregated and "fisticuffs" often broke out, the game brought together the future leaders of Grant County. Both Anglo and Hispanic men remembered their days on the diamond as important exchanges, encounters that helped to assuage ethnic tensions heightened by occupational and residential segregation practices in the company town. Some players, especially the Chicanos, viewed baseball as an opportunity to showcase their superior (or at least equal) skills. Girls and women, on the other hand, played in softball leagues (see figure 4.34). In the end, recreational baseball and softball games brought Hispanic and Anglo community members together by the 1940s and 1950s, similar to integration at the schools and the mine works.[58]

The cultural features of Santa Rita discussed above illustrate the steady maturation of a southwestern community. The layering of educational, religious, and recreational activities formed a cohesive community seemingly destined for an endless run. But several factors had already begun to intervene to dramatically transform the town and the relationships of its inhabitants. Some of these changes came about as a result of the exigencies of mining itself. The expansion of the open pit forced a reordering of and then final removal of the town altogether because of blasting, digging, and dumping. Other changes resulted from the impact of national and international events, such as the labor movement and World War II. The growth of big labor along with patriotic military service inspired Chicanos to question and protest paternalistic company-town practices. Mainly through Local 890 of Mine Mill, they disputed occupational and residential segregation, limits to promotions at the mine works, and obstacles to political empowerment. Hispanic Santa Ritans believed they deserved the full rights of American citizenship like the Anglos of the copper town. So they actively sought change as a result. Of the sixteen Santa Ritans who gave their lives in the world conflagration in the 1940s for their community and the nation, for example, fourteen were Hispanic. Of those wounded, thirteen of twenty-one were Chicano. Those who returned home realized that they were being treated as second-class citizens in their own home country and many of them were no longer willing to put up with it.

Beginning in the 1930s, two features of this transformative world had already visited the copper town. First, company officials decided to begin to remove some of Santa Rita's homes to make room for the expansion of the open pit. Everyone knew, of course, that the enlargement of the massive human-made canyon would ultimately lead to the demise of their beloved town. This inexorable process began by the early 1930s. In 1932, 1933, and 1934, in fact, the company moved forty-nine houses out of Santa Rita. Although many folks who were required to leave Santa Rita altogether or take up new residences elsewhere in town were saddened by the change, they realized they had no choice. The economic benefits were far too great to effectively question the emerging policy. Consequently, many moved to nearby towns like Turnerville, Bayard, Vanadium, Hanover, and Fierro.[59]

Second, company officials targeted labor organizers and sympathizers for eviction beginning in 1934.

Workers' angst over the removals turned to anger because the company began evicting employees who had voted in September 1934 to join the International Union of Mine, Mill, and Smelter Workers. The Kennecott Copper Corporation had already fired the employees whom general manager Rone Tempest had placed on a blacklist and then made them leave town. In 1935 and 1936 the company handed out sixty-one eviction notices to those fired for union organizing and then forced them to remove or tear down their homes. In these two years Santa Ritans were forced to move or destroy 175 homes. The National Labor Relations Board later ruled that Kennecott had acted illegally in firing the employees. Many of those who were blacklisted got their jobs back. Still, their homes were gone for good. The dust from this labor-management fracas did not settle until the US Supreme Court ruled in 1942 in favor of the workers. This legal decision did condemn Kennecott's duplicitous actions against activist workers and influenced the corporation's decision about a decade later to sell its company town.[60]

The continued enlargement of the pit, however, could not preclude the inevitable removal of the town's homes and businesses in the 1950s and 1960s. The drama

FIGURE 4.32. *Santa Rita, Ball Park neighborhood, 1937. The mining company built Anglo homes next to the baseball field beginning in 1926. In this photograph four of the eventual seven rows of homes appear. These three-, four-, and five-room homes sported indoor plumbing and rented for $14 to $22 a month. Note the benches of the Chino Pit in the background.*
COURTESY OF TERRY HUMBLE.

FIGURE 4.33. *Santa Rita baseball game, ca. 1920s. This rare photograph of a baseball game in early Santa Rita reveals the locals' strategy, still used today in Grant County, for seating by parking their cars on the first- and third-base sides of the diamond. Note the benches of the open pit and the Kneeling Nun in the background.*
COURTESY OF DON TURNER.

of Santa Rita's removal because of mining actually began to unfold during World War II. Around-the-clock wartime production maximized by advances in blasting, digging, hauling, and processing technologies meant that the open pit was growing and the town's foundation dwindling. Seemingly simple changes in the operations, such as the implementation of more efficient blasting gelatins, increases in the capacity of the shovel dippers, and construction of double rail lines in the pit to accelerate ore removal, resulted in the exponential growth of the mine. These measures also expedited the fate of the town. In 1942, Movietone News filmed an explosion that dislodged 450,000 tons of ore in a single moment. Blasts like this one rapidly ate away the sides of the pit to speed up the demise of the downtown "island" and other sections of Santa Rita. Electric shovel crews were digging up one of North America's oldest mining towns. By 1957 the pit was so gargantuan that the company built a lookout on the west rim that attracted more than 2,000 vis-

itors from forty-five states and ten foreign countries in its first year. Locals had already come to terms with the death of Santa Rita. The *Silver City Enterprise* reported in April 1951 that "the basic reason for the removal is that a portion of the village and some of the shops used in connection with the mine operations are sitting on ore considered vital to continued operation. Kennecott shovels are ready to bite into the remaining areas on that side of the mine whenever the buildings are removed." This unprecedented pace also required mining engineers to recalculate the projected age of the mine and begin long-term plans for the exodus from Santa Rita.[61]

By the mid-1960s, shovel crews had dug up most of the area of what was the old town of Santa Rita. For many former residents the dismantling of the copper town was distressing even though they realized that the mine was the source of their very livelihood. Paul Jones lamented that "a way of life was gone forever, and a deep sentimental attachment has developed among the

FIGURE 4.34. *Santa Rita Girls' Softball Team, 1952. Young Chicanas enjoyed the game of softball in the post–World War II era.* Left to right: *Ramona Legarda (Gray), Mary Lou Esquibel (Ojinaga), Erlinda Puentes (Portillo), Virginia Esquibel (Rodriguez), Coach Sara Jimenez (Puentes), Lucille Acosta (Chavez), Irene Arzaga (Valenzuela), Rosa Hurtado (Cordero), and Helen Esquibel (Carreon).*

COURTESY OF SILVER CITY MUSEUM.

ex-residents who remember the good and bad times of their lives there . . . with a feeling of great loss and emptiness." Olga Salce's mother delayed her home's removal to the very end, declaring that "they're going to have to stick dynamite under me. I'm not going to move." Not even a stubborn grandmother of dozens and mother of eight could stop the company's plans. María Martinez remembered that "most people stayed as long as possible before leaving . . . It was a sad time for the people of Santa Rita because they thought that Santa Rita would be there forever." When Frank Blair returned to Santa Rita in 1958 after many years away, he was devastated that the town was gone. As he and his family drove into the Santa Rita Basin on their homecoming, he proudly told his wife, "'when we get over this little rise here you're going to see where I was born and raised,' and there was nothing there but a hole in the ground. Tears came to my eyes and I cried." It was "heartbreaking," Mary Lou

Lopez recalled, "when they were told they would have to move off the land." These and other such personal stories inspired the formation of the Society for People Born in Space.[62]

The projected costs for Kennecott to remove the company-owned housing also instigated a rethinking of how best to go forward with the inevitable. The logistical and financial realities of this transformation eventually forced the company's hand and officials decided in the early 1950s to sell its residential buildings to the John W. Galbreath & Company firm out of Columbus, Ohio. From 1947 to 1965 the real estate business bought defunct industrial towns all over the American West. In the months leading up to Kennecott's sale of its Chino properties, locals could not believe the mining corporation could sell the homes of their much-loved communities. Longtime resident and Chino employee Claude Dannelley, for instance, recalled that "rumors were out that the Company was going to sell the townsite [of Hurley], which people doubted. It was just not believed possible!" It happened, however, and only three months after the opening of Cobre High School (to replace Hurley High) in nearby Bayard, Kennecott announced in December 1955 the sale of its company towns of Santa Rita and Hurley. Galbreath paid nearly $5 million for these and six other towns at the copper corporation's western divisions, with about $1.7 million going for the purchase of the New Mexico properties. Kennecott relinquished ownership of the towns and then over the next few years gave up control of the hospital, the store, and the schools. The powerful corporation's officials were coming to terms with the company's new status in surrendering its paternalistic hold over its workers in the post–World War II period.[63]

Although many of the Santa Rita–Hurley homes had to be moved or were being moved at the time, Galbreath gave tenants first choice to buy the onetime rentals. The new landlord allowed residents who owned their homes to lease the lot over the next decade or until the mine expanded to their location. Employees had the option of buying homes at 5.5 percent interest with a down payment of at least $400 on the purchase price, which ranged from $1,000 to $4,500. Galbreath sold only to Chino workers in the first year, then planned to open the sale to the general public. By the end of August 1956, however, the firm had sold every house to a Chino employee. Between the middle of May when the Richard family bought the first home north of Hurley and August, the Ohio-based company sold 234 houses in Santa Rita and 444 in Hurley.[64]

Over the next two years Kennecott deeded the remaining community buildings, valued at $1.1 million, to the people of Santa Rita and Hurley. The mining company, for example, gave the recreation hall to the Catholic church, the parsonage to the Community Church, and the Chino and Casino Clubs and Masonic Lodge to their members. Yet relinquishment of these and other properties did not end Kennecott's involvement in the community. When General Manager William H. Goodrich announced in September 1957 the donation of the remaining properties, he also proclaimed that the decision to transfer the community buildings was symbolic of the end of the company's parental role in Grant County. The handing over of company property to the towns was made to "get away from the paternalism of the old-fashioned

company town," Goodrich declared. "When the company first acquired or built this property, it was necessary for nearly every western mining company to provide facilities for its employees. The mining camps were in isolated spots and transportation was poor. Times have changed and we feel that the towns should be turned over to the people who live in them."[65] Mine Mill, of course, had already successfully replaced the company as the principal representative of the workers. Kennecott realized it could no longer intimidate employees as its predecessors had during the pre–World War II period. Chino's founding general manager, John Sully, would not have recognized this new corporate acceptance and reformulated philosophy. Kennecott had acted in good faith a few years earlier when in 1952 it replaced the company store. The new Santa Rita store, a 20,000-square-foot structure with ten departments, welcomed customers from all over the county. Rather than paying in boletas, customers made purchases in cash, though Chino employees still had the option of buying on credit. Kennecott also financed in 1955–1956 nearly the entire $900,000 cost of the new Cobre High School in Bayard.[66]

The natural and human landscapes of Santa Rita had changed dramatically during the 1940s, 1950s, and 1960s. Most obviously, the activities at the mine, mill, and smelter altered the natural terrain. Less obvious were the changes in the human landscape. As the drama unfolded after the 1950s, the natural landscape continued to be molded to meet the needs of production and profits while the employees' world on the human landscape evolved in relationship to monumental changes in the nation. No longer were the workers cast in a social hierarchy that placed them at the mercy of the company in an isolated and segregated corner of the world. Now they drove their cars back and forth from home to work and elsewhere. They were like most Americans living in postwar America. Their sphere was becoming more personalized as technological and social advances created a more individualized experience. At the same time, those identical factors worked to broaden their vision of their place in Grant County, in the Southwest generally, and in the United States as a whole. These changes in an indirect way contributed to the rise to power of Local 890 of Mine Mill, itself an international union. The human landscape was becoming a freer place, too. Santa Ritans could choose where to live and where to get medical treatment. Hispanic children no longer had to endure segregated classrooms in the early grades and separate restrooms. Leaders among their parents could expect to freely participate in local government for the first time since the inception of the company towns. Chino's workers could demand better wages, fairer opportunities for advancement, and the economic expectations of the American Dream.

The Kennecott Copper Corporation can take some of the credit for these positive shifts in life in Santa Rita and elsewhere in Grant County. Sustained efficiency, perennial productivity, recognition of the workers' rights, and the good fortune of owning millions of tons of copper ensured a key place for the company in this new world order. The mining company's resources translated into a prosperous future for the southwestern community. Yet without the intervention of the union as well as new state and federal laws, the transformation could have been delayed in large part because the corporation had been used to unfettered domination of the community. How Kennecott and Mine Mill negotiated these changes with the implementation of larger-scale technology and the rise of big labor in this fascinating story is the focus of chapter 5.

NOTES

1. Initially they established one criterion for membership in this loosely organized group, to have been born in the Santa Rita Hospital. Soon they discovered, though, that many Santa Ritans, especially Hispanics, had been born in their homes under the supervision of midwives. As a result, anyone born in Santa Rita automatically became a member of the society. Ironically, Schmitt became an astronaut, making him the only man to have been born in "space" and to have traveled into space; see Sheila L. Steinberg, "Santa Rita, New Mexico, Community Report," *Community Studies Series Report No. 4* (Dept. of Sociology, Humboldt State University, March 2003), 6–7.

2. Among the informants were four Hispanics (one woman and three men) and three Anglos (all men), all of whom had a deeper knowledge of the history of Santa Rita, having lived there an average of 58.5 years; Steinberg, "Santa Rita," 9.

3. These quotes come from Steinberg, "Santa Rita," 14–15; interview of Jesus "Tuti" Chacón, Santa Rita Project, 1992.

4. Steinberg, "Santa Rita," 23–24.

5. See Ackerly, *A History of the Kneeling Nun*; Calvin, ed., *Lieutenant*; Raymond, *Statistics of*, 337; *Weekly New Mexican*, January 30, 1877; *Grant County Herald*, May 28, 1881; *Silver City Enterprise*, June 26, 1885, May 6, 1887; Wellman, "Above a Great

Copper Mine the Kneeling Nun Still Prays for Mercy," *Kansas City Star*, April 18, 1937; KCC, *Chino World*, August 16, 1976; Muñoz, *The Kneeling Nun*; and Topmiller, "Making the Kneeling Nun" are just a few of the sources for the Kneeling Nun.

6. Steinberg, "Santa Rita," 10, 17.

7. Ibid., 10.

8. Ibid., 10, 17; interview of Daniel C. Marrujo, Santa Rita Project, 1992.

9. "The Chino Copper Company," *The Mogollon Times* (1913), 33.

10. David Robertson, *Hard as the Rock Itself: Place and Identity in the American Mining Town* (Boulder: University Press of Colorado, 2006), 82–84; also see Leland M. Roth, "Company Towns in the Western United States," in John S. Garner, ed., *The Company Town, Architecture and Society in the Early Industrial Age* (New York: Oxford University Press, 1992), 173–205.

11. Chino Copper Company (CCC), *First Annual Report*, June 1, 1910, Monthly Report #10, September 15, 1910, Report, March 25, 1911; *Silver City Enterprise*, March 25, September 30, 1910, January 13, 1911; *Silver City Independent*, March 29, April 19, 1910. The numbering system was known among Santa Ritans. Also see *Map of Buildings and Water Lines on Company Property*, Kennecott Copper Corporation, Chino Mines Division, Santa Rita, New Mexico, 1957.

12. Upper Santa Rita Hill was also known as Iron Hill and Grandview Heights.

13. House listings, Kennecott records, 1920 Company Census.

14. Interviews of Ramón Arzola, Jesus "Tuti" Chacón, Mary Lou Lopez, and Guadalupe Fletcher, Santa Rita Project, 1992.

15. Vargas, *Labor Rights*, 185–187, shows that Chicano workers often struggled with lower wages and therefore were often forced to live in inferior housing in segregated parts of towns in the Southwest.

16. Paul M. Jones, *Memories of Santa Rita* (Silver City, NM: Silver City Daily Press and Independent, 1985), 33; Claude A. Dannelley, "Memories of Hurley," 11–12, Special Collections, Miller Library, Western New Mexico University; interview of María Martinez, Santa Rita Project, 1992.

17. See Sarah Deutsch, *No Separate Refuge: Culture, Class, and Gender on an Anglo-Hispanic Frontier in the American Southwest, 1880–1940* (New York: Oxford University Press, 1987), 89–91. Deutsch contends that mining corporations designed company towns in northern New Mexico and southern Colorado to destroy the village/communal atmosphere of traditional Hispanic plazas to maximize authorities' abilities to control the workers and to instill loyalty and foster obedience in them. This strategy may have been implemented in Hurley, a brand-new company town. Santa Rita already had a downtown area based on a grid pattern. An exception to this rule was Tyrone, New Mexico; see Christopher J. Huggard, "Reading the

Landscape: Phelps Dodge's Tyrone, New Mexico, in Time and Space," *Journal of the West* 35:4 (October 1996): 29–39.

18. An examination of photographs in the Chino Copper Company's *Annual Report* (1941) reveals these housing patterns; also see Jones, *Memories*, 27, 29; Dannelley, "Memories," 12–20.

19. CCC, *Annual Report*, June 1, 1909, *Report*, March 25, 1911; *Silver City Enterprise*, March 25, April 8, 19, May 19, July 27, September 30, October 28, November 11, December 23, 1910, January 6, 1911, June 21, September 27, 1912, August 28, 1914, January 24, 1919; *Silver City Independent*, March 1, 1910, May 16, December 19, 1911; F. J. Anderson to Sully, correspondence, September 12, 1917, June 29, 1918; interview of Daniel Marrujo, Santa Rita Project, March 1992; F. J. Anderson to Horace Moses, correspondence, January 1, 1918.

20. CCC, Monthly Reports, 1916.

21. *Silver City Independent*, November 9, December 21, 1909; *Silver City Enterprise*, March 1, 22, 1910, August 30, 1919, January 2, 9, 1931; *Silver City Independent*, July 21, 1931; Thorne to Sully, February 5, 1932; Goodrich to Moses, June 6, July 7, 1941, December 6, 1941; interview of Elvira Moreno Augustine, Santa Rita Project, 1992.

22. *Silver City Enterprise*, September 5, December 12, 1913, December 6, 1918, January 13, August 8, 1919, January 16, 23, 1920, May 2, December 26, 1924, September 25, 1925, February 24, 1928; *Silver City Independent*, August 12, October 14, November 25, 1913, March 11, 18, May 20, June 3, December 30, 1924, February 1, 1927, February 14, 1928; Thorne to Sully, *Reports*, January 6, August 14, October 6, 1925, February 20, September 5, December 6, 1928; also see interview of Elvira Chacón by Josephine Peña, Santa Rita Project, 1992.

23. Jones, *Memories*, 27, 87–89; Dannelley, "Memories," 7.

24. Interviews of Ramón Arzola and Robert Gardner, Santa Rita Project, 1992. Arzola claimed that the Chicano boys were fined by law enforcement officers; interview of Terry Humble by Chris Huggard via telephone, August 10, 2010.

25. Interview of Josephina B. Enriquez, Santa Rita Project, 1992.

26. Jones, "Memories," 67.

27. Interview of Frank Blair, Santa Rita Project, 1992; interview of Rafael Kirker by Joe E. Kirker, Santa Rita Project, 1992.

28. See Terry Humble, "The Chino Bandits," *Quarterly of the National Association for Outlaw and Lawman History* 29, no. 1 (January–June 2005): 8–14, for a detailed account; *Silver City Enterprise*, August 11, 1911, January 5, April 12, May 10, 1912; *Silver City Independent*, August 15, 1911, January 9, April 16, 1912.

29. *Silver City Enterprise*, March 27, 1914, June 4, 1915, April 8, 15, 1921, December 1, 1933; *Silver City Independent*, February 24, 1914, April 5, 12, 1921, December 4, 1933.

30. *Silver City Independent*, September 30, October 14, 1930, August 4, 1931; *Silver City Enterprise*, October 3, 10, December 5, 19, 1930.

31. George De Luna, "The Mexican-American: Two Generations in Santa Rita and Kennecott" (paper presented to Dr. Dale Giese for History 300, Western New Mexico University, December 1972), 4. De Luna interviewed German and Joaquin De Luna, Anne Westover, Moises Ruiz, and Herminio Gonzales for this paper; also see Vargas, *Labor*, 284–286, who shows that Chicanos complained of inordinate harassment by law enforcement officials throughout the Southwest in the twentieth century because of their ethnicity, similar to treatment of Hispanics in Santa Rita.

32. Sully to New Mexico Secretary of State, correspondence, May 2, 1917.

33. Dannelley, "Memories," 8, 16; Goodrich, General Superintendent of Mines, *Annual Report* (1942), 21.

34. James B. Allen, *The Company Town in the American West* (Norman: University of Oklahoma Press, 1966), 128–136; Jones, "Memories," 47; interview of Robert Gallardo by Tracy Muñiz, Santa Rita Project, 1992; interview of Bill Wood, Santa Rita Project, 1992.

35. Jones, "Memories," 47–49.

36. Robertson, *Hard*, 83.

37. Teacher's Monthly Reports, September 1915 to April 1916, September to April 1917; *Silver City Independent*, February 1, 1916, September 15, 1931; *Silver City Enterprise*, September 14, 1921, March 3, 1922, May 21, 1953; Thorne to Sully, September 1, 1925, September 6, 1926; Thorne to Tempest, September 5, 1933, September 3, 1935, September 7, 1936; Thorne to Moses, October 5, 1940; Goodrich to Moses, September 8, 1941, October 7, 1947; Slover to Goodrich, September 7, 1949.

38. *Silver City Independent*, May 21, 1918; *Silver City Enterprise*, December 27, 1929, June 5, 1931, June 10, 1932.

39. This Spanish term could also have meant "mouse trap" or "little rat house." Clearly, the name had a negative connotation for many of the Hispanic students, who felt "trapped" in the school.

40. Interview of Olga Viola Chavez, March 1992; interview of Jesus "Tuti" P. Chacón by Dorina Chacón, Santa Rita Project, November 1992; interview of Elvira Moreno Chacón, Santa Rita Project, 1992.

41. Jones, "Memories," 35.

42. Interview of Lilly Durán Delgado by Lilly Durán, Santa Rita Project, 1992; Baker, "On Strike," 38–39.

43. Baker, "On Strike," 105–113, 267; *Silver City Enterprise*, May 30, 1940, May 20, 23, 1941, May 21, 1942, May 22, 27, 1952, April 30, 1953, March 4, 1954; *Silver City Daily Press*, May 15, 1945, May 16, 1947, May 24, 1949, May 24, 1950.

44. Interview of Lilly Durán Delgado by Lilly Durán, Santa Rita Project, 1992.

45. Interview of Rafael and Nicanora Kirker by Joe Kirker, Santa Rita Project, October 1992.

46. *Silver City Enterprise*, December 19, 1910, October 25, November 1, 1918, June 12, 1931, July 28, 1933, February 12, 1948; *Silver City Independent*, November 28, 1911; CCC, Monthly Reports, April 25, May 18, October 28, 1911; Sully to Jackling, correspondence, May 18, 1912; Thorne to Sully, Report, September 25, 1924; Santa Rita Hospital records, 1910–1954; Jones, *Memories*, 63–65, 99; interview of Daniel Marrujo, March 1992.

47. Sully to Jackling correspondence, July 29, 1910; Sully to Wilson (CCC attorney), correspondence, December 7, 1910; interview of Mary Lou Lopez, Santa Rita Project, 1992; *Silver City Enterprise*, February 18, 23, 1910, April 25, 1919, July 30, September 24, October 1, 1937, January 16, 1942; interview of Soccoro Herrera Lopez by Victoria Madrid, Santa Rita Project, November 1992, Goodrich to Moses, correspondence, February 6, March 6, 1942.

48. *Silver City Enterprise*, March 13, April 3, June 12, August 21, 1925, December 6, 1929, April 6, August 31, 1934; *Silver City Independent*, December 10, 1929, March 27, April 10, 1934; Gerald Ballmer to Reverend Christiansen, correspondence, August 1929.

49. *Silver City Enterprise*, March 21, May 2, June 6, 13, September 26, November 14, 1930, May 8, 22, June 5, July 10, November 20, 1931, March 4, April 8, May 13, June 10, July 29, September 2, 16, 23, December 2, 1932, May 12, June 16, 23, July 28, December 1, 1933, July 26, August 23, 1935, February 21, May 8, 15, October 16, 1936, March 20, 27, 1936, April 21, May 5, July 7, October 6, 1939, June 4, 1953; *Silver City Independent*, May 19, 1931, June 28, September 6, 1932; *Silver City Daily Press*, February 11, April 9, 1936, February 5, March 3, 17, 23, 1937, July 19, 26, November 24, 1939, January 26, February 1, 12, 1940, December 21, 1954, September 6, 1955, November 7, 1956; Frank Thurber, *Report on the Annual Boy Scout Camp* (held at the Black Canyon of the Gila, June 4–16, 1932).

50. *Silver City Enterprise*, December 6, 13, 1929, February 13, 27, March 20, October 9, November 17, 1931; *Silver City Independent*, May 20, September 12, October 31, December 16, 1930, June 9, 1931, February 16, April 27, May 11, June 1, 22, 29, July 27, September 14, 28, December 28, 1934, December 19, 1940; Thorne to Tempest, January 3, 1935.

51. *Silver City Independent*, April 5, 1921; CCC, Monthly Report #4, April 11, 1921; Harry Thorne to Sully, Reports, April–December 1921; CCC, Monthly Report #10, December 10, 1921; for Sully quote from a comment he made in 1910, see *Silver City Daily Press*, March 31, 1951.

52. *Silver City Independent*, January 13, 1931, May 10, 1932; *Silver City Enterprise*, October 9, 1931, January 22, July 22, 1932, November 30, December 14, 1934, August 23, 30, 1935; Thorne to Sully, *Reports*, June 5, July 5, 1932; Thorne to Tempest, Monthly Reports (1934); Tempest to Boyd, Monthly Reports

(1934), December 7, 1935, January 7, 1936; Thorne to Tempest, *Reports*, March 4, April 3, May 7, June 3, November 4, December 4, 9, 1935; Tempest to Jackling, *Reports*, February 5, 1935, August 7, September 7, 1935.

53. *Silver City Independent*, July 9, 1901, March 4, 1913, April 7, May 19, September 25, October 2, 1914; *Silver City Enterprise*, June 14, 28, July 5, 1901, August 1, 15, 1913; March 6, April 3, 10, May 2, 23, 30, June 5, September 22, 1914, March 26, 1926; Sully to MacNeill, correspondence, September 4, 1913.

54. *Silver City Independent*, April 9, 1918.

55. *Silver City Enterprise*, March 1, 29, 1918, February 26, March 26, April 9, 1926, June 10, 1927; *Silver City Independent*, March 19, April 9, 1918, April 6, 1926, June 14, 1927.

56. *Silver City Enterprise*, September 16, 1927, July 18, 25, August 1, 15, 29, September 26, 1930.

57. See Ralph Kiner with Joe Gergen, *Kiner's Korner, at Bat and on the Air: My Forty Years in Baseball* (New York: Arbor House, 1987); Ralph Kiner, Danny Peary, and Tom Seaver, *Baseball Forever: Reflections on Sixty Years in the Game* (Chicago: Triumph Books, 2004); and http://www.ralphkiner.com/ for information on Ralph Kiner's baseball career.

58. Jones, *Memories*, 121.

59. Interview of Bill Wood, Santa Rita Project, 1992.

60. Thorne to Tempest, Monthly Reports (1935–1936); Tempest to Boyd, correspondence, November 7, December 7, 1935, January 7, 1936; pamphlet, "Society for People Born in Space, Santa Rita, New Mexico, 1804–1965" (Dedicated to Those Who Served).

61. Y. S. Leong et al., *Technology, Employment, and Output per Man in Copper Mining*, Report No. E-12, Department of the Interior, Bureau of Mines (Philadelphia: Works Project Administration, 1940), 44–45; Goodrich, general superintendent of mines, Chino Division, *Annual Reports* (1941, 1942); Moses, general manager, *Annual Reports* (1941, 1942); Ballmer, general superintendent of mines, *Annual Report* (1951, 1955), 12; *Silver City Enterprise*, February 6, 1942, April 5, 1951, April 1, 1954, June 13, July 18, 1957; Kennecott, *Annual Reports* (1951, 1954, 1955).

62. Jones, *Memories*, 1; interviews of Olga Chavez and Frank Blair, Santa Rita Project, March 1992; interview of María Martinez, Santa Rita Project, 1992; interview of Mary Lou Lopez, Santa Rita Project, 1992.

63. James B. Allen, *The Company Town in the American West* (Norman: University of Oklahoma Press, 1966), 48–49, 52, 73–74, 140–142; Dannelley, "Memories," 19; *Silver City Enterprise*, December 15, 1955.

64. *Silver City Enterprise*, December 29, 1955, January 19, February 19, August 9, 1956.

65. Goodrich, general manager, *Annual Report, The Chino Mines Division, Kennecott Copper Corporation* (1952); *Silver City Enterprise*, December 4, 1952, September 26, 1957.

66. *Silver City Enterprise*, September 26, 1957.

V

The Kennecott Era

MODERN TECHNOLOGY AND BIG LABOR

n 1955 the Chino Mines Division of the Kennecott Copper Corporation premiered *Chinorama*. Modeled after the multinational corporation's *Kennescope*,[1] the colorful glossy monthly magazine replaced the company towns at the Santa Rita–Hurley complex in New Mexico as the symbol of a maturing industrial "family." Clearly, officials hoped to retain the paternalistic overtone of the company-town era, especially after the sale of Kennecott's property to the real

estate firm of John W. Galbreath & Company in this mining-milling-smelting corridor of Grant County, New Mexico. They were attempting to use *Chinorama* to perpetuate that patriarchal tradition. The popular publication peppered with photographs highlighted general managers' commentaries, departmental features, household advice, women employees, old-timer insets, copper and strike news, medical and retirement plans, and other information.

Still, officials must have realized things were different now. The early volumes, in fact, reveal the company's public relations plans to generate a positive image among its workers and the general public. They sensed that they had to adjust to the changing world. So the features disclose an emerging acceptance of big labor

and the budding Chicano leadership in Local 890 of the International Union of Mine, Mill, and Smelter Workers (Mine Mill). In their "greetings" to *Chinorama*'s readers in the first volume, Charles Cox, president of Kennecott; J. P. Caulfield, general manager of the Western Mining Division; and William H. Goodrich, general manager of Chino, aired their hopes for "modern" communications, "unity within the whole organization," and recognition of "worthwhile accomplishments" of employees, respectively.[2] Another new era had dawned at Santa Rita.

The first edition's cover featured the heart of the operations, the Chino Pit, after a winter storm had dusted the southwestern landscape with a snowy powder (see figure 5.1). Locomotive No. 54's ore cars are receiving ore from the No. 11 shovel in the electricity-generated

FIGURE 5.1. *Cover photograph of first issue of* Chinorama *1, No. 1 (May 1955), showing the Electric Shovel No. 11 loading waste rock into railcars pulled by Loco No. 54 in the Chino Pit after a snowstorm.*
COURTESY OF TERRY HUMBLE.

operation. The ascending thirty-foot benches look like a stairway for the ever-present Kneeling Nun, who devotedly "prays" above the mining scene before the "altar" at the north end of Ben Moore Mountain. The company mailed the flashy free magazine—indicative of the financial success of Kennecott in Cold War America—to all of its employees and retirees. Undoubtedly, *Chinorama* marked the coming of age of one of the key properties of the world's top-producing copper corporation.

The combined success of the copper enterprise and the rise of activist unionism created a new atmosphere in postwar Grant County. Though Kennecott's efforts at Chino started slowly—with the Great Depression and the union certification crisis (1934–1942)—economic and working conditions improved dramatically during World War II, but especially in the postwar period. Chino's production soared to nearly 130 million pounds in 1940 and rarely dropped below that figure for the remainder of the century. This profit potential gave Kennecott the funds necessary to steadily improve technology. With these resources, for instance, the company regularly upgraded smelter technology. It introduced giant diesel haulage trucks. Its engineers perfected copper precipitation techniques and expanded Chino's water conservation program. At the same time, they devised strategies to extract molybdenum, gold, and silver from the porphyry ores of the Chino Mine; for much of the postwar period, moreover, the proceeds from these minerals paid for operational expenses. Simultaneously, engineers carefully studied the level of efficiency at the operations. Their efforts resulted in the introduction of more sophisticated mechanization processes. The copper company even paid employees for cost-saving innovations as part of a new potentially lucrative (although controversial) "Suggestion System."[3]

Chino's workers also reaped the rewards of the rise of radical unionism. Under the leadership of Mine Mill officials, especially after the arrival in 1947 of international representative Clinton Jencks and his activist wife, Virginia, Chicano labor leaders like Arturo "Art" Flores, Juan "Johnny" Chacón, Severiano "Chano" Merino, and other union men flourished from the late 1940s through the 1960s. At the height of Kennecott's ability to produce untold amounts of copper, big labor thwarted the traditional corporate hegemony over the workers, especially Hispanic ones. This newfound power brought results for the beleaguered employees as well. Local 890 regularly

won higher wages, better health benefits, substantial retirement packages, access to skilled training, and other profit-sharing gains. To achieve these new rewards the international union overcame company obstructions, the imposition of Taft-Hartley's anti-communist affidavits, expulsion from the Congress of Industrial Organizations (CIO), and raiding efforts by the United Steel Workers of America (USWA).[4] Working conditions changed, in effect, because of effective Hispanic leadership under the guidance of the international and regional offices of Mine Mill. In the process, Chino's grassroots labor leaders led the way in the fight for political and civil rights in the modern Southwest as well.

Kennecott's successes combined with societal changes emboldened the workers to stand up for their rightful portion of the profits at the New Mexico property. The AFL and CIO locals also effectively unified into the Chino Unity Council to solidify their position in collective bargaining negotiations with Kennecott. The gains of Chino's employees also encouraged other workers represented by Local 890, most notably those of the Empire Zinc Company, to struggle for fifteen long months (1950–1952) in the famed Salt of the Earth strike. In the end, Local 890 officials used their experience in dealing with Kennecott to embolden their brothers at the Empire Zinc property and eventually achieve similar financial packages and occupational rights. Their boldness, in fact, manifested as leading roles for Juan Chacón, Joe T. Morales, and others in the famed *Salt of the Earth* movie (1954). Although banned by federal authorities until the 1970s as "communist," the inspiring film brought national attention to Grant County's labor movement and is still celebrated today as a daring attempt to thwart anti-communist red baiting at the height of the Cold War. These labor efforts got results. By the mid-1950s, Kennecott led all the copper corporations—Phelps Dodge, Anaconda, and ASARCO included—in negotiating with its workers, setting the pattern until the early 1980s for advantageous contracts in the industry. This development represents the fulfillment of labor's rise to power. It also signifies the demise of unfettered corporate hegemony over the workers at the New Mexico enterprise and elsewhere in the United States.

Kennecott's origins center on the early twentieth-century formation of the Alaska Syndicate, a partnership of J. P. Morgan and Daniel Guggenheim. First organized to exploit rich underground deposits near Kennicott

Glacier, the Kennecott Mines Company[5] developed the copper mine discovered by mining engineer Stephen Birch, who had won the financial backing of sugar magnate Henry Osborne Havemeyer of New York City. More significantly, Daniel Guggenheim hoped to discover and develop copper mines via the Guggenheim Exploration Company (Guggenex) to feed the ore and concentrate needs of the family's principal mineral resource enterprise, the American Smelting & Refining Company (ASARCO). By 1915 the syndicate realized that the Alaska venture had limited potential because of rapidly dwindling ore supplies and exorbitant freighting costs. So the investors incorporated the Kennecott Copper Corporation that year and began exploration efforts to discover and then develop mines to sate ASARCO's appetite for copper concentrates.

Their efforts started with the acquisition of a 25 percent interest in the Utah Copper Company. Under the guidance of Daniel Cowan Jackling,[6] a soon-to-be-famous mining engineer trained at the Missouri School of Mines, the copper company had implemented mass mining strategies to develop the "richest hole on earth," Bingham Canyon, just outside Salt Lake City. During the next two decades the emerging copper giant acquired several open-pit projects in Nevada, Arizona, and New Mexico to go with their properties in Chile. By the 1930s, Kennecott began constructing its own smelters, vertically integrating its operations from mining and refining to the manufacturing of copper, while also phasing out its long-standing contractual agreements with ASARCO.

Kennecott purchased the Chino Mine in 1933 just three years before gaining full ownership of Utah Copper's Bingham Canyon Mine. By the beginning of World War II the international mining corporation owned two additional US copper holdings at Ray, Arizona, and Ely, Nevada, that along with the Utah and New Mexico pits formed Kennecott's Western Mining Division. These properties strengthened Kennecott's position in the international copper market, which was already secured because of the corporation's investments at Braden and Chuquicamata, Chile. These South American properties, combined with its prolific operations in the American West, placed Kennecott by the 1930s at the top of the copper industry hierarchy.

The copper corporation also gained control of the smelter at Garfield, Utah, and then constructed others at Hurley, New Mexico, and Ray, Arizona. These refining investments freed the corporation from its obligations to ASARCO. Like many mineral resource firms in the United States, Kennecott had initiated a vertical integration strategy, acquiring two wire-manufacturing operations—Chase Companies Incorporated (later Chase Brass & Copper Company) in 1929 and the American Electrical Works (later Kennecott Wire & Cable Company) in 1935. With these extraction and refining properties Kennecott produced nearly 6 billion pounds of copper, earned $23 million, and had accumulated assets of $460 million during World War II. With no funded debt, the company decided to invest in the development of its numerous properties in the western United States and South America.[7] At Chino, Kennecott invested $9 million in the late 1930s and early 1940s to revamp the aging mining and milling works at Santa Rita and Hurley. The jewel of the company's rejuvenation program in New Mexico, of course, was the newly constructed smelter, whose 501-foot stack majestically rose above the mill town of Hurley (see figure 3.33).

Despite a wartime 12-cent limit on copper imposed by the US Office of Price Administration, Chino Mines produced in 1942 a record 146 million pounds of the red metal. As a result, the division generated $1.2 million in profits that year from its output of copper along with molybdenum (2.6 million pounds), gold (3,672 ounces), and silver (53,672 ounces).[8] This production trend continued throughout the war and with few exceptions in the postwar period. Although Chino grappled with labor shortages during the war, the federal government assisted Kennecott with its housing deficit by constructing 200 units at Hanover (see Map 5.1). Despite these obstacles, the corporation's New Mexico property continued to produce copper at a near-record pace to assist in building the nation's stockpile of the strategic metal. Kennecott did extend 90-day contracts to furloughed soldiers during the war as well to facilitate high production levels (see figure 5.2).[9]

On the other hand, Chino Mines did struggle with labor shortages during the war. Its force dropped by more than 200 employees, the majority quitting to enlist in the Armed Services (see Santa Rita Workforce in appendices). The dramatic effect on labor because of the war would also have a profound lasting impact on Grant County, the American West, and the nation as a whole. Patriotic service, for example, inspired Hispanic veterans

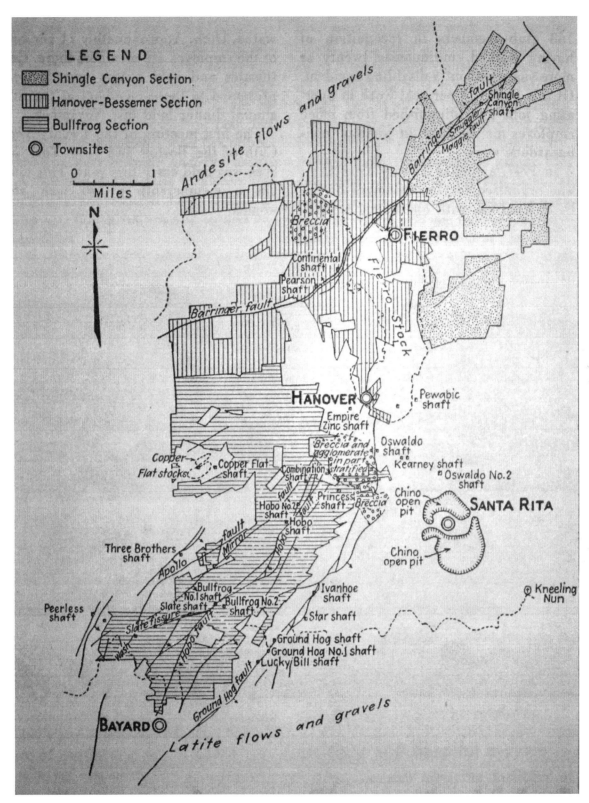

MAP 5.1. Geology and Mining Practice at the Bayard, N. Mex., Property, *by Leo H. Duriez and James V. Neuman Jr., Mining and Metallurgy (October 1948): 559–566. This map reveals the contours of the Chino Mine in the late 1940s, as well as the immense production of hardrock ores elsewhere in New Mexico's prolific Central Mining District. Note Hanover (center), site of the famed Salt of the Earth strike against the Empire Zinc Company in the early 1950s.*

FIGURE 5.2. *World War II Honor Roll Board, March 1943. The Kennecott Copper Corporation erected this kiosk in front of the Pinder shops to recognize workers who entered the Armed Services during World War II. The last name on the board is Adelino I. Grijalva Jr., who enlisted on March 11, 1943, and was honorably discharged on January 6, 1946. Numerous additional names were added later. At this point in the war, the monument lists the names of 168 Chino employees. After serving principally in the Pacific Theater (some of Chino's men died in the Bataan Death March), many of the patriotic Chicano employees were inspired to fight for their equality at work and in the community at large in the postwar period. By this time, for example, the company contracted with the unions to retain the servicemen's seniority status while they served.*
COURTESY OF TERRY HUMBLE.

on their return to their homeland to fight for full citizenship, an expectation that Mine Mill effectively worked toward in labor negotiations and political activism in the postwar period. Minority men who served all over the globe for their nation came to the realization that they were "Americans first." Not surprisingly, they came to believe that they deserved the rights promised in the US Constitution. Patriotism ran deep in Grant County. More than 10,000 locals, in fact, attended a Desert Command Force exhibition at the Santa Rita ballpark on August 12, 1943. The event featured tank, anti-aircraft, and machine-gun demonstrations as well as nationalistic speeches, one

given by Private Bill Gillaspy, a hero of the New Guinea campaign. Soon after the war Mine Mill Local 63 led the patriotic effort to fund a veterans' memorial building by pledging a half-day's wages per member, a strategy adopted by several other CIO- and AFL-affiliated locals. It was not a coincidence that three World War II veterans—Albert Vigil, Albert Muñoz, and Art Flores—conceived of and then volunteered to lead the fund-raiser. They also insisted that the fund drive air the "need to eliminate racial discrimination" so that they could "acquire more public sympathy" for their plight in postwar New Mexico.[10]

FIGURE 5.3. *Female track gang members pose on rail equipment car, ca. 1945. Chino Mines began hiring women in 1944 because of the labor shortage during World War II. Mostly Chicanas, these workers proved they could do "men's work."* Left to right: *Rosa Hernandez, Rufina Rodriguez (standing), Delores Gonzales (with cap), Dora Gutierrez, unknown, Manuela Betancourt (behind her is an unknown woman), and Juanita Hernandez (holding post).*

COURTESY OF SILVER CITY MUSEUM.

Those men who did continue to work at Chino during the war also willingly "kept the peace." They accepted annual labor contracts as well as arbitration decisions of the Nonferrous Metals Commission rather than go on strike. The company offered all the unions successive one-year deals that included vacation and holiday pay and a recognition of seniority rights. Kennecott volunteered in 1942 to give a 25-cent-a-day raise, and the union, with strong support from the War Labor Board, secured an additional 50-cent bump retroactive to January 16. The unions won wage increases in 1943, 1944, and 1945. More significantly, they earned two-week paid vacations for employees with five or more years of service, hourly shift pay differentials (an additional 4 cents an hour for the afternoon shift and 8 cents more for the night one), and rim-to-rim pay for time spent getting to their jobs in the pit and on the tailings dumps, which were often several miles from the operation's entrance. Kennecott also agreed to end discrimination based on "race, creed, or color" despite its well-known reputation

for poor treatment of Hispanic and Native American workers.[11] Officials consented to the formation of a neutral grievance committee to settle management-worker disputes. In 1946 the company accepted the union's demands for pay for the hour each day that workers spent getting to and returning from their job sites dating back to March 1942. Chino Mines then acquired two one-ton Ford stake bed trucks and modified them with benches to ensure ample portal-to-portal transportation.[12] Soon thereafter the company acquired buses to accommodate on-property travel. Not until the postwar period and the imposition of the Taft-Hartley Act would Mine Mill threaten militant action in the effort to achieve better wages, fairer lines of promotion, improved working conditions, and other benefits.[13]

Likewise, the labor shortage for the first time afforded women, Navajos, and teenage boys employment opportunities at Chino beginning in the 1940s. By the end of the war Kennecott had hired more than 100 females (mostly Chicanas). They worked in various capacities at the smelter and mill but principally on track gangs in the mine (see figure 5.3). Although the copper company in 1946 laid off most of the women workers, several stayed on the payroll, inspiring others to seek employment at the mine works in years to come. Guadalupe Fletcher, a divorcée with a son, for example, left domestic work in 1942 for employment in the track department. Despite the men's superstitions, especially general mine foreman Harold "Tuff" Moses, about women working with them, Fletcher persevered. Only twenty-two when she started, she endured the difficulties of working on a track gang. "We had to carry the ties on our shoulders. Not like the men who had to carry them by themselves, we had three of us . . . Some of the women were real good at spiking. They had a bunch of just girls, in gangs. We got along. There were gangs of men and gangs of girls and we worked side by side. We did the same sorts of work the men did." But "they only had men foremen. José Morales[14] was one . . . [and so was] Lucero Gonzales." Later she worked as a "janitress." She earned a promotion to truck driver in the 1950s, staying on in that capacity until 1983, when she retired. In July 1943, Chino began to employee Navajo men. The first thirty-eight of them pitched tent houses in a segregated enclave called "Indian Village" at Vanadium near the precipitation plant. By war's end Kennecott had hired nearly 400 Navajos, among them the well-known Begay family who lived in the temporary segregated community.[15]

Kennecott fully welcomed the Chino Mines Division into its corporate fold in 1942 by eliminating all the Nevada Consolidated Copper Company logos on buildings, machinery, and signage at the New Mexico operations. This expensive gesture, which included repainting equipment as well as purchasing new letterhead, signified that Chino would be a key division in the corporation's copper enterprises. Changes in Chino's management in the 1940s also marked the end to the company-town era with the retirement of Harry Thorne, mine superintendent and an employee at Santa Rita since 1906. His replacement, William Goodrich, part of the new guard, served in that position until 1947, when he succeeded Horace Moses, another original member of Sully's staff, as general manager. Gerald Ballmer moved into the position of mine superintendent and Walter Herkenhoff, formerly general blasting foreman, replaced Harold "Tuff" Moses as general mine foreman.

Under their leadership Chino completed its transition to an all-electric pit operation. Kennecott invested in two additional Marion Model 4161 electric shovels, the Nos. 16 and 17 (for $81,000 apiece), to go along with the five 4161s already in service. These shovels sported five-yard buckets. They joined Chino's first electric shovel, the 1924 Model 350, equipped with an eight-yard scoop. The company retired the remaining steam shovels. Likewise, officials parked all of the steam locomotives, sold the operable ones, and purchased three additional General Electric 85-ton units to complete the fleet of thirteen electric engines and finalize Chino's wartime pit machinery[16] (see figure 5.4).

The remainder of this chapter is devoted to the evolution, from the late 1940s to about 1970, of the mining, precipitating, and smelting of copper *and* the history of labor achievements under the leadership of Mine Mill Local 890. Chino's dedication to continued and improved mining as well as haulage techniques epitomizes the rise of modern industrialization. The processing of precipitation copper at the "p-plant" signifies the long-term future of copper production especially in the twenty-first century. The history of smelting at the New Mexico operations represents the role of vertical integration for Kennecott and the future struggles with new environmental regulations. Engineers fueled the mechanization of the operations to reduce costs and maximize efficiency.

Local 890's role in the history of Chino illustrates the grassroots emergence of exceptional leadership among the county's Chicanos as well as the realization of big labor's successes throughout the United States in the postwar period. Ironically, Kennecott's implementation of new technology and Mine Mill's partnership in carrying out these modernization plans translated into the disappearance of Santa Rita, the copper camp of the Spanish, Mexicans, and frontier Americans and the generations-long company town for thousands of Chino's workers. Nothing could stop the inevitable digging up of the community, which by 1970 no longer existed except in memory as "space."

Nothing much changed in the mining process at Chino in the three decades after World War II. Drillers still drilled blast holes. Shovel runners still dug up the blasted rock. And locomotive engineers still hauled trainloads of ore or waste rock to the crushers or the dumps. The technology, the machinery itself used to mine and haul materials, however, did change. By the mid-1950s, for example, Chino officials decided to replace churn drills with new rotary models while also beginning the transition from train to truck haulage, a gradual process that lasted through 1980.

Before replacing the churn drills, Kennecott had modified the gasoline-powered rail versions by adding caterpillar tracks and electricity generation in the 1930s to 1950s (see figure 5.5). These more efficient machines drilled 2.5 times more footage in a shift. Engineers discovered by 1955, however, that rotary drills were even more efficient and cost-effective. Used since the inception of open-pit mining, churn drills made blast holes based on

the "shattering impact" method. As a *Chinorama* report described, "The heavy drill bit [of the churn drills] lifted and dropped, lifted and dropped, lifted and dropped until the rock was broken into bits and could be lifted out as a mud when mixed with water."[17] This noisy, expensive process required large amounts of water, a scarce resource in the Desert Southwest.

Previously used for oil drilling, rotary drills could dig far deeper and with far less fuel consumption. Yet they still brought the powdered rock to the surface in a mud mixture, requiring a substantial water source, a commodity Kennecott had devised elaborate conservation measures to secure for milling processes at Hurley.

FIGURE 5.4. *Marion Model 4161 (ES-16) loads overburden into railcars, ca. 1955. Purchased in 1940 for $80,700, this shovel sported a 5-yard bucket. Here the shovel runner is loading materials into 40-yard pneumatic railcars that unloaded automatically at the waste dumps. Note the Kneeling Nun in the background.*
COURTESY OF TERRY HUMBLE.

FIGURE 5.5. *Churn Drill No. 2, February 1956. Note the crawler tracks, the heavy electric cable (to the left), and the 12-inch drill bit (hanging from the cable of the boom).*
COURTESY OF TERRY HUMBLE.

Concerned with water conservation, Chino's drillers and engineers worked with the Joy Manufacturing Company to design rotary drills that brought the pulverized material to the surface using an air exhaust system. By early 1957, Chino purchased four of the new models (see figure 5.6). Drillers excitedly initiated the electricity-generated drills (the No. 22 the first to be used in May 1955), delighted at the prospect of making blast holes three to five times more quickly than before. On a good day a single churn-drill crew, for example, knocked out sixty feet of hole. Depending on the ground, a two-man rotary drill team bored from 200 to 400 feet in an eight-hour shift. Driller Byron "Pete" Cody and his helper, E. L. Porter, set the standard on February 20, 1958, with a record 520 feet of hole. The combination of 60,000 pounds of hydraulic pressure and the spinning of three hard steel rotating bits (with a 12-inch rather than 9-inch diameter) on a wheel greatly enhanced drillers' abilities to prepare the ground for blasting and with far less annoying vibration and noisy pounding. Drillers also could now more easily and quickly make blast holes sixty feet deep to "break the bottom" of the ore body and improve shovel engineers' ability to dig up rock at a faster pace. In addition, the two-man crew did the work of four or more, resulting in reduced labor costs. By the mid-1960s, Kennecott began using angle hole rotary drills to break the rock into even more manageable sizes for mining and hauling. About the same time, Chino drillers, such as long-timers J. "Dewey" McLehaney, George "Frank" Coldwell, Bill Dooley, Jimmie Smith, and their helpers, like Louis S. Montoya, Pete Lardizabel, and Luis V. Flores, welcomed the more mobile truck-mounted units.[18]

Once the drillers prepared the ground for blasting, powder crews filled the sixty-foot holes with dynamite. Exclusively Hispanics, the blasting crews risked their lives on a daily basis, literally pouring powdered dynamite into the holes by hand (see figure 5.7). Tragically, five men lost their lives in 1954 when a powder truck exploded, throwing bodies and debris hundreds of feet. The Powder Department foremen—among them Herminio Gonzales, Julián Vargas, Andres F. Rodriguez, José M. Gonzales, Gaspar Ramirez, and Gavino Gallardo—took accidents and safety seriously. To alleviate the inherent dangers of working with explosives Chino began in 1955 to use carbamite, a mixture of nitrate (dynamite) and diesel fuel. The new explosive proved to be more efficient at breaking rock at a much lower cost. By the early 1960s the

FIGURE 5.6. *Rotary Drill No. 24 (Joy 60-BH), April 1963. Purchased in 1959, this electric-generated rotary drill outperformed the churns as much as 5 to 1, making blasting a far more efficient enterprise. It also used far less water by air blowing the stemming material (drill cuttings) that powdermen later used to fill the holes after loading carbamite. Note the outriggers on both ends of the drill that stabilized the machine to ensure level vertical holes.*
COURTESY OF TERRY HUMBLE.

department improved safety by loading the blast holes with hoses connected to a truck filled with carbamite (see figure 5.8). From this time on, Santa Ritans rarely heard and felt the blasts, which sometimes involved tens of thousands of pounds of explosives, further convincing officials of the efficacy of the new technique.[19]

Once blast crews blew the materials into manageable sizes, the shovel runners and their helpers loaded the ore or waste rock into railcars, as had been the case since

FIGURE 5.7. *Powderman pours dynamite into blast hole at Chino Pit, ca. 1955. These men daily risked their lives working with explosives. After the 1954 accident in which five men lost their lives, safety became an obsession among Kennecott's blasting crews.*

COURTESY OF NEW MEXICO STATE UNIVERSITY LIBRARY, ARCHIVES
AND SPECIAL COLLECTIONS (IMAGE #03200168).

1910. Changes in shovel technology had already begun in 1924 with the introduction of the first electric shovel, known as ES1 (Electric Shovel Number 1) and described in chapter 3. Shovel engineers continued to operate the Marion 4161s purchased in the 1940s. Kennecott decided to add in 1957 a Pauling & Harnischfeger (P&H) 1800; the ES20, with an 8-yard bucket (see figure 5.9) and two massive Bucyrus-Erie (B-E) 280Bs; the ES21 in 1969; and the ES22 in 1970, with 15-yard scoops (see figure 5.10). During the 1980s and 1990s, Chino Mines purchased a se-

FIGURE 5.8. *Workers load carbamite from Ireco powder truck into blast holes, June 25, 1968. Filling blast holes with the wet explosives minimized risks for the workers at Chino.*

ries of P&H and B-E electric shovels with bucket capacities ranging from 17 yards to a whopping 56 yards.

The electric shovels and locomotives consumed most of the electricity generated at the Hurley power plant (see figure 5.11). To meet these machines' digging and hauling needs of about 8 million tons of ore and 14.5 million tons of waste rock a year in the 1950s, Chino Mines developed an extensive network of power lines. These pervasive wires carried the 24,000 volts of alternating current required to run the drills, locomotives, and shovels scattered throughout the growing pit, which by 1960 gaped a mile wide and descended 1,000 feet. The hum of electricity could be heard and felt by the workers at the various substations in the pit and at the precipita-

tion plant. Two substations—one on the rim of the north pit and the other west of the pit—delivered 750 volts of electricity to the locomotives through transformers and rectifiers. Four additional substations strategically placed around the pit transformed the entering 24,000 volts of current to 4,000 for use by the shovels, which were plugged into numerous electric feeder stations. Drillers plugged in their rotary drills "by means of long cables, kind of super extension cords," to these same units but through transformers that dropped the voltage to 440.

The Mine Electrical Department maintained these electric generation stations from the 1940s through the 1970s. Electricians in this department also erected more than fifty miles of power lines to "juice" the trolley wires that carried the power to the locomotives (see figures 5.12 and 5.13). This expansive electrical operation greatly reduced fuel costs from the 1940s on while also generating more power per unit. To take advantage of these conditions, Chino purchased in the 1950s two additional 85-ton and four new 125-ton General Electric locomotives that facilitated increased production at a much lower per-unit cost. Steam power had become a thing of the past. "Nowadays," the *Chinorama* reported in 1962, "a ton of coal is about as useful around the mine as a horse collar." But more importantly, the transition to electricity

eliminated the intrusive nuisances of steam power. The winds of Santa Rita no longer carried the clanking, sputtering clatter of steam boilers and screeching cylinders and the billowing coal dust that sometimes shrouded the Kneeling Nun and irritated locals' respiratory systems. Unfortunately, Hurley's residents could not claim the same clean air because of the smelter emissions.[20]

The Track Department repaired, maintained, moved, and constructed rail lines at Chino. Still known as "track gangs," these almost exclusively Hispanic crews implemented new technology to facilitate completion of their

FIGURE 5.9. *Shovel engineer of the Electric Shovel No. 20 loads blasted materials into a 40-ton Euclid haulage truck, July 1961. Equipped with an 8-yard bucket, this ES-20 came on board at Chino in 1957 as part of the transition from rail to truck haulage. The two "Eucs" sported two side-by-side Cummins diesel engines without mufflers. They were so noisy that not only could residents in Bayard six miles away hear the raucous machines but many of Chino's early truck drivers suffered hearing loss.*

FIGURE 5.10. *Chino bull gang completes the assembly of the Electric Shovel No. 21, a Bucyrus-Erie 280B, February 1969. Purchased for $600,000, this massive piece of machinery arrived unassembled on twelve railroad cars in December 1968. Note the 15-yard bucket, which cost an additional $50,000, and the workers standing next to it.*
COURTESY OF TERRY HUMBLE.

duties along the fifty miles of rail in the pit and at the dumps. Although the men still assembled sections of rail by hand (see figure 5.14), they benefitted by the 1950s from new technology, namely bulldozers and cranes and modifications to these pieces of machinery. As a result, they could relocate rail sections, especially at the ever-growing dumps (see figure 5.15). Relocation eliminated the need to rebuild new sections, from scratch. Now the

FIGURE 5.11. *Electric Shovel No. 12 loads railcars pulled by the No. 101, a 125-ton General Electric locomotive, September 1957. The 125-ton designation marked the weight of the loco, which could haul from eight to ten 85-ton ore cars. Note the extensive power and trolley lines that "juiced" the shovels and locos. Careful examination reveals a switch-tender shack with electricians working on top of a line car next to it, slightly above and to the right of the No. 101 in the photograph. The electricity-generated operation proved more cost-efficient by producing more per-unit power than had the steam machinery. Even more efficient truck haulage would soon replace the electric rail system. Electric shovels, though, still dominate loading strategies at Chino and elsewhere in the open-pit mining world.*
COURTESY OF TERRY HUMBLE.

crews moved reconditioned units during the re-laying process.

On these well-kept lines locomotive engineers pulled their trains of ore and waste along a 2 percent grade. From the bottom of the pit in the late 1950s and 1960s the trains climbed the 800 to 1,000 feet to the rim on a series of switchbacks or zigzag inclines to reduce the length of the average trip. As the *Chinorama* reported in January 1958, "these inclines are the more or less permanent tracks on which all the trains 'go out' . . . In other words, contrary to the way it appears to a casual observer, trains do not circle round and round the pit and spiral their way to the top."[21]

FIGURE 5.12 (ABOVE). *Santa Rita electrical foremen, October 1957. These men supervised the electrical crews that could be found throughout the pit and in the maintenance shops on any given day. Left to right: L. C. Benavidez, Jesse Blair (son of James Blair, former head of security for Chino), Tom Welsh, and Charlie Slout.*
COURTESY OF TERRY HUMBLE.

FIGURE 5.13 (BELOW). *Electric line car with unnamed electricians working on trolley line in Chino Pit, ca. October 1958. These mobile "shops" carried equipment and supplies needed to maintain and repair fifty miles of electric lines. Note the hydraulic lift and platform that raised the electricians for work on high wires as well as trolley and power poles. From* Chinorama 4, no. 9 (October 1958): 10.
COURTESY OF TERRY HUMBLE.

FIGURE 5.14 (ABOVE). *Track gang replacing ties on one of Chino's main railroad lines, November 19, 1960. Here workers put in new 130-pound rail. Note the trackman with the sledgehammer pounding in a rail spike. The specially designed sledge head was longer on one side to allow men to drive spikes over the rail without breaking the handle.*

COURTESY OF NEW MEXICO STATE UNIVERSITY LIBRARY, ARCHIVES AND SPECIAL COLLECTIONS (IMAGE #03200456).

FIGURE 5.15 (BELOW). *Track crew shifts rail with bulldozers, ca. September 1956. Using bulldozers with customized "cranes" by the early 1950s, the track gangs could more easily reconfigure the rail lines in the pit to provide access to the shovels for the haulage trains. Augustin Chavez is the track gang member in the center of the photograph. The bulldozer operators are unnamed. From* Chinorama 2, no. 9 (September 1956): 14.

COURTESY OF TERRY HUMBLE.

FIGURE 5.16. *Lloyd Grinslade calls another employee from the lookout to coordinate locomotive traffic in the Chino Pit, ca. 1959. The Central Traffic Control system initiated in 1953 allowed lookout men to monitor all rail traffic in the pit. These skilled workers had telephone lines backed up by two-way radios to communicate with locomotive engineers, switch-tenders, and other workers throughout the pit operations. They were also important for safety, serving as "eyes in the sky."*
COURTESY OF NEW MEXICO STATE UNIVERSITY LIBRARY, ARCHIVES
AND SPECIAL COLLECTIONS (IMAGE #03200030).

Beginning in 1953, locomotive engineers no longer counted solely on switchmen to redirect rail lines in the pit. Rather, the Central Traffic Control (CTC) monitored and controlled rail traffic. "Lookout men" radioed switch-tenders with instructions on where to direct approaching trains from a glassed-in tower on the east edge of the pit (see figure 5.16). CTC operators like Clarence A. Anderson and Elwood Fleming in the 1950s monitored train traffic via flashing lights on a board. When a locomotive approached a switch, the CTC men in the lookout no longer depended on switchmen to change the direction of the train in most sections of the pit. They just

flipped a switch on the control board. In case of malfunctions or accidents, the CTC operators also had direct radio access to locomotive engineers. This system proved particularly useful in guiding locomotive engineers through the switchbacks on the 5900, 5942, 5984, 6050, and 6100 levels of the rising benches of the pit. To ensure maximum visibility for the "eyes in the sky," Chino periodically relocated the lookout to strategic spots along the rim of the pit.[22]

Always with a mind to reducing operating costs, Chino's officials decided in 1952 to introduce truck haulage after five years of research into the efficacy of this

FIGURE 5.17. *Northwest diesel shovel with 2.5-yard bucket loads a LeTourneau scraper, 1957. In 1952, Chino Mines began using haulage trucks to remove overburden or waste rock. Note the two 25-ton Euclid trucks, which proved to be so efficient and versatile that Kennecott decided to increase its truck fleet every few years with even larger "Eucs." The scraper proved to be inefficient and was sold two years later.*

COURTESY OF TERRY HUMBLE.

new strategy. In 1947, Chino Mines contracted with the Isbell Construction Company of Reno, Nevada, to remove more than 8 million tons of overburden (waste material) from the Niagara Pit as part of the mine expansion plan. Over the next six years Isbell worked with Chino shovel crews to strip more than 40 million tons of material to prepare the ground for mining. What intrigued Chino officials was the efficiency of the Nevada-based company's truck haulage system in expediting the process. Consequently, they decided to initiate a five-year study to determine the effectiveness of truck haulage, even though the eleven Euclid dump trucks Isbell employed had only 10- and 15-ton capacities. William Goodrich, at the time assistant general manager, and

Abe Morris, assistant mine superintendent, first examined the possibility of changing to the new technology in January 1947. Although nothing materialized after their ten-day exploratory trip to pits in Arizona and Nevada, Chino's officials kept the possibility on the back burner until 1952.

In March of that year Gerald Ballmer, mine superintendent (Goodrich was now general manager), and Dan Thorne, mine maintenance general foreman, made another investigative excursion, visiting several pits in Arizona. They were convinced that hard-to-reach sections of the Chino Pit could be more easily mined using trucks. Within a few weeks Kennecott ordered seven Euclid Model 36TD 25-ton haul trucks (see figure 5.17).

This purchase marked the beginning of the end of rail haulage, even though locomotives still freighted fully 98 percent of the ore in 1952 and the electric trolley system would be in place for another three decades in the higher reaches of the pit. Within two years Chino Mines purchased three more 25-ton Euclids, making the transition inevitable.[23]

Chino initially used the trucks in new mining areas of the pit. "On the outer edges of the Chino pit," the *Chinorama* reported in February 1958, "and in spots where the curves are sharp and the grades steep, the 'Eucs' are blazing trails as wide as a superhighway into new areas of the Santa Rita mine." The Truck Department had just received several new "big Eucs" with 40-ton capacities to facilitate the new strategy. "You might call this job [of using the trucks] development work," Euclid haulage foreman Bill Sessions reported. "New mainline tracks are going on the bench below." The millions of dollars of investment in electric train freighting clearly precluded an immediate end to locomotive service. Nevertheless, the effectiveness of the trucks was apparent. Maintenance of the waste dumps, for example, required far fewer workers as well as less equipment. Still, the danger of unloading too close to the edge of the dumps concerned drivers like Jesus "Chuy" Martinez Jr., a haulage truck driver since 1953 and one of the first to pilot the 40-tonners. "You have to watch the ground to be sure it isn't cracking. Otherwise you might go over the bank."[24] Despite this new safety concern, the cost savings and flexibility of the truck system impressed mine officials.

The gradual transition also affected workers. Many Track Department jobs were eliminated, requiring retraining for employment in the Truck Department. Over the next decade or so, many of Chino's workers found themselves in the position of either moving on to jobs elsewhere or seeking training at the Kennecott property to fill new posts at Chino. Among the most important new positions along with truck driving were truck mechanics and oilers. In addition to making repairs, these men inspected the trucks after 100 driving hours and then performed major checkups every 500 hours. "One of our biggest jobs is repairing tires," foreman Dennis Murphy pointed out in February 1958. "We change tires with a crane. It takes two men to roll a tire [21 inches wide and 5 feet in diameter] and rim into the shop and maybe five or six men to lift it from a flat position." Oilers made the

FIGURE 5.18. *Truck and dozer mechanics and foreman pose for photograph in truck shop, August 1965. Field repair general foreman Don Schmidt,* right, *presents certificates for completion of a 500-hour diesel mechanic correspondence course to,* left to right, *Arturo Arias, Sam Saenz, and Jim King. Most truck shop mechanics took the yearlong course to prepare them to repair Chino's heavy equipment. After Kennecott initiated the truck and dozer apprenticeship program in 1963, all apprentices had to take this course as part of their four-year training program.*
COURTESY OF TERRY HUMBLE.

rounds with a grease truck, lubricating each unit on a need basis on location. By the late 1960s, mechanics inspected the haul trucks on a daily basis in the "oil house" or truck maintenance shops during fuel refills and lubrication checkups, a 20-minute chore that on completion gave the drivers greater confidence in their machines as they headed back into the pit.[25]

New management slots also opened. The changes, though, resulted in the need for more highly skilled workers—at the expense of laborers' positions—who would earn higher pay (see figures 5.18 and 5.19). With the transition to new technologies and greater mechanization, mine superintendent John M. Haivala and other Chino officials regularly encouraged workers to "keep up." "The employees themselves must meet their part of the challenge," Haivala contended, "by improving their skills and adapting themselves to changing job conditions." If they adjusted, the company could reduce costs

FIGURE 5.19. *Chino truck and dozer oilers pose for photograph, March 1960. They are,* left to right, *Felipe Valerio, Pete Quevedo, Ramón Arzola, Sammy Gonzales, Mariano Grijalva, and Sirildo Durán. Valerio and Quevedo lubricated dozers and the other four serviced trucks. Note the brand-new service truck equipped with engine, transmission, and hydraulic oil, grease, antifreeze, and air in the background.*
COURTESY OF TERRY HUMBLE.

and increase efficiency at a time when ore values were in decline and labor and equipment costs on the rise. Under these conditions Chino officials recruited Frank Dominguez and Gilbert Martinez to serve as driver instructors during the transition from rail to truck haulage. Those employees laid off in the Track Department had the option of seeking retraining based on seniority. Some men chose to enter a new apprenticeship program for truck and dozer mechanics.[26]

From 1958 to 1964, Chino mine department officials formulated a strategic plan to open new low-grade ore bodies (of about 0.9 percent copper). They initially targeted the Niagara and Lee Hill sections on the north end of the pit and the "island" at its center. To implement this strategy, beginning in the early 1960s Kennecott purchased even larger haul trucks with 40- to 100-ton capacities. In 1963 alone the company invested $2 million for sixteen 65-ton diesel-powered trucks, ten KW Dart units (see figure 5.20), and six LeTourneau-Westinghouse (Haul Pak) versions. The company also purchased three 85-ton Lectra Haul rigs with the first electric drives, a technological feature that improved the trucks' ability to climb steep grades (up to 7 percent grade as compared to rail at 2 percent) and make sharp turns (see figure 5.21). In 1965, Chino acquired its first 100-ton Lectra Haul unit and then in 1974 the division brought on board the first 150-ton rig (see figure 5.22). In recent years Chino has employed 320-ton units.[27]

FIGURE 5.20. *Forty-ton Euclid, left, and 65-ton KW Dart haulage trucks, Chino Pit, August, 6, 1963. Note the dual stacked mufflers on the new Dart. Drivers and other employees greatly appreciated the Dart's quieter exhaust system (as compared to the raucous straight exhausts of the Eucs).*

COURTESY OF TERRY HUMBLE.

FIGURE 5.22. *Haulage trucks, Chino Mine, 1974. Left to right: 150-ton Unit Rig (also known as Lectra Haul), 100-ton Unit Rig, 85-ton Unit Rig, and 65-ton KW Dart. Note the modified ladders placed at the front of these units.*

COURTESY OF TERRY HUMBLE.

FIGURE 5.21. *Eighty-five-ton Lectra haul truck, Chino Pit, August 1963. Capable of a far greater haulage capacity with their 700-horsepower Cummins 12-cylinder Turbo diesel engines, these trucks also sported the first electric drives that made climbing 7 percent grades (as compared to 2 percent with locomotives) a possibility in the mine. This unit, the No. 80, was the first one built by the Unit Rig Company of Tulsa, Oklahoma. Note the steep vertical climb to the cab up the ladder for the driver. Chino later modified the awkward ladder after several "climbing" accidents, a modification the Unit Rig Company adopted on all its future trucks.*

COURTESY OF TERRY HUMBLE.

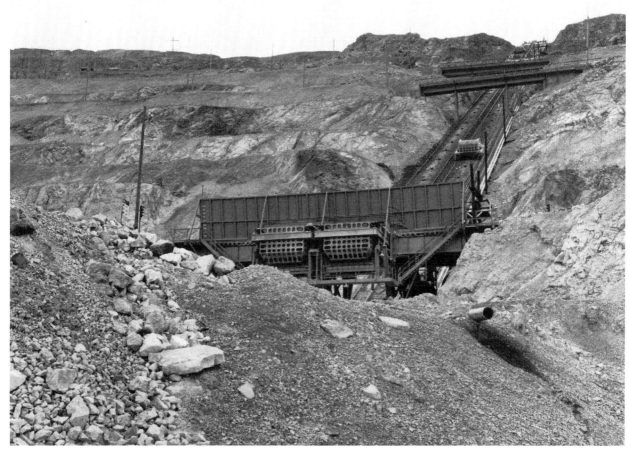

FIGURE 5.23. *Chino's new skip haulage system, October 1962. Haulage trucks dumped ore and at times waste into the gondolas at the base of the 1,400-foot skip. Despite the $2 million price tag for the skip, it became obsolete within ten years.*
COURTESY OF TERRY HUMBLE.

Kennecott also decided in 1958 to construct a $2.25 million skip haulage system as the second major technological feature of the mine expansion plan. "A skip is a carriage on wheels," *Chinorama* announced in March 1959, "running up and down the side of the pit on rails. When one loaded skip is going up the rail an empty one will be coming down." Engineers had calculated that continued rail haulage was cost-prohibitive because the pit had "become so deep and the curves of the benches so 'tight'" that only trucks could access the area that lay some 1,000 feet below the rim. During 1959–1961, Chino bulldozer operators carved the 45-foot-wide slot for the skip into the side of the pit, making way for the National

Iron Company's (a Minnesota firm that had built similar skips for the iron mines of the Mesabi Range) construction of the 1,400-foot tracks (see figure 5.23). In January 1962, truck drivers began unloading ore and waste rock into the 40-ton "gondolas" at the foot of the incline. Using electric power, the skip took the gondolas in about one minute to the ore dock at the 5950 level and about 175 feet below the rim of the pit for unloading into railcars (see figure 5.24). Driven by a 2,000-horsepower electric motor and with a capacity of about 1,000 tons per hour, the skip facilitated the transition from rail to truck haulage. By the end of 1962, in fact, track gangs had removed all of the rail and trolley lines in the pit below the

FIGURE 5.24. *Top of the skip haulage system with 40-ton gondola loading a rail waste car, February 1962. The limited capacity of the gondolas eventually motivated Chino Mines to abandon the skip, which the company had hoped to use for decades.*
COURTESY OF TERRY HUMBLE.

6,000-foot level. Considered a decades-long solution to deep pit haulage, the skip actually lasted only ten years. It became obsolete because of the expense of upkeep and its inability to carry the much larger loads of the 65- to 150-ton trucks.[28]

By spring 1963, mine superintendent and future general manager Frank G. Woodruff led a team of engineers in completing the 1959 expansion plan. On this talented team were Robert W. Shilling, pit operations superintendent; Dan Thorne, mine maintenance superintendent; and D. D. McNaughton and Young Kim, mine engineers. Targeting the Niagara and Lee Hill ore bodies as well as the "island," Santa Rita's former downtown, this group

of engineers celebrated the successful implementation of the large haul trucks and the skip system. Yet they understood like every other Kennecott employee that the achievement also meant the demise of the town of Santa Rita. That announcement came in the March–April 1965 issue of *Chinorama*. "Pit expansion will need Santa Rita townsite," the headline read. "Operations have now reached a point," officials declared, "where another program for the removal of houses and buildings from the Santa Rita townsite is necessary." From January 1966 to January 1969 what remained on the north rim of the pit of the longtime copper town—the neighborhoods of Ball Park and Booth Hill and the barrios of

Espinaso del Diablo, de la Iglesia, del Cañon, and de Los Osos—succumbed to the giant electric shovels (see figure 5.25). Officials rationalized the inevitable demise of the town for *Chinorama*'s readers. "Open pit mining by its very nature must be an expanding operation if it is to prosper and provide jobs over the years. Actually the relatively long time which families have occupied many of the houses in Santa Rita is remarkably long for a 'typical' mining town. It reflects the sound planning done years ago by Chino management."[29] By 1970, Santa Rita had disappeared only to be remembered decades later as the "town in space."

Chino's future operations became dependent on the expansion of the pit. Yet it was the piling-up of the mountain-size dumps that would eventually ensure the continuation of the works, especially in the twenty-first century. This reordering of the landscape—degradation via mining and amassing via dumping—reflected the historic importance of the precipitation of copper, a process that precluded concentrating or milling of the ore.

The Chino Copper Company first produced precipitate copper in 1916. Knowing that copper-laden waters would precipitate the red metal onto pieces of iron, workers constructed wood flumes to transport the "impregnated waters" from the bottom of the pits into wood boxes filled with scrap metal, such as old boiler tubes and blasting powder cans. Over several days the copper in the water literally dissolved the iron, leaving an earthy red mud of 70 to 90 percent purity. After this process,

workers shoveled the sludge into railcars for transport to ASARCO's El Paso smelter. In that first year of using precipitation methods, the company produced a substantial 250,000 pounds of "precip" copper. Targeting the hard-to-mill oxide ores, Chino officials decided in 1917 to build a precipitation plant (known as the p-plant). Within five years engineers realized that Whitewater Creek was funneling green copper water from the waste dumps and local watershed near the plant. To capture this rich copper-laden runoff, crews constructed dams or catchment basins. The company then installed pumps and pipes at the "leach ponds" or water reservoirs to transport the precious liquid to the p-plant. Famous engineer T. A. Rickard in 1924 viewed this scene of "winning the copper." "When I stood on the big dump of waste, representing the stripping and other rock too poor to be treated as ore, I looked down from a height of 125 ft. upon the gully through which runs Whitewater Creek. The blue of the sky was reflected in the clear water, which was fringed by a band of green copper salts. Two hundred yards from the big dump was a dam from which the water that comes through the dump is led to launders in which the copper is precipitated on scrap iron." At this early juncture, Chino directed natural flowing waters from the creek as well as those that seeped through some of the dumps to the precipitation works.[30]

During the 1920s and 1930s the new facility increased the company's overall production by just 2 percent. Consequently, officials decided in 1936 to begin pump-

FIGURE 5.25. *"Proposed Schedule for Vacating Santa Rita Townsite," April 1965. This diagram reveals Chino's plans for removing homes, shops, and other buildings from Santa Rita from 1966 to 1969. By 1970, Kennecott removed all company buildings and any homes not yet removed by their owners. Santa Rita no longer existed! From* Chinorama *11, no. 2 (March–April 1965): 4.*
COURTESY OF TERRY HUMBLE.

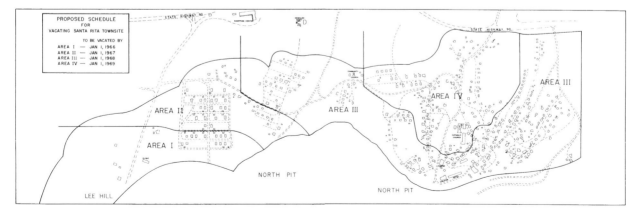

ing larger quantities of water from pit drainage and seasonal runoff directly onto the dumps so that it could seep through the massive piles of rock and vastly increase the flow of copper-laden waters to the leach ponds. This new process of "leaching" increased the flow of impregnated waters to the p-plant to about 1,000 gallons per minute; the company increased that flow to 1,500 soon thereafter when engineers discovered that tailings waters could be effectively used in leaching. The immediate success of leaching convinced officials, after engineers' chemical analyses projected large outputs, to construct in 1939 a new "modern" precipitation plant equipped with two six-cell concrete units. The results were phenomenal. So workers added three additional six-cell units in 1941, 1948, and 1950, successively, to increase the plant's capacity to a total flow of 3,500 gallons per minute (see figure 5.26). By the late 1940s, precipitate copper made up at least 20 percent of annual production, a trend that lasted for the remainder of the century, even though a relatively small number of employees (twenty-three in 1956) ran the highly profitable department. Precipitate copper, for example, made up 37 million pounds of a total output of 143 million (26 percent) in 1948 and 38 million of 141 million (27 percent) in 1951.[31]

These kinds of returns motivated Chino officials to expand leaching and precipitation. In 1958, engineer Frank Colbert, in fact, earned the job of studying the state of the precipitation plant with an eye to expand the highly profitable production strategy. Colbert's supervisors charged him with coming up with "a plan to make the operation more efficient, more productive and more profitable." Familiar with similar operations "in other parts of the world," the engineer brought international experience to the project. In "a few months, he had combined his local observations, the information in technical publications and his experience into a proposed plan," a strategy his supervisors decided to implement at Chino.[32]

Colbert's priority for expanding the precipitation of copper centered on water. Always a limited resource, water had been the focus since 1909 of the corporation's conservation program. During the intervening fifty years, first Chino then Kennecott acquired water rights on more than 300,000 acres of land in Grant County. Initially sources at Apache Tejo, the B Ranch, the 700, and other wells met the water needs of the operations. By mid-century, however, the increase in milling capac-

ity at the Hurley concentrator combined with growing needs of the power and the precipitation plants meant the corporation needed to find more water in the arid climate. The crux of Colbert's plan involved capturing summer rains (although the company also procured groundwater access on the Stark Ranch). These floodwaters had been allowed to seep into the local desert landscape for much of the history of the operations; even after the construction of the water flume in 1927, much of the water was still "lost." Colbert designed several dams with piping systems to direct the bulk of the rainwaters into Reservoir No. 5. "Instead of allowing the heavy summer downpours to rush madly down Santa Rita creek and into the dry sand to be lost forever," the *Chinorama* reported in 1961, "the dams will stop and hold the floodwater." Although some runoff had been collected in wells via the natural hydrologic cycle (despite the above comment that it would be lost forever), the company had no choice in the 1960s and later but to capture rainwater to expand its precipitation as well as milling operations. In the end, Chino built six new dams with a storage capacity of 400 million gallons of water, allowing for an increase in water flow to the p-plant from 3,500 gallons per minute to 7,500. This massive water program gave Chino access to nearly 1 billion gallons of water a year and allowed for the expansion of the leach ponds on the dumps (see figure 5.27). This supply also inspired the sprinkling system still used today at Chino (see figure 5.28). "What does the leaching expansion program mean to Chino?" officials in 1961 asked rhetorically. "Happily, it is good insurance for the division's security. The rich copper cement 'sweetens' the feed [with a much higher-grade concentrate than even milled ore] going into the reverberatory furnace [at the smelter] and has a favorable effect on the cost of producing Chino's copper."[33]

In 1964–1965, Chino Mines added seven "cones" to increase the production of precipitated copper. Instead of placing the iron—now in the form of shredded tin cans, hence the metaphor "Tin Can Mine" to describe the p-plant—in concrete tanks, crane operators, such as Frank Madlock and Luther Beilue, filled the top portion of cones with scrap metal. Under foreman Pete Perea's supervision cone operators like Rudy Madrid monitored the flow of the impregnated water into the cones, instructing the cranemen to add more shredded cans after "the green water consumes them." Washers, such as Fidel Lucero, Luis Lopez, Ray Leon, and Adolf Lopez,

FIGURE 5.26. *Chino precipitation plant, ca. 1959. Fully 20 percent of Chino's copper production came from precipitated copper beginning in the late 1940s.*
COURTESY OF TERRY HUMBLE.

had "the most spectacular job." They got into vats with the red copper mud with powerful hoses and washed the rich concentrates into decant basins, where the water was drained off to begin drying. Pumpmen then operated the pump house to recirculate the used water through a massive piping system (see figure 5.29) to the top of the leach dumps, where bulldozer operators had plowed out 300-foot squares for ponds. Eventually this water-chemical process would be improved with the introduction of the solution-extraction electro-winning (SX-EW) plant in 1988 to ensure long-term "mining" at Chino well into the twenty-first century.[34]

Because of the huge new demands for water, company officials implemented the Water Development Program in 1964. The expansion of precipitation and milling required the "capture" of 8.8 million new gallons of water a day at the operations. Investing $1.5 million in the program, Chino engineers devised new conservation strategies to increase water flows as well as to reduce water waste. The principal strategy involved drilling new water wells like the 2-C ones, spaced out at great distances, to minimize the drop in the water table in the underground aquifer. To initiate this plan, workers under Herb McGrath's supervision put in ten more

FIGURE 5.27. *Aerial photograph of the Chino leach ponds, February 13, 1957. Kennecott leached nearly every ounce of copper out of the waste dumps beginning in the 1940s. This technique, still in use today (using solution-extraction electro-winning, or SX-EW), extended the life of the mine works by decades.*
COURTESY OF NEW MEXICO STATE UNIVERSITY LIBRARY, ARCHIVES AND SPECIAL COLLECTIONS (IMAGE #03200231).

FIGURE 5.28. *Sprinkling system on leach dumps at Chino Mine, December 1967. Kennecott engineers discovered that sprinkling sulfuric acid and water over the dumps leached copper from the very low-grade waste rock. The resultant "impregnated water" then passed through the p-plant to produce a very high-grade copper concentrate. This process forced Chino Mines to modify its already efficient water conservation program to ensure that enough of the precious resource was available for this processing strategy.*

miles of pipelines that converged at the old Apache Tejo pumping station and other newly installed pump houses. At the Tejo site, water department employees put in a new 150,000-gallon steel tank; they also built an additional 150,000-gallon tank in the pipeline system itself at a new "booster" station. Called "underground storage" facilities, these pipeline reservoirs greatly reduced evaporation rates. Water foreman William L. "Shorty" Smith, who worked in the conservation program for more than thirty years, witnessed the rise of an "oasis" in the desert land operations from the 1920s to the 1960s.[35]

This water system allowed Chino Mines to produce millions of tons of concentrates for refining in the smelter at Hurley. Kennecott had invested $5 million in 1938 to construct the smelter and, thus, eliminated the costs for freighting concentrates to the ASARCO refinery in El Paso. Chino officials "blew in" the smelter on May 2, 1939. Just a few weeks after initiating the new plant, assistant editor John B. Huttl of the *Engineering and Mining Journal* toured the new operation, which was one feature of Kennecott's vertical integration plan in the postwar period.[36] "The visitor here sees an example of the most

FIGURE 5.29. *P-plant or "Tin Can Mine" with water lines climbing mountain in the background, 1964. Kennecott devised an elaborate water transportation system to feed the precious liquid to the leach ponds, the sprinkling system, various shops, and the p-plant at Santa Rita. Engineers also designed reservoirs and catchment basins to feed the Hurley concentrator's and smelter's water needs.*

modern design," Huttl wrote, "coupled with an adequate and intelligent expenditure of money. Built for permanence, economical upkeep, and low operating costs, the operators have a plant to be proud of." The engineer's observation proved correct. The main features of the smelter—the 130- by 32-foot reverberatory furnace and the two 30- by 13-foot converters—produced the bulk of Chino's matte and blister copper for many years to come. In 1941 1942, Kennecott added another "reverb" unit and a converter along with a new fire-refining furnace.

The addition of this equipment, combined with other innovations, resulted in the production of 2 million tons of smelted copper at Chino from 1939 to 1969.[37]

Feedermen like Al Gaines and H. V. Talavera in the 1950s fed the dried precipitation mud mixed with milled concentrates into the reverberatory furnace at the smelter to melt the material into matte (see figure 5.30). This combination of the two concentrated types of copper did "sweeten" the ore feed into the massive 130-foot furnace. Still, most of the copper produced at Chino came from the traditional method of crushing then milling the ores mined at the giant pit. In the late 1930s, in fact, the company dismantled the primary crusher plant at Santa Rita and rebuilt a new version next to the Hurley concentrator. After that, millmen processed ores using the same techniques implemented at the beginning of the operations in the first decade of the century. The principal distinctions in crushing involved newer, larger jaw and cone crushers as well as ball and rod mills that reduced the rock from a coarse to a fine grade of ore. At the mill the company continued to use flotation cells (or roughers) but with more sophisticated air injection systems and more efficient chemical combinations (pine oil and raconite for copper separation and fuel oil and alcohol for molybdenum). This system ensured greater extraction of copper as well as molybdenite from the ore.[38] These refined techniques, using the same crushing and milling principles of 1910, combined with the increase in the production of precipitation mud, ensured a steady supply of concentrate to Chino's smelter at Hurley.

To begin smelting, feedermen loaded the reverbs with about 1,000 tons of material from conveyors through the roof of the gas-heated unit. On reaching 2,800 degrees Fahrenheit under 100-foot natural gas flames, the four-foot-thick liquid "bath" separated into two layers—copper matte on the bottom and slag on top (about 50 percent iron along with heavier rock and other "earthy material"). Skimmers or leadmen, such as C. B. Gonzales and J. F. Ribald in the 1950s and 1960s, then periodically skimmed the top portion of the bath for disposal on the slag dumps. Slagmen like E. C. Baca then transported ladles of slag—which, once cooled, were used for rail and roadbeds—to the dumps. These same workers supervised tappers, who tapped the matte (about 35 percent copper) into 12-ton ladles hooked to a crane. Cranemen then directed by sight the molten ladles full of matte (about six per bath) to the converters (see figure 5.31).

FIGURE 5.30. *Feederman Al Gaines "feeds" concentrates into the reverberatory furnace at the Hurley smelter, ca. March 1958. Note the gas mask designed to protect the feedermen from poisonous gases emitted from the firing "reverb." Here Gaines also wears protective gloves, clothing, steel-toe boots, and a hard hat. From* Chinorama *4, no. 3 (March 1958): 10.*
COURTESY OF TERRY HUMBLE.

Converters oxidized the matte by blowing large quantities of air into the molten bath through tuyeres or air lines. This oxygen injection turned sulfur and other impurities into gases that rose as emissions into the 501-foot smokestack. At the same time, the air infusion worked to maintain the heat of the bath, eliminating the need for gas burners in this step. Leadmen then added silica as a flux to combine with iron to create more slag material. Periodically slagmen skimmed off slag into ladles for reintroduction into the reverb furnace. After three or more skimmings, utility men like J. A. Sandoval adjusted the blowing of air through the tuyeres, a process done by hand until the late 1950s, when mechanical punchers were installed. After about twelve hours of blowing, leadmen tilted the converters to pour 99.5 percent blister copper into ladles. Cranemen then directed the nearly pure molten copper into the fire-refining furnace to remove more of the impurities.

A much smaller furnace, the fire-refining unit, took a 260-ton charge and oxidized it with air injections, similar to the converters. Workers added flux to the bath to remove traces of lead, selenium, and tellurium, achieving

FIGURE 5.31. *Reverb crew posing in Hurley smelter, ca. March 1958. Standing,* left to right: *E. C. Baca, slagman; G. J. Marín, larryman; N. J. Abreu, motorman; William Garton, foreman; P. E. Pecotte, tapper; Pat Chester, utility man; and J. F. Ribald, skimmer. Kneeling,* from left to right: *Al Gaines, feederman; S. N. Bustamentes, utility man; V. V. Villegas, reverb laborer; and H. V. Talavera, feederman. From* Chinorama 4, no. 3 (March 1958): 13.

99.85 percent purity. Prior to 1958, smeltermen "poled" the furnaces (reverbs and converters) with pine logs to supply carbon to burn the oxygen out of the molten metal bath (see figure 5.32). After that they used silica injections to burn off the impurities. Tappers then tapped this last furnace to pour 1,000-pound wedge cakes. The "chippers" then smoothed the edges of the giant ingots after cooling for stacking and then shipment to fabricators around the country. By the late 1960s, tappers used molds to produce 50-pound ingot bars that were cooled by dunking in a trough of water. They were then stacked and strapped into 4,000-pound bundles.[39]

The Chino smelter crews introduced several innovations into the refining process that Kennecott used at other facilities in its international copper enterprise. As early as the 1940s, for example, E. A. Slover, future general manager at Chino, and a Mr. Mossman designed the Chino-type waste heat boiler to siphon steam from

the smelting process to generate electric power at the Hurley Power Plant. Chino's engineers also modified the converters' automatic tuyeres, first used in 1954 at the Nevada Mines Division, so effectively that they became known as the "Chino automatic mechanical punchers." Smelting engineers at Chino also designed "stack fluxing" into the converters in 1954, introducing crushed silica into the bath to form iron slag and reduce the time it took to blow the matte into blister copper.

In the 1960s, Chino engineers made several improvements that revolutionized smelting. First, metallurgists discovered in 1966 that adding oxygen to the normal air injection through the tuyeres removed the need to melt the concentrate in reverberatory furnaces. Called "dry smelting," this new technique saved the company labor costs as well as time in producing refined copper. That same year, Chino Mines installed mechanical tapping machines in the reverbs to improve efficiency in tapping

FIGURE 5.32. *Refinery crew "stages a fireworks display" while "poling" the refinery furnace at the Hurley smelter, ca. March 1957. This five-man crew, with each member wearing fireproof clothing and a face shield, puts pine logs into the refinery to burn off impurities. Using a chain block hoist, they plunged the pole into the molten copper to burn off the oxygen and create a purer product. At night, residents of Grant County could see this fiery process from miles away. From* Chinorama *3, no. 3 (March 1957): 3.*
COURTESY OF TERRY HUMBLE.

matte and to improve safety for the tappers. Kennecott also increased the length of service of the reverberatory furnaces in the 1960s by designing refractory brick interiors to take more heat for longer periods, increasing the time between refurbishing the units from one to three or more years (see figure 5.33). Construction in 1967–1968 of a second smokestack—626 feet tall, 48 feet in diameter at the base, and 24 feet in diameter at the top—allowed Kennecott to continue producing smelted copper during the reconditioning of the old stack. The new "chim-

ney" allowed the company to add a fourth converter to increase production (see figure 5.34). Chino Mines also implemented the use of dust collectors at the base of the smelter stacks. At the original stack, engineers designed an electrostatic precipitator to remove solid particles from the emissions using an electric charge. Once the charge was turned off, the dust fell from plates inserted in the stack into bins for retrieval and reintroduction into the smelter. This unit also reduced particulate matter emitted into the air, a growing environmental concern

FIGURE 5.33. *Photograph of the inside of the reverberatory furnace, 1950. The trough at the bottom of the furnace, known as the crucible, held the 2,700-degree molten bath of matte forty inches deep. The concentrate feed entered the furnace through the openings in the sides of the reverb walls. The gas burners, which cast a flame of about thirty-five feet, were located at the far end of the furnace. Workers replaced the different types of brick on an annual basis until the early 1960s, when new material and stacking techniques extended the length of the furnace's service to three or more years. From* Chinorama *8, no. 2 (March–April): 5.*

COURTESY OF TERRY HUMBLE.

after Congress passed new stringent emissions standards with the 1970 Clean Air Act. Smelter crews installed a mechanical-type dust collector at the new stack that removed dust and smoke by cyclonic action, achieving the same result as the electrostatic precipitator.[40]

The massive mining, milling, and smelting operations at Chino required the employment of more than 2,500 workers by the early 1950s at Kennecott's New Mexico property. Although these numbers gradually declined from the 1950s to 1980s to about 1,800 with increased mechanization, these conditions did not deter big labor from playing a key role in the corporation's production successes. Soon after World War II, AFL- and CIO-affiliated locals at Chino Mines formed consolidated units. The Chino Metal Trades Council was composed of the Brotherhood of Locomotive Firemen & Enginemen,

Local 902; the Brotherhood of Railroad Trainmen, Local 323; and the International Association of Machinists, Local 1563. Bayard Amalgamated Local 890 represented Mine Mill Locals 63 of Santa Rita and 69 of Hurley as well as several other locals negotiating with Empire Zinc; ASARCO; United States Smelting, Refining & Mining Company; and Phelps Dodge. By the mid-1950s the AFL and CIO units merged locally into the Chino Unity Council. This alliance among the locals improved their bargaining power in numbers. In the meantime, all Mine Mill locals with jurisdictions in the Western Mining Division had already unified into the Kennecott Council. This region-wide solidarity among all units facilitated advantageous negotiations with the most powerful copper corporation in the world. This collectivization effort reflected the workers' desires to unify to counter traditional

FIGURE 5.34. *Two emission stacks at the Hurley smelter, March 1976. To the left is the original 501-foot stack with "Kennecott" painted on it. In 1967–1968, Chino Mines constructed a second, 626-foot stack to maximize smelter production and cast emissions farther from Hurley. These two stacks, which were taken down by Freeport-McMoRan in 2007, served as a symbol of Kennecott's mining, milling, and smelting operations in Grant County for forty years.*

COURTESY OF TERRY HUMBLE.

corporate hegemony in the industry. The partnership also empowered the copper workers to fight congressional impositions placed on labor during the early years of the Cold War.[41]

Known as the Taft-Hartley or Labor-Management Relations Act of 1947, the new federal law struck directly at the heart of unions' abilities nationwide to win new wage increases and other benefits. The new "right-to-work" law encouraged workers to turn down membership in unions and increased judges' power to impose injunctions against strikers. The most controversial feature of the anti-labor law required union officials to sign anticommunist affidavits. Later declared unconstitutional, the affidavits forced labor leaders to deny membership in the Communist Party or relinquish collective bargaining rights under NLRB jurisdiction earlier guaranteed by the Wagner Act of 1935. Consequently, the first major obstacle in gaining better wages, improved working conditions, and other benefits in postwar America centered on overcoming this congressionally backed effort to weaken the union movement. Emboldened by the new federal law, Kennecott, ASARCO, USSR&M, and the rest of Grant County's major mining companies refused to negotiate with union officials until they signed the affidavits. Still, Mine Mill officials resisted, using the unjust law to embolden their members and fight the corporate-government collusion against labor.

Clinton Jencks was the key figure in the early response of labor to Taft-Hartley at Chino and other mining properties. Arriving in Grant County in 1946 from Denver, the native Coloradan led the amalgamation movement in the southwestern New Mexico mine fields one year after his appearance on the scene. He took on the dual role of southwestern representative of the International Union of Mine, Mill, and Smelter Workers (Mine Mill) and president of Local 890. Soon to be known as "El Palomino" (yellow-haired stallion) for his flowing, thick blonde hair, Jencks earned the love and affection of the Chicano workers and their families. His courageous leadership in the face of sometimes abusive treatment by local law enforcement, company officials, and even the Federal Bureau of Investigation impressed upon them his willingness to sacrifice for them.[42] Along with his activist wife, Virginia, whom Clinton had met at the University of Colorado during his studies in the 1930s for an economics degree, Jencks chose a leadership style that required him to "live among" the Hispanic workers. Once he gained their trust he was able to train them to take up the mantle of leadership for their own cause. His principal goal, he imparted to the union men, was to "put himself out of a job." Soon after his arrival, Jencks began working with grassroots leaders, such as Joe T. Morales (see figure 5.35), Brijido Provencio, Albert Muñoz, Art Flores, and many others. Together they negotiated fair contracts that empowered them at work and in the community. He cultivated their desire to develop leaders especially through training at national and regional Mine Mill meetings as well as at local workshops, tapping into the international's resources (see figure 5.36). These endeavors endeared the Jenckses to many native Santa Ritans. Virginia Chacón, women's auxiliary activist in the 1950s, claimed that most Chicanos embraced the Anglo unionists "because they didn't act like they were *americanos*. They acted like one of us. That's the kind of feeling they always gave, and you can ask a lot of people here in Grant County about it." Soon after their arrival, in fact, the Jenckses demonstrated the seriousness of their mission, beginning a crash course in Spanish, achieving fluency in the years to come.[43]

Chacón's sentiments were shared by hundreds of Chicano families in the area. Decades after Jencks had left Grant County, local organizers invited him to speak at a reunion. Titled "Bringing 'Salt of the Earth' Home: A 50th Anniversary Symposium,"[44] the special commemoration took place in May 2004 at Cobre High School in Bayard. Among several speakers at the four-day event, the octogenarian Jencks approached the microphone on the stage in the auditorium and announced, "Yo soy El Palomino" (I am the Palomino). The mostly Hispanic crowd erupted into an animated five-minute standing ovation, even though many in the audience did not know him personally. He had become a hero of mythological proportions in the oral tradition of previous generations. Mexican American fathers and mothers had told stories to their sons and daughters about El Palomino's struggle in their cause for justice. He was an Anglo man who had fought for *their* occupational, political, and civil rights in the "early years" of the labor battles.[45]

Jencks understood that the Chicano leadership needed basic skills along with bold strategies to counter the traditional strength of the company, especially in the context of company-town life. Consequently he offered Mine Mill leadership training in English *and* Spanish. He initiated role-playing exercises to anticipate grievances

FIGURE 5.35. *Joe T. Morales speaks at a party in honor of long-time employees at the Murray Hotel in Silver City, New Mexico, March 26, 1958. Each year Kennecott feted its employees who had completed twenty, thirty, or forty years of service at a banquet. Mill Labor Department foreman Morales, who had already won fame for his role in the controversial* Salt of the Earth *movie, spoke on behalf of the twenty-year class. One of the first Hispanic foremen at Chino Mines, Morales served as president of Local 69 and as a shop steward before being promoted to foreman during the heady years of the early grassroots labor movement in Grant County. He retired in 1970 with more than thirty years' service at Chino.*
COURTESY OF NEW MEXICO STATE UNIVERSITY LIBRARY, ARCHIVES
AND SPECIAL COLLECTIONS (IMAGE #03200333).

and how best to proceed. He taught the stewards and officers how to use mimeograph machines, design leaflets, take out newspaper advertisements, air bilingual radio programs, and other logistical tactics devised to com-

municate with the rank and file as well as the general public. In dealing with grievances Jencks insisted that workers take their complaints directly to foremen "face-to-face" rather than through an anonymous paper trail.

FIGURE 5.36. *Officials of the International Union of Mine, Mill, and Smelter Workers (Mine Mill), Local 890, pose for this group photograph at the 1953 regional meeting of Mine Mill in Denver, 1953. This distinguished group laid the foundation for the rise of big labor in Grant County. Noted for their extraordinary efforts during the Salt of the Earth strike against the Empire Zinc Company, most of Local 890's officials worked at Chino for Kennecott. The stability of the Chino operations, combined with the emergence of a strong labor presence, made Chino's employees prime candidates for leadership roles in Local 890. Standing, left to right: Robert Kirker (Local 890 officer), Juan Chacón (Local 890 president, 1953–1956, 1959–1962, 1973–1974, and lead actor in Salt of the Earth), Virginia Jencks, Maclovio Barrazas (International representative), Virginia Chacón, Clorinda Alderette, and Frank Alderette (Local 890 secretary). Kneeling, left to right: Nick Castillo (Local 890 officer), Art Flores (Local 890 president, 1970–1972), Ernest Velasquez (Local 890 president, 1952), Clinton Jencks (International representative, Local 890 president, 1948–1951), and Clinton Jencks Jr.*

This strategy altered the traditional deferential behaviors common to Anglo-Hispanic employee relations. He also instilled in them a great appreciation for solidarity, a kind of collective democratization of the workplace. Labor activist Lorenzo Torrez recalled Jencks asserting that the steward "always defends the union's policies even if he didn't vote for them. He always defends the democratic wishes of the union members . . . He never says anything anywhere that can be used to weaken or disrupt

his union." Furthermore, they learned, "an injury to one is an injury to all." The blonde activist also introduced them to networking strategies that, according to future Local 890 president Art Flores, taught the men to "talk to the guy next to us . . . and then have that guy talk to others."[46]

Flores's experience reflects the transition of Chino's workers from tacitly accepting discrimination to becoming advocates for change. A stonemason with extensive experience as well as four years' service (1942–1946) in the US Navy, Flores discovered that Kennecott officials profiled workers by name even if their resumés revealed exceptional skills. "They didn't ask for a résumé . . . They just asked for my first name and my last name, and my last name determined where I was going to work. No matter what." So he accepted the labor position of ash shoveler on a coal-powered locomotive. Incensed at this discrimination as well as an incident on his return from overseas after World War II when he was told that "we don't serve Mexicans here" at a saloon in Dallas, Texas, Flores had a transformative experience under Jencks's tutelage. Soon he took a leading role in Local 63 and served as vice president in the late 1940s before eventually becoming president of Amalgamated Local 890 in the early 1970s. One of his first assignments must have emboldened him as well. He accompanied US senator Carl Hatch, a Democrat from New Mexico, on a visit in 1947 with President Harry S Truman at the White House to lobby against the Taft-Hartley Act. Although disappointed when Congress overrode Truman's veto of the anti-labor bill, Flores gained invaluable experience early in his tenure at Chino Mines. This formative training motivated him to take a leading role in the late 1940s and 1950s in the local labor movement.[47]

Guadalupe Fletcher recalled the major changes brought about because of World War II and the rise of Mine Mill as well. Before the war, "we wouldn't defend ourselves, the Anglos would just walk all over us, and they had the power to do it. People like my mom and dad, they were from Mexico. When they came to the states, the states were supposed to give them a lot of money. It was supposed to be a good living up here." The Anglos, she claimed, took advantage of them and limited their opportunities at work. But with World War II, she reflected, "things changed. That's when it started picking up. In some departments, there were just Anglos, just Chicanos. Some didn't want the Chicanos because they figured they didn't know how to do the job." Their success as soldiers, though, inspired a change in the Hispanics' attitudes: "We can do the job out there, we can do the job here too." The company also had to respond. Before the war, there were separate showers, "one for the Chicanos and one for the Anglos. And they changed after [the war and with the emergence of Local 890]. There was one shower for everybody."[48] Postwar changes, combined with the emergence of Mine Mill at

Chino, began a transformation of life at work and in the community for Santa Rita's Hispanics.[49]

Jencks and other Mine Mill trainers fueled the union members' desires for equality in the context of a long-time class struggle. They introduced them to Marxist ideology that portrayed them as the oppressed who had to fight to achieve new rights and benefits at work while also pursuing the larger goals of civil and political justice in the broader community. This theoretical instruction resonated with the Chicano workers, who had lived with this sentiment for generations and now understood it in historical terms. They were inspired as a result. Union members aired their work and civil rights concerns in their Spanish-language pamphlet *El Reportero*, which to a certain degree countered the company's publication *Chinorama*. In addition, Local 890 officers often reflected by the mid-1950s on the achievements of the previous decade. Cipriano Montoya, Local 890 president in the 1950s, announced to his brothers at a national Mine Mill convention in Spokane, Washington, in March 1950 that Local 890 "dealt with the question of civil rights . . . [and] has been outstanding in defense of civil rights. But in some locals there is weakness, even to the point of some brothers being afraid to assume positions of leadership in their locals." Joe Ramirez reminded the members on August 14, 1955, that their collective actions garnered better wages, health and life insurance, pensions, a grievance system, and other benefits "for our families and us." Two weeks later he announced that credit for these successes should go to "the Militant guys . . . those are the ones that help the union most." Their risky radicalism in the 1940s reaped rewards in the 1950s.[50]

Vitalized by their newfound power, they also became politically assertive. In 1948, while in a heated battle with Kennecott officials concerning Mine Mill's refusal to sign the Taft-Hartley anti-communist affidavits, for example, Local 890 announced in March its support for Henry Wallace, Progressive Party candidate for president of the United States. "Our union members [who voted unanimously], almost 1,500 strong," the *Silver City Daily Press* quoted from Local 890's endorsement letter, "see clearly that the two old parties, Democratic and Republican, are in the hands of the employers. These parties, champions of billion-dollar corporations, talk for the poor and act for the rich . . . Mexican-American workers see monopoly capital used to keep discrimination alive."[51] Wallace's platform advocating an end to segregation, full voting rights for people of color, and universal government-funded health insurance reflected many of Local 890's hopes for its members in twentieth-century America. By the early 1960s the union was regularly inviting political candidates to their meetings to address their membership, which usually endorsed their invitees.

This activism introduced a new dynamic to labor-management relations at Chino in the postwar period. Despite congressional passage in 1947 of Taft-Hartley, Jencks, Flores, Muñoz, and Morales agreed that they would not sign the anti-communist affidavits while negotiating for a new contract that same year. Moreover, they collectively conceived of a whole new set of demands as part of Jencks's and the International's strategy to seek to redress inequities while also strengthening their position as sole bargaining agent for all the county's CIO locals. In 1947, for instance, Local 890 proposed eleven new demands: (1) a $2.00-a-day wage increase; (2) adjustments in wages for inequities at Chino itself and for geographical discrepancies (Utah, Nevada, and Arizona workers earned higher wages for the same jobs); (3) a closed shop; (4) double-time pay for work over ten hours and on Sundays; (5) severance pay; (6) triple-time pay for holiday work; (7) sick leave; (8) guaranteed annual pay; (9) increase in pay for shift differentials; (10) a health and accident fund; and (11) removal of the no-strike clause in contracts with the company.[52] These daring demands, most of which would be achieved over the next decade, floored company officials, who ignored all of them, willing only to consider a new, much lower wage increase. In the new atmosphere the union sought to flex its new muscles under Jencks's leadership, an audacious approach not apparent at Chino during the certification crisis from 1934 to 1942 and for the duration of the war. Still, it was going to be an uphill battle for the union.

Almost simultaneous with these new demands, Congress passed the Taft-Hartley Act. The restrictive provisions placed on labor as a result emboldened Kennecott officials to counter these new demands with no response. But things had changed with the new Mine Mill leadership. The results were an extraordinary stand-off that ended with the company giving in to a higher wage increase (though less than requested) and a new stronger bargaining strategy for the unions, the use of the threat of a strike. Jencks knew how to use the print media to air the union's concerns *and* seek public sympathy. Consequently, he taught the early grassroots leaders—like Albert Muñoz, president of Local 63—to make themselves available to reporters of the Silver City newspapers, the *Daily Press* and the *Enterprise*. The day after negotiations between Kennecott and Local 890 officials broke off on June 23, 1947, for example, Muñoz gave an interview with the *Daily Press*, suggesting that if the company did not comply with the new requests, there would be "no contract, [and] no work." The negotiations stalled as a consequence of Mine Mill's insistence that a higher wage was deserved by their members. Even after F. G. Broome, secretary of the AFL's Chino Metal Trades Council, successfully contracted for an 11.5-cent wage increase, five paid holidays, job re-ratings and reclassifications, and the promise that Kennecott would not raise utility costs and house rents, Local 890 held out for more. Jencks, Muñoz, Morales, and Flores had already earned a vote of confidence from their members. They voted in favor of a strike. They also knew their brothers at the Utah and Nevada divisions were willing to man picket lines in their jurisdictions if Local 890's officers could not negotiate an end to the deadlock. Chino Mines officials began to realize their domination over their workers was being compromised. This realization may have been a key reason for general manager Horace Moses's retirement[53] in 1947, which resulted in William Goodrich's rise to that position.[54]

A month later Jencks used the media to announce to Grant County that a "deadlock" required Local 890 to request a federal conciliator to negotiate a "reasonable settlement." Using rhetorical tactics, El Palomino warned that if all the mining companies—ASARCO, USSR&M, Empire Zinc, and Kennecott—refused to negotiate, "they will force the unions to definite action." "We are making every effort," he continued, "to settle our differences by amicable means. We are still hoping for the best, but we are preparing for the worst." He, along with other Local 890 officials and members, had already begun preparations for a strike. They had formed finance, food, and picket committees, giving the impression that the rank and file was ready to go on strike. "We feel the company offers are wholly inadequate," Jencks told the *Daily Press* reporter, "and not fair in light of their present-day profits and in light of the recent wage increases and contract improvements in comparable industries . . . The proposed increase is not sufficient to give the men a living wage . . . The company desires to maintain our district as a cheap labor area."[55]

Jencks was using the media to air the workers' concerns. He also cleverly instilled Marxist principles in the rank and file, such as profit sharing (surplus value concept), a living wage, and equity pay, without making his source for these goals obvious to the workers and the newspaper's readers. By comparing Chino employees' wages with those of other jurisdictions, he also appealed to the general public's sense of fairness. He then concluded by placing the blame for the deadlock on the company. "It is hard to bargain against flat refusals, and when the company will not make counter-offers." Three days later he made a veiled threat that the union members were ready to strike. "The men are becoming impatient and feel that the wage increase we are asking is overdue. So far the company has made no offer adequate for settlement." By early August, Albert Muñoz and Joe T. Morales, president of Local 69, penned a news release with Jencks stating that they had exhausted "all possibilities of arriving at an agreement without the necessity of resorting to strike action," implying that a walkout was imminent. They further refused to negotiate as long as company officials insisted on including a no-strike clause in the contract.[56]

At regular meetings of Local 63, officers warned the rank and file that Kennecott planned to use the Taft-Hartley anti-communist affidavits after the August 23 implementation date to intimidate union members. "Keep tonight['s] discussion [of July 31, 1947, meeting] quiet and reveal it to no one. [The] company is about to break up our organization. If we keep our negociations [sic] quiet we will have the Co. where we want it. So far we have brought up the best contract in the Southwest . . . Our point is to put more power daily into breaking the Co. bluff." They remembered Kennecott's attempt to block union certification thirteen years earlier. They hoped to preempt any effort to destroy their nascent local labor movement. At the next meeting on August 7, Local 63 officials further strategized that they should complete the current negotiations before Taft-Hartley became law. "Will the Taft Hartley Slave Labor Bill . . . hurt us if we strike? No, it can't hurt us if we strike before it goes into full effect."[57]

This stance seemed to have worked when Kennecott agreed to a new contract before the labor law went into effect. On August 11 Kennecott officials agreed to give Mine Mill employees a 12-cent-per-hour raise, twenty-five job reclassifications (which meant as much as a 30-

cent raise for some jobs), and six paid holidays and double-time rates for working those days. Although the CIO union failed to win all the demands it proposed, Local 890 set the standard for negotiating with Kennecott and the other mining companies of Grant County. Kennecott, in fact, renegotiated contracts with its AFL affiliates, resulting in a wage increase from 11.5 to 12 cents, retroactive to the date of the initial contract signings several weeks earlier.[58] Soon union reps throughout the county looked to Local 890 to set the pattern in contracting with the major mining corporations. Composed principally of Chino employees, Mine Mill officials inspired members who worked for Empire Zinc to hold out, beginning in 1949, for better wages and improved working conditions in large part because of their successes at the Kennecott property. This countywide solidarity eventually motivated the celebrated Salt of the Earth strike.

Despite the emerging solidarity of Grant County's unions, Kennecott did interject in 1948–1949 the anti-communist affidavit controversy into negotiations with Local 890. As talks for the 1948 contract approached, general manager Goodrich refused to meet at the bargaining table. As long as Local 890's officials chose not to sign the affidavits, even though the law did not require their signatures to negotiate (though unions could not request NLRB intervention or arbitration during negotiations if their officials had not signed), Goodrich remained recalcitrant. Soon after AFL local officials signed the affidavits Kennecott did agree to new contracts with the Chino Metals Trade Council. However, Goodrich sent a letter in May 1948 to Local 890 officials to declare that he would not contract with Mine Mill as long as the affidavits, remained unsigned. In the letter he raised "questions of [Mine Mill's right to] representation" and that they "should become good Americans" and comply with Taft-Hartley. "We consider that such action [signing of affidavits] is necessary," Goodrich wrote, "so that democratic principles, unhampered by concepts foreign to the American way, would prevail in bargaining between labor and management." The Kennecott Council, which met in Denver in June, voted for a new policy of "no signing of affidavits" to support their brothers at Chino and throughout the American West. Furthermore, local labor officials proclaimed that "the company has made submission to Taft-Hartley their big gun because it is their way to break the union . . . This is a matter of personal freedom. The suppression of any minority means the death of democracy. People in this area know this lesson well."[59] Labor leaders were boldly threatening corporate hegemony in the copper industry.

Local Mine Mill officials Flores, Provencio, Morales, and Jencks replied to Goodrich on July 10 with their own letter. They claimed that Kennecott's position in May on the "yellow-dog affidavit issue" was a "smoke screen to prevent contract negotiations." In June "the Company continued in its failure to bargain," the return letter claimed, because of "the supposed excuse . . . that differing with [their] . . . point of view in [sic] un-American." This response elicited from the union on June 11 a unanimous strike vote and this retort from Local 890: "We have complied with every essential provision of the Taft-Hartley Act in spite of our hatred for this unjust law designed to weaken, divide and destroy . . . trade unions in America . . . Your company is in violation of the law . . . when you [Goodrich and other Kennecott officials] attempt to force our union into subservient acceptance of those purely voluntary provisions . . . You are in violation—not us! You are undemocratic—not us!" In the conclusion of the communiqué the union officials countered the company's red-baiting accusations. "If, in spite of this democratic expression, you persist in this arbitrary refusal to bargain with our Union, we shall withhold our labor and your profits. We shall further call upon every local in the International to support our membership. Our patience is worn thin."[60]

Chino Mines officials objected to this aggressive position. They complained of Local 890's refusal to sign the affidavits as well as the union's "propaganda" campaign during the stalemate. Goodrich criticized the union-sponsored "parades with banners and signs emphasizing the company's 'unreasonable' position." He further complained of their use of "the radio [with] . . . recorded programs which obviously originated from the union's national headquarters . . . [that] highlight the company's profits and salaries of higher officials" as unfair obstructions to contract negotiations. He also objected, despite refusing to budge on the affidavit issue, to a "staged demonstration" on June 9 by a group of 150 men made up "entirely of Spanish-Americans" outside the administration building at Chino. Throughout the summer, Goodrich and Jencks exchanged barbs in letters and interviews printed in the local media. On June 30 Jencks threatened that "we may strike at any time." Goodrich responded on July 11 that he "remained firm in his re-

quest that the filing of these affidavits" take place before negotiations go forward. The next day, Chino's CIO workers voted overwhelmingly to strike. Management and labor were at a stalemate.[61]

Local 890 staged a six-hour strike at Chino on August 7 despite efforts in the previous week by company, labor, and federal mediation officials to negotiate a settlement at meetings in Washington, DC, and Salt Lake City, Utah. Although the Kennecott Council voted with Local 890 in favor of the strike, only Chino employees staged the protest, setting up pickets at the entrances to the mine, mill, and smelter. At the Hurley Power Plant several strikers "barricaded themselves in the powerhouse with cots and a kettle of beans," planning a lengthy holdout. Tempers flared when a Kennecott official in his automobile sped through one of the picket lines and "nicked" Art Flores, injuring his wrist. The picketers soon after this incident "barred" entry into the company gates. They also brandished signs with phrases such as "Stop Acting Like Hitler, Kennicott. Bargain on a New Contract" and "For Your Home and Mine, Don't Cross the Picket Line" (see figure 5.37). These bold public statements juxtaposed the workers' views of the corporation's autocratic bargaining ploy (reference to Hitler) against their plea for Kennecott officials to consider the consequences of the deadlock for their employees' families. Like labor leaders elsewhere, they were employing a strategy that countered corporate accusations of union "communism" with appeals to company officials' "democratic" sense of justice. Goodrich later reflected that although the "pickets themselves were not numerous, a large group of men sitting on the sidelines, who were ostensibly not pickets, were there in case trouble developed." He also lamented the "loss" of 17,000 tons of ore production, because the men shut off the mill feed two nights earlier, in anticipation of the walkout.[62]

This first major strike, however brief, also reveals the important role of the women's auxiliary in supporting Mine Mill actions. As many as ten wives of the workers made 800 sandwiches to feed the picketers. This "family" solidarity emboldened women to play a significant role in Mine Mill actions from this time forward and especially during the Salt of the Earth strike. This precedent forewarned of the willingness of women to take a proactive role often against their husbands' and boyfriends' wishes in the 1950s and 1960s labor and civil rights movements in Grant County.[63] Non-striking CIO members also committed 25 percent of their wages to share in the costs of the strike. This cooperative effort impressed Local 890 president Jencks, who stated to the press, "We felt the first show of strength was a huge success." Clearly, Mine Mill was preparing its rank and file for future labor battles with America's most powerful copper corporation.[64]

The following night, federal mediator D. C. Goodman and Utah state industrial commissioner Daniel Edwards announced a settlement. Kennecott agreed to recognize Mine Mill as the sole negotiator for their membership while claiming it had not questioned the union's legitimacy to this right. The corporation also announced that the CIO workers would be "accorded" the "same wage treatment" as the AFL employees.[65] The strike action got results. Big labor was successfully putting a dent

in traditional corporate domination over workers. Using the threat of a work stoppage set the stage for labor-management relations in the 1950s and 1960s at Kennecott's Chino Mines Division.

The following October, Local 890 officials finally grudgingly signed the Taft-Hartley affidavits two months after International officers of Mine Mill had given in to red-baiting pressures. This action, however, did not deter the radical union's officials from moving forward with new demands despite facing expulsion in 1950 from the CIO. If anything, conditions of the Red Scare of the early 1950s inspired even bolder actions by Local 890. Chicano leaders, although still in training under Jencks's tutelage, put forward new demands in 1949–1950. They included a call for a new health and welfare plan, a company-financed pension plan with a generous $100-a-month sti-

FIGURE 5.37. *Mine Mill Local 890 strikers carry signs on August 7, 1948, in Santa Rita. The one-day strike led to a wage increase at the Chino Mines Division of the Kennecott Copper Corporation. Foreground* (left to right)*: Albert "Chino" Muñoz, Clinton Jencks, Salvador "Hueso" Espinosa, Greg Mesa. Mine Mill's militancy inspired other future strike actions, such as the Salt of the Earth strike against the Empire Zinc Company in the early 1950s in Grant County.*
COURTESY OF TOM MONTEZ.

pend for twenty-year employees of sixty years of age or older, new shift differentials, and fifty-two new job reclassifications. The company responded to these lofty goals with "proposals [that] were minor in scope . . . and the source of considerable misunderstanding," since they ignored all of them. In the end, the company stonewalled the union on most of the demands, offering annual contracts with modest raises through 1953.

Still, Local 890 initiated a nationwide two-week strike in the summer of 1951 as well as several walkouts and sit-downs during the remainder of the decade. On July 17, 1952, for example, pit workers left their jobs to protest a contractor's use of trucks in the mine. In August of that year underground miners initiated a sit-down to object to unsafe working conditions. At the height of Local 890's efforts in the Salt of the Earth strike against Empire Zinc, 800 Chino employees sat on the job on September 24, 1950. They were protesting the ninety-day jail sentence for contempt of court imposed on their officers who had ignored the injunction to stop the strike. Among those incarcerated were Jencks, Cipriano Montoya, Ernest Velasquez, Vicente Becerra, Pablo Montoya, and Fred Barreras, all Local 890 officials. Precipitation plant workers "struck" on March 2, 1953, disputing a foreman's complaints about the men taking too long in the change room. Track gang members sat on the job for several hours on September 23, 1954, objecting to management's poor treatment of one of their Hispanic foremen. The entire Chino workforce struck March 4–9, 1955, for the same reason. An industry-wide forty-seven-day strike in July and August 1955 also inspired the formation of the Chino Unity Council, which informally consolidated all of the district's craft and industrial locals. Twenty men participated in a sit-down at the smelter on August 23, 1955, to safeguard a forklift operator's right to drive his machinery.[66] These actions in the 1950s reveal the growing authority of the unions in earning new gains at the workplace as well as asserting their rights in the early phase of the Civil Rights Movement in the United States. The most telling strike prior to 1959 occurred from August 16 to September 2, 1954, to disapprove of Kennecott's new hospital plan.[67]

In August 1952, the NLRB certified the Santa Rita Hospital employees to be represented by Local 890 of Mine Mill. Similar to Jackling's negative response in 1934 to the formation of the first locals, Kennecott decided in January 1953 to close the hospital. The company then replaced it with two clinics—one in Santa Rita and the other in Hurley—and began offering health insurance in place of company medical care. The tense yearlong negotiations concerning the company's Hospitalization Insurance Plan, however, ended in August in a stalemate. Local 890 officials responded in late summer 1954 with a two-week strike. After the walkout and a promise from company officials to resolve the medical benefits question, the parties returned to the negotiating table in late 1954, with Kennecott reaching an agreement with all of the unions. Individual fees were to be $1.00 a month and family plans $3.50. The powerful copper corporation then contracted to pay all medical costs beyond these fees.

On October 27, 1954, Anna Marquez gave birth to Rose Mary, the last native Santa Ritan to be born in the company hospital. Marquez recalled later that Kennecott workers literally removed the bed next to her sometime before she and her new daughter were discharged on October 30 as the last patients. On November 1, Chino officials closed the longtime health facility. Another symbol of the company town was gone.[68]

Soon Kennecott replaced its defunct hospital with the Chino Insurance Office in dealing directly with worker health. This administrative unit managed the employees' claims for medical as well as death benefits. In the 1950s, employees' services administrator Lewis Presnall and his two assistants, Charles Evins and Joe Sandoval, processed dozens of claims a month. Chino's workers and their dependents commonly made claims to cover the costs of surgery, hospital stays, doctors' visits, and various other treatments. The office also often compensated mothers for pregnancy care up to $100 for normal births while also processing claims for "premature terminations of pregnancies." By the early 1960s, Mine Mill officials, such as presidents Vicente Becerra and Juan Chacón, had negotiated extraordinary medical and health benefits. These "fringes" included major medical, surgical, obstetrical, and hospital benefits that by this time cost employees with individual plans about $3 a month and those with families about $6. Even more compelling, Local 890 had negotiated benefits for retired, sick, and disabled employees for up to two years. Any discharged employee continued to receive benefits until his or her case was settled. During "stoppage of work due to a trade dispute" all workers continued to get every contracted benefit as long as they paid their monthly fees.[69]

Kennecott addressed less serious medical needs at its two clinics. Under the administration of chief surgeon Dr. E. A. Rygh, the clinic doctors and nurses treated Chino's employees for "more off-the-job accident[s] . . . than on-the-job injuries by three to one." The clinics offered vaccinations, X-ray examinations, treatment for everyday ailments, and minor surgical procedures. For more serious health concerns Chino's employees went to Silver City. Over the next fifteen years Kennecott encouraged its employees to seek medical treatment at the Hillcrest Hospital. The corporation regularly helped to fund upgrades at the facility, which by 1971 sported seventy-eight rooms, specialized physical- and inhalation-therapy experts and equipment, and a whirlpool to treat burn, arthritic, circulatory, and other conditions. The large number of Chino employees who sought specialized medical attention at Hillcrest also facilitated Kennecott's financing of the hospital's coronary and intensive-care unit.[70]

When Kennecott agreed to pay most of the medical costs of its employees, safety became extremely important to company and labor officials. Chino established a Safety Department that by the mid-1950s included forty full-time employees under director Paul Hunter. Emphasizing plant safety procedures, industrial hygiene, fire control, and general security, the division regularly trained and inspired employees of all stripes to follow new and evolving safety policies and procedures. By the late 1950s and 1960s these efforts brought excellent results in the form of National Safety Council and Joseph A. Holmes[71] safety awards in recognition of the millions of man-hours of work without serious accidents. *Chinorama* also regularly featured articles on the importance of safety by offering tips for on-the-job and at-home accident prevention. Company officials posted Sammy Safety icons—a cartoon character sporting a hard hat and "green cross for safety"—on departmental bulletin boards and in *Chinorama*. In 1962, Kennecott sponsored a safety slogan contest that elicited more than 300 entries. Among the featured phrases were Salvador Puente's "If you are safe today, you will be here tomorrow"; Mrs. Ernesto Vargas's "Courtesy pays, safety saves"; and David M. Castañon's "Men must be alerted so that accidents can be averted."[72]

Beginning in the late 1950s, Chino Mines also held safety contests between the pit workers, called the Ridgerunners, and the millmen, named the Flatlanders. The winners—the department with the fewest lost man-

hours due to accidents—were to receive hams, turkeys, and other holiday "bonuses" in December at the end of each year's contest. Individual workers who followed safety procedures and experienced "near misses" by wearing proper eyewear, steel-toe boots, and hard hats earned special awards recognitions. Those with shattered safety lenses won entrance into the Wise Owl club. Others saved by their hard hats earned slots in the Turtle club. Workers avoiding foot injuries by wearing proper industrial shoes gained entrée into the Golden Shoe club. Miguel G. Garza Jr. won both Owl and Turtle club awards within a week's time. When a flying rock broke his safety glasses on May 2, 1960, he won the former. Seven days later when a piece of metal punctured his hard hat without harmful consequences, he was awarded the latter. Foremen and other officials feted these safety awardees with certificates at company-funded luncheons.

Safety officials at Chino were already regularly celebrating excellent safety records. In spring 1957, Chino's safety director, Paul Hunter, acknowledged several workers for their extraordinary performance. Blacksmith shop foreman Hose Gill won recognition for his crew's 2,675 days and 278,030 man-hours without a lost-time injury. Dump crew foreman David B. Saenz earned distinction for his crew having worked 1,165 days and more than 287,000 man-hours without lost work time. Shovel department foreman J. L. Archer's men also received accolades for having worked 429 days and 185,830 man-hours without injury. Longtime labor leader Art Flores earned special recognition in the summer 1960 issue of *Chinorama* for ten years of accident-free truck driving. These and other extraordinary safety records brought state and national attention to Chino. On July 16, 1969, for example, New Mexico governor David Cargo presented to Chino's general manager Frank Woodruff the Award of Honor, the National Safety Council's highest mark of distinction. That same year the US Bureau of Mines awarded three of Chino's departments Awards of Merit and four others Certificates of Commendation in safety. Labor leaders worked closely with company officials to ensure that all employees emphasized safety and then won recognition when they got results.[73]

Despite Kennecott's obsessive focus on safety, the same year labor and management agreed to the new medical plan, the Chino family experienced the worst tragedy in the history of mining in Grant County. On March 31, 1954, five workers lost their lives in an explosion. During the midnight shift the powder crew had transported 800 pounds of dynamite to a pit bench to fill churn-drill holes to blast for ore removal when without warning it exploded. Four powdermen—Enrique Pedraza, 27; Andres Gonzales, 48; José Portillo, 48; and Magdaleno Chavez, 43—died in the tragic accident. Two of the men's bodies were never recovered and two others were blown 175 and 250 feet from ground zero only to die soon thereafter. A fifth victim—Chesley Chappell, a 35-year-old dozer operator—lost his life when a wheel from the powder truck flew about 100 feet into the unsuspecting worker's cab, decapitating him. Investigators found debris from the blast more than 600 feet from the site. Their months-long examination, however, never uncovered the exact cause of the premature explosion, leaving many locals to wonder whether foul play was involved, especially in light of the obsessive safety procedures followed by the powder crew. By this time Local 890 had negotiated death and pension benefits as well that, although they could not alleviate all the suffering of the four widows and their eleven children, did give life insurance and a settlement award to the deceased men's families. Labor's gains brought relief not available to workers under similar circumstances prior to the union movement in Grant County.[74]

Mine Mill continued to fight for greater worker benefits in the late 1950s and 1960s. Two major strikes in 1959 and 1967–1968 reveal the willingness of union members at Chino to unify for wage laborers' rights. From August 7 to December 27, 1959, Chino's workers struck Kennecott for the longest tenure to date to set the stage for the 1960s. Union officials of Mine Mill as well as the AFL-CIO affiliates each held out for better wages, resulting in benefits increases of about 25 cents per day per worker. They won the first of several pay adjustments known as Southwest differentials, which incrementally increased Chino workers' wages over the next three years. Under the 1962 contract, in fact, Chino Mines employees finally received the last pay increase to remove the differential and establish pay grades equal to those at Ray, Arizona; Ely, Nevada; and Bingham, Utah (see figure 5.38). As part of this equity pay plan, however, Kennecott also initiated the Job Evaluation Program, which both legitimized pay grades for evolving positions at Chino and assessed the need for certain jobs. Ironically, this administrative strategy, combined with Kennecott's Suggestion System, worked to eliminate many jobs over the next

two decades, resulting in a decline in the number of employees from about 2,500 in 1959 to fewer than 2,000 by the end of the 1970s. This decline in the workforce did allow Kennecott to increase each employee's benefits in the 1960s. Still, Mine Mill officials often wondered aloud at their meetings whether Kennecott's strategy involved efforts to eliminate jobs and reduce costs at the expense of its membership.[75]

Mine Mill and the other unions continued to negotiate in the 1960s for increases in benefits for their members. The 1964 contract, which reflects the unions' emerging strategy to negotiate company-wide, multi-year contracts and increased wages, while also augmenting health, retirement, and vacation benefits. Kennecott, for example, increased payouts for doctors' visits and maternity expenses and assumed all medical costs without deductions. The company was even willing to increase hospital services from 30 to 365 days a year for "mental" and "nervous" disorders. Big labor had come of age and the workers at Chino Mines reaped the rewards.[76]

Mine Mill's extraordinary achievements in behalf of its members emboldened the International union in the mid-1960s to consolidate its authority in negotiating with Kennecott and other companies. Yet fiscal concerns, as well as competition from the United Steel Workers of America, led Mine Mill's International leadership to decide to relinquish its longtime position in industrial unionism in favor of a merger with the United Steel Workers of America. President Al Skinner, in fact, approached his counterpart, I. W. Abel, of the USWA with the idea of merging, during their joint effort in early 1966 to see through new federal mine safety legislation. By the end of August the two union presidents and their executive committees had brokered a merger. When both men gave speeches justifying their actions at a special convention in Tucson, Arizona, in January 1967, they lamented the long-standing animosity and "bitter hostility" between the two most powerful unions in the nonferrous metals industry. At the same time, both of the loyal union leaders found common ground in their institutions' histories. Skinner reflected on Mine Mill's role in forming the CIO in 1936 and then the difficulties of being expelled from that organization fourteen years later. Abel commiserated with Skinner and then in his speech imparted to the audience the shared history of the two unions in fighting corporate domination, referring to their antecedent unions and some of their

FIGURE 5.38. *AFL Craft Union representatives pose during signing of new two-year contract at Santa Rita, summer 1962. By the early 1960s, AFL-CIO unions negotiated generous wage, health and life insurance, pension, severance, vacation, and other benefits at Chino and the other Kennecott properties. The collective union reps from each of the divisions joined in the Cooperative Committee to work for company-wide agreements. Standing, left to right: Frank Anderson (International Association of Machinists, Chino), Bill Smith (Brotherhood of Railroad Trainmen, Chino), Alton L. Matthews (Chino Metal Trades Council), Jesse Wysong (Brotherhood of Locomotive Firemen & Enginemen), three unknown representatives, and Henry A. Billings (Santa Rita foreman and a member of the company negotiating committee). Seated, left to right: E. F. Brehany (Brotherhood of Locomotive Firemen & Enginemen, Grand Lodge), R. J. Cantrill (BRT, Grand Lodge), L. F. Keegan (IAM, Grand Lodge), M. E. Doxford (White Pine Metal Trades Council, Nevada), F. A. Yagmin (industrial relations director, Ray), J. W. Shuster (industrial relations director, Chino), and Max Blackham (industrial relations director, Nevada). From* Chinorama *8, no. 4 (July–August 1962): 2.*

struggles. From the Homestead Strike of 1892 to Big Bill Haywood's and Joe Hill's "martyrdoms" to the Steel Workers Strike of 1919, Abel contended, they shared a common fight to improve workers' lives on the job and at home.

Skinner concluded his speech by arguing that Mine Mill had always believed in Bill Haywood's "One Big Union" strategy to counter corporate hegemony. He understood the power in numbers a merger would bring to a mega-union, as well as its historic significance. "We shall regard this merger . . . not as the end of a great and wonderful union, but as a new challenge and a new opportunity to contribute our vitality, our experience and our militancy to a new phase of history that now compels us to recognize that disunity, division, rivalry and raids can no longer be tolerated among unions in the nonferrous metals industry in North America." Abel responded similarly at the end of his talk. "There will be no playing off one union against another. There will be no undermining of one another's bargaining position. There will be no race to see who can settle first at the expense of the rest." They were now "freed of rivalry" and "have made a great decision" to join together. "Let us join our hearts, our minds and our energies so we can move onward, together, to build a greater and more united union, a better American trade union movement, and a better America."[77] The consolidation of Mine Mill and the USWA would prove nearly unbeatable in formulating industry-wide agreements for the next decade.

The immediate result at Chino was the longest strike in Kennecott's history. From July 14, 1967, until March 30, 1968, the company and the union were at loggerheads. Although several old-timers lamented the merger because of their experiences in thwarting USWA raiding efforts at Chino since the early 1950s, consolidation of the unions gave Local 890 still greater bargaining power. Kennecott officials were incensed at developments, especially when the USWA leadership initiated the industry-wide strike. In the end the standoff lasted 260 days. Local 890 president Severiano "Chano" Merino headed the local effort to "man" the picket lines, and each of Chino's eight other unions[78] stood with their brothers and sisters even though it was a USWA action. Although local bread-and-butter issues concerned USWA negotiators, they hoped to eliminate industry-wide distinctions, such as pay differentials and incongruent health insurance benefits between the copper industry of the Southwest and the steel industry of the Upper Midwest. The principal goal was industry-wide contracting. They hoped to use their power in numbers of nearly 1.2 million members to force Kennecott and other powerful nonferrous metals corporations to accept across-the-board wage increases; still better health, retirement, and pension plans; and other "fringes."[79]

Chino general manager Frank Woodruff's response to the USWA's strategy reveals the argument Kennecott officials would use to combat the growing authority of big labor. At the same time, his attention to labor issues shows how far the union movement had come at the New Mexico property in the post–World War II era. In a news release, "A New Twist to an Old Question," printed in part in the local press, Woodruff and his staff recognized the efficacy of the labor movement while appealing to the workers' and the public's economic concerns during the nine-month stalemate. "Unions are necessary," they acknowledged. "Where would the working man be without them?" This admission in support of unions, even if insincere, reveals how Kennecott's executives had to act under the realization that the public accepted big labor.

Still, Kennecott officials were already formulating in this open letter an economic hardship argument against this and future strikes. "Who pays the wages at Chino?

Who pays for most of the fringe benefits? Who puts up the money to keep the business going? Who buys the shovels, the trucks, the crushers and the furnaces that we use to make a living? Who finds the markets and sells the copper we produce? Where would the working man be without his employer? Where would the unions be?" Company officials also argued that Kennecott had set the pattern for advantageous contracts over the previous decade and deserved more respect from the USWA. Three weeks after the parties finally settled the labor dispute, Chino Mines released the "costs" of the strike to the public. The 1,200 strikers, the company publicized, "lost" $5.6 million in wages (about $3,600 per employee) and Kennecott $19 million in profits. State and local revenues were reduced by $2 million. This report was the last of many released by Chino Mines during and just after the nine-month ordeal. Ironically, Kennecott officials would use a similar economic hardship argument in the 1970s and 1980s to delay compliance with new federal and state clean air standards at the Hurley smelter. Nevertheless, Kennecott's public relations plea failed to halt the threat to corporate hegemony, which the labor and environmental movements compromised in the 1960s and later.[80]

The strike finally ended after President Lyndon B. Johnson called the combatants to the White House in mid-March 1968. Local 890 sent president Chano Merino and representative Art Flores to the high-level meeting. The president's summit got results. A few days after the meeting, in fact, Merino knew the deal was nearing completion on March 25 when he stated to local reporters that "we have a very good settlement. I don't think we'll have any trouble getting it ratified [by the locals]. We hope to be able to resolve future issues without the necessity of a strike." After four days of negotiating on local issues, among them a compromise on "archaic" lines of promotion, a revision of the job evaluation process, and back pay and wage increases for truck drivers, "ratification day" came on March 30. The copper industry's longest strike was over. New Mexico governor David Cargo telephoned Local 890 that day with congratulations. Kennecott, which had fired up the smelter furnaces several days earlier, published in the Silver City Daily Press a "Notice! Chino Employees" announcing the end of the strike and listing the telephone numbers of each department so that employees could "avoid confusion" and find out when they were scheduled to return to work.[81]

The end of the 1967–1968 strike marked a new era in labor-management relations at Chino. Local 890 continued to lead all the unions in militantly standing by its members. Much had changed since the heady days of the late 1940s and early 1950s when Albert Muñoz, Joe T. Morales, and Clinton Jencks first coordinated the rise of big labor in Grant County. The bold strategies of these men had worked to change labor-management relations at Santa Rita and Hurley. At the same time, labor's gains translated into political and social change as well. Although discrimination was still a major concern for Local 890 leadership, things were different. Chicanos and Anglos earned high wages compared to the rest of New Mexico's workers. Many Hispanics had been promoted to foreman, some to general foreman, and others to white-collar positions as engineers. Numerous Kennecott employees owned their own homes even if they could no longer live in Santa Rita, which by 1970 no longer existed. Chicanos for the first time began to hold town council slots, judgeships, and other political positions in Hurley and Bayard and Grant County, generally. The copper corporation's investment in modern technology translated into generations-long jobs for southwestern New Mexico's citizens while also providing the context for the ascent of big labor and the decline in discrimination. And even though Local 890 and the other unions had often faced stiff resistance from Kennecott officials as well as the problematic Taft-Hartley Act, the workers' fortitude and perseverance had paid off. Clearly, Chino's employees and their families, although branded "working class," could now be considered working middle-class. Many of the area's social and civic leaders, some of whom received Kennecott college tuition scholarships, either worked at Chino or were the children of folks who worked there.[82]

Lost in the excitement of the strike, of course, was the demise of Santa Rita. The mythological Spanish presidio, its successive manifestations under the Mexicans and the frontier Americans, and then its status as a company town were now just memories. By 1970 the only remnants of the nearly 200-year-old town were three cemeteries and lots of remembrances. The copper town was no more. Still, Santa Rita's story, so inextricably tied to the mining and processing of copper, lives on but now as space. For those who live in Grant County, reminders of the operations and the town will remain for centuries in the form of the massive mile-and-a-half-wide pit, the giant waste dumps, and the ever-present Kneeling Nun.

NOTES

1. *Kennescope* was the first of Kennecott's divisional publications in this case to highlight the activities and employees of the Utah Copper Division.

2. *Chinorama* 1, no. 1 (May 1955): 2–3.

3. Some of Local 890's leaders were skeptical of the Suggestion System, believing it was Kennecott's attempt to eliminate jobs, especially manual labor positions held almost exclusively by Chicanos, and, therefore, to reduce operating costs as well as the power of the union.

4. Mine Mill did finally merge with the United Steel Workers of America in 1967, fully twenty years after the first efforts of the USWA to raid Mine Mill.

5. Kennecott became the name of the company rather than Kennicott because of a clerical error; see www.reference forbusiness.com/history2/97/Kennecott-Corporation.html.

6. For more on Jackling's biography and role in developing porphyry copper in the American West, see Hildebrand and Mangum, *Capital and Labor*, 81–86.

7. See Hildebrand and Mangum, *Capital and Labor*, 71–86; Kennecott Copper Corporation, *All about Kennecott: The Story of Kennecott Copper Corporation* (1961); Charles K. Hyde, *Copper for America: The United States Copper Industry from Colonial Times to the 1990s* (Tucson: University of Arizona Press, 1998), 137–142, 164–165, 196–197; Leonard J. Arrington and Gary B. Hansen, *The Richest Hole on Earth: A History of the Bingham Copper Mine* (Logan: Utah State University Press, 1961).

8. Kennecott Copper Corporation (hereafter KCC), *Annual Report for 1942*; Chino Mines Division, Monthly Reports for 1942.

9. Goodrich to Moses, Reports, September 6, October 6, December 7, 1943, January 7, 1944.

10. "Local 63 Pledges Half-Day's Pay to Vets Memorial," *Silver City Daily Press*, August 24, 1946; most of the employees served in the Pacific Theater, also inspiring Grant County officials later to name a local park the Bataan Memorial Recreational Park; also see *Record Book*, "Minutes of Regular Meeting," International Union of Mine, Mill, and Smelter Workers (hereafter IUMMSW), Local 63, August 15, 1946, Box 864, Western History Collection (WHC), Archives, University of Colorado, Boulder.

11. See Memorandum: "Report on Field Trip to Silver City, New Mexico" of Carlos E. Castañeda, Assistant to the Chairman, Fair Employment Practices Commission, to Will Maslow, Director of Field Operations, March 3, 1944, in Box 1, Jack Cargill's Empire Zinc Strike Papers, Special Collections, Western New Mexico University, Silver City, New Mexico. Castañeda claimed in his report that of seven mining compa-

nies with federal contracts in Grant County, Kennecott was the "worst offender" in discriminating against "Mexicans" and "Indians." According to Castañeda, Kennecott practiced three specific types of discrimination: improper classification, refusal to upgrade, and a differential wage scale.

12. *Silver City Enterprise*, June 11, July 16, 1942; the CIO workers received retroactive pay from June 12, 1943, and the AFL employees from July 16, 1943, when Kennecott agreed to rim-to-rim pay in late 1945; General Manager, *Annual Report*, Chino Mines Division, February 2, 1946; by November 1943, Chino purchased a forty-passenger truck to transport workers to their stations; Moses to Kinnear, Monthly Reports, 1946, Kennecott paid out $188,000 for this retroactive rim-to-rim pay.

13. Chino Mines Division, Monthly Reports, 1942; Moses to Boyd, March 11, May 13, June 13, July 13, August 13, 1942, February 7, 1944; *Silver City Daily Press*, June 6, November 30, 1944; Moses to Boyd, Monthly Reports, 1944; General Manager, *Annual Report*, Chino Mines Division, February 2, 1946.

14. Joe T. Morales.

15. Kennecott failed in its attempt to entice Mexican workers of the *bracero* program to work at Chino; Goodrich to Moses, Monthly Reports, September 6, October 6, December 7, 1943, January 7, 1944, 1945, 1946; Moses to Boyd, Monthly Reports, 1944, 1945; Interview of Guadalupe Fletcher, Santa Rita Project, 1992.

16. KCC, *Annual Report*, 1942; *Silver City Enterprise*, April 2, May 28, 1942, December 4, 1947; Goodrich to Moses, Monthly Reports, February 7, 1942, April 8, 1948; Moses to Boyd, Monthly Reports, 1942; General Manager, *Annual Report*, February 14, 1948; *Silver City Daily Press*, December 3, 1947.

17. "Rotary Drilling," *Chinorama* 5, no. 12 (May 1958): 10–11.

18. Goodrich to Caulfield, General Manager's Report, March 6, 1953; Goodrich to C. D. Michaelson, *Annual Reports*, 1955, 1956; Ballmer to Goodrich, Monthly Reports, 1952, 1956, *Mining Operations Report*, January 12, 1956; *Chinorama* 4, no. 5 (May 1958): 10–13; 8, no. 3 (May–June 1961): 11; 10, no. 2 (March–April 1963): 12; 11, no. 4 (July–August 1964): 9.

19. Goodrich to C. D. Michaelson, *Annual Report, 1955*, January 23, 1956; Ballmer to Goodrich, *Mining Operations Report, 1955*, January 12, 1956; *Chinorama* 3, no. 11 (November 1957): 18–19.

20. *Chinorama* 3, no. 6 (June 1957): 12–15; 4, no. 10 (October 1958): 8–11; 8, no. 5 (September–October 1962): 5.

21. *Chinorama* 2, no. 9 (September 1956): 12–15; 4, no. 1 (January 1958): 10–13.

22. Goodrich to Caulfield, Reports, April 11, August 11, December 11, 1953; *Chinorama* 1, no. 12 (December 1955): 12–15; 4, no. 1 (January 1958): 10–11.

23. Goodrich to Moses, October 7, 1947; Moses to Kinnear, November 1947; Horace Moses, *General Manager's Annual Report, 1947*, February 14, 1948; Ballmer to Goodrich, Monthly Reports, 1952, *Mine Operations Report, 1953*, January 8, 1954; Goodrich to Caulfield, Monthly Reports, 1953, Monthly Report, March 1954, April 12, 1954; *Silver City Daily Press*, March 31, April 1, 2, 3, 5, 1954; *Silver City Enterprise*, April 1, 8, 1954.

24. *Chinorama* 4, no. 2 (February 1958): 11.

25. Ibid., 10–13.

26. Ibid., 10–13; 6, no. 6 (November–December 1960): 1.

27. *Chinorama* 9, no. 2 (March–April 1963): 2–7; 9, no. 3 (May–June 1963): 2–5; 9, no. 4 (July–August 1963): 14–15; 9, no. 5 (September–October 1963): 4–7; 10, no. 3 (May–June 1964): 10–15; 15, no. 4 (Fourth Quarter, 1969): 25–29.

28. *Chinorama* 5, no. 3 (March 1959): inside cover; 7, no. 1 (January–February 1961): 12–15; 7, no. 6 (November–December 1961): 10–11, back inside cover; 8, no. 2 (March–April 1962): 12–15.

29. *Chinorama* 11, no. 2 (March–April 1965): 2–5.

30. Moses to Sully, February 11, 1916; CCC, *Monthly, Quarterly, Annual Reports*, 1916; CCC, *Annual Report*, 1918; US Bureau of Mines, *Minerals Yearbook* (Washington, DC: GPO, 1920–1960); T. A. Rickard, "The Chino Enterprise—V, Concentrating the Ore at Hurley and Smelting at El Paso," *Engineering and Mining Journal-Press* 117, no. 1 (January 5, 1924): 15–16; A. B. Parsons, *The Porphyry Coppers in 1956*, 126–127.

31. *Chinorama* 7, no. 2 (March–April 1961): 14–17.

32. Ibid., 17–18.

33. Ibid.

34. *Chinorama* 12, no. 1 (January–February 1966): 12–15; 15, no. 1 (First Quarter, 1969): 4–8.

35. *Chinorama* 10, no. 4 (July–August 1964): 11–14.

36. By 1960, Kennecott had constructed the Ray, Arizona, smelter and had purchased the Garfield smelter in Utah; see *Chinorama* 6, no. 6 (June 1960): 5–6.

37. John B. Huttl, "Chino Today," *Engineering and Mining Journal* 140, no. 9 (September 1939): 29–33; Colin J. Bury, "How Copper Is Smelted," *Chinorama* 2, no. 15 (Second Quarter, 1969): 14.

38. For discussions of crushing and milling of ores at Chino in the 1950s and 1960s, see *Chinorama* 4, no. 4 (July–August 1958): 12–15; 3, no. 5 (March 1959): 8–11; 1, no. 12 (January–February 1966): 2–7.

39. *Chinorama* 2, no. 11 (November 1956): 24; 4, no. 3 (March 1958): 10–13; 4, no. 4 (April 1958): 10–13; 2, no. 15 (Second Quarter, 1969): 14–19.

40. *Chinorama* 2, no. 15 (Second Quarter, 1969): 14–19.

41. "Kennecott Council," Folder 9, IUMMSW Collection, Box 870, WHC, Archives, University of Colorado, Boulder.

42. "Sheriff, Local CIO Leader Tangle, One-Punch Fight Ensues over Talk about Fierro Riot," *Silver City Daily Press*, May 5, 1949, this article reveals that Sheriff Bartley McDonald punched Jencks at the county courthouse for allegedly calling the lawman a "liar" while he was attempting to find out the

details surrounding the arrest of several union men after a bar brawl in Fierro. On April 20, 1953, Jencks was arrested at his Santa Rita home in front of his young son in his front yard when FBI agents "roared up and jumped out" to tell him he was under arrest. When he asked to go inside to get shoes and socks, they refused and treated him like an "armed fugitive"; see Baker, *On Strike*, 238–239.

43. Quoted from Baker, *On Strike*, 74.

44. This event was sponsored by the Latinos Unidas, the Silver City Museum, Western New Mexico University, and the WNMU International Film Society and was partially funded by the New Mexico Endowment for the Humanities; see *Silver City Sun News*, May 5, 7, 2004, and *Silver City Daily Press*, May 1, 10, 2004.

45. See Baker, *On Strike*, 71–80.

46. Quotes from Baker, *On Strike*, 82, 76. Baker offers an extensive discussion of how Local 890 and Mine Mill in the 1940s and 1950s combined the struggle for workers' rights with the fight for civil and political rights.

47. "CIO-Kennecott Negotiations Are in Deadlock," *Silver City Daily Press*, June 18, 1947; Baker, *On Strike*, 78–79.

48. Interview of Guadalupe Fletcher, Santa Rita Project, 1992.

49. Ramón Arzola claimed that discriminatory practices began to change in the postwar period principally because of the rise of Mine Mill; see interview of Ramón Arzola, Santa Rita Project, 1992.

50. Shop Stewards Council Meeting, March 24, 1955, 2, IUMMSW Collection, Roll Book, Box 869, WHC, Archives, University of Colorado, Boulder; Special Amalgamated Meeting, August 14, 1955, 185–189, and Special Meeting of Southwest Locals, August 28, 1955, 195–201, IUMMSW Collection, Roll Book, Box 864, WHC, Archives, University of Colorado, Boulder.

51. "Local Mine Union Backs Wallace," *Silver City Daily Press*, March 3, 1948.

52. Horace Moses, Chino Mines general manager, to Kinnear, general manager Western Mining Division, "Labor Developments," April 1947.

53. Moses had been anti-labor for more than fifteen years, as exemplified by his strong-arm approach to the workers' strike of Gallup American Coal Company (a subsidiary of Chino Copper Company) in 1933 when he was serving as general manager of that company. He may have realized that officers of Local 890 were emboldened by New Deal legislation and conditions brought about by World War II; see Vargas, *Labor Rights*, 96–97, 100–103, 108–109.

54. "Labor Contract Talks Postponed Until Thursday," *Silver City Daily Press*, June 24, 1947, "Local Unions Vote to Strike in Backing Demands," July 7, 1947, "AF of L Union Ratifies Contract with Kennecott," July 10, 1947; "Santa Rita and Hurley Miners Take Strike Vote," *Silver City Enterprise*, July 2, 1947.

55. "Unions Ready for Strike," *Silver City Daily Press*, July 21, 1947, and "Conciliator for Mine Dispute May Arrive This Week," *Silver City Daily Press*, July 24, 1947.

56. *Record Book*, Mine Mill Local 63, Regular Meeting Minutes, July 31, August 7, 1947, Box 864, IUMMSW Collection, WHC, Archives, University of Colorado, Boulder.

57. Ibid.

58. Moses to Kinnear, "Labor Developments," August 1947.

59. "Labor Leaders' Compliance with Taft-Hartley Act Asked by Kennecott," *Silver City Daily Press*, June 2, 1948; "Union Attacks Chino Letter to Workers," *Silver City Daily Press*, June 7, 1948.

60. Letter from Arthur Flores, B. G. Provencio, Jose T. Morales, and Clinton Jencks, Local 890, to W. H. Goodrich, Superintendent, Chino Mines Division, July 10, 1948, Box 864, IUMMSW Collection, WHC, Archives, University of Colorado, Boulder.

61. Goodrich to Kinnear, "Labor Developments," July 15, 1948; "Mine-Union Zero Hour Near," *Silver City Daily Press*, June 25, 1948; "Union Holding All-Day Meet as U.S., Chino Parleys Stall," *Silver City Daily Press*, July 11, 1948; "Chino Workers OK Strike Vote," *Silver City Daily Press*, July 12, 1948.

62. "No Agreement Reached Yet," *Silver City Daily Press*, August 7, 1948; Goodrich to Kinnear, "Labor Developments," September 15, 1948.

63. In 1951, when a court injunction precluded Empire Zinc employees of Local 890 from picketing in the Salt of the Earth strike, their wives took to the picket lines at first against the wishes of their husbands, who eventually accepted their wives' important role in that famous show of union and "family" solidarity. See Baker, *On Strike*; Jack Cargill, "Empire and Opposition: 'Salt of the Earth' Strike," in Robert Kern, ed., *Labor in New Mexico: Unions, Strikes, and Social History since 1881* (Albuquerque: University of New Mexico Press, 1983), 183–267; and James J. Lorence, *The Suppression of* Salt of the Earth: *How Hollywood, Big Labor, and Politicians Blacklisted a Movie in Cold War America* (Albuquerque: University of New Mexico Press, 1999).

64. "No Agreement Reached Yet," *Silver City Daily Press*, August 7, 1948; Goodrich to Kinnear, "Labor Developments," September 15, 1948.

65. "Union Says Walkout Off 'Indefinitely'; Raise Retroactive," *Silver City Daily Press*, August 9, 1948; "Chino-CIO Settlement Explained," *Silver City Daily Press*, August 10, 1948.

66. Four of the men who participated in the smelter sit-down were suspended for their actions. Nick Castillo and Wally Garcia received thirty-day and David Ronquillo and Alfredo Abalos ten-day suspensions; see *Chinorama* 1, no. 10 (October: 1955): inside back cover.

67. Goodrich to Kinnear, "Labor," September 21, 1949, November 11, 1949, December 19, 1949; Goodrich to Buch-

man, Monthly Reports, 1952; Goodrich to Caulfield, *Annual Report*, 1952, Reports, April 11, September 1953; *Silver City Enterprise*, July 24, 1952, August 25, October 20, 1955; *Silver City Daily Press*, August 10, 12, 1953, August 9, October 17, 1955.

68. Santa Rita Hospital Record #28980, October 27–30, 1954.

69. *Chinorama* 3, no. 3 (March 1957): 22–23; 8, no. 5 (July–August 1962): 4–9.

70. Ballmer to Goodrich, Monthly Reports, 1954; Goodrich to Caulfield, Monthly Reports, 1954; *Silver City Daily Press*, December 28, 1954; *Chinorama* 3, no. 7 (July 1957): 12–15; 17, no. 2 (Summer 1971): 6–9.

71. The US Bureau of Mines gave the Holmes awards in honor of its first director, Dr. Joseph A. Holmes, who emphasized safety in the mining industry while he was head of that federal agency.

72. "No One Can Afford an Accident!" *Chinorama* 8, no. 5 (September–October 1962): inside cover.

73. For example, see "I've Been Working on the Railroad—Safely," *Chinorama* 2, no. 9 (October 1956): 10–11; "The Safety Department," *Chinorama* 3, no. 1 (January 1957): 12–15; "Protective Equipment Policy," *Chinorama* 3, no. 12 (December 1957): 19; "Safety Needs Teamwork," *Chinorama* 3, no. 4 (April 1957): 11; "Sammy Safety's New Contest," *Chinorama* 4, no. 4 (July–August 1957): 5; "Various Safety Awards Presented," *Chinorama* 5, no. 5 (May 1959): 3; "Who Will Win?" *Chinorama* 6, no. 4 (July–August 1960): 15; "Owls, Turtles and Gold," *Chinorama* 6, no. 5 (September–October 1960): 16; "Is Your Job Safe?" *Chinorama* 10, no. 2 (March–April 1964): 10–13; 10, no. 5 (September–October 1964): 12–13; 15, no. 3 (Third Quarter, 1969): 16–17.

74. Ballmer to Goodrich, Monthly Reports, April 9, September 9, 1954. Mary Pedraza, widowed when her husband, Enrique, was killed in the 1954 accident, used her life insurance (not available through the company before the 1940s) in 1955 to finance the construction of the last house built in Santa Rita. In 1964, Mary moved the home to Bayard during the evacuation of Santa Rita. Ironically, Terry Humble bought the house and still resides in it with his wife, Micha Humble.

75. "Chino Strike Over," *Chinorama* 6, no. 1 (January–February 1960): 1–7; 8, no. 4 (July–August 1962): 2–3; as early as November 1957, famed Local 890 president Juan Chacón wondered aloud in a Smelter Unit meeting of Mine Mill whether the Suggestion System, publicized as a cost-savings plan by Kennecott, "is doing us more harm than good since we are giveing [*sic*] the company ideas as to how to eliminate more

jobs causing curtailment of men"; see Smelter Unit Meeting, November 29, 1957, 2, Box 869, IUMMSW Collection, WHC, Archives, University of Colorado, Boulder.

76. *Chinorama* 10, no. 4 (July–August 1964): 3, 22.

77. See Letter of IUMMSW president A. C. Skinner to All Local Unions, September 1, 1966; "Report of President A. C. Skinner," Special Convention, Pioneer International Hotel, Tucson, Arizona, January 16, 1967; and "Address, I. W. Abel, President, United Steel Workers of America," Special Convention, January 18, 1967, Box 866, Book 3, IUMMSW Collection, WHC, Archives, University of Colorado, Boulder.

78. They included the International Association of Machinists, Local 1563; the Office & Professional Employees International Union, Local 62; the Brotherhood of Railroad Trainmen, Local 323; the Brotherhood of Locomotive Firemen & Enginemen, Local 902; the International Brotherhood of Electrical Workers, Local 496; the Plumbers & Pipefitters, Local 586; the Carpenters Union, Local 987; and the Boilermakers Union, Local 632.

79. For complaints about USWA raiding efforts at Chino, see comments of Cipriano Montoya and Joe Campos in Minutes of Meeting of Kennecott Local Delegates, Denver, CO, March 27, 1951, Box 870, IUMMSW Collection, WHC, Archives, University of Colorado, Boulder; "Special Strike Issue," *Chinorama* 14, no. 1 (January 31, 1968): 4–5; "Union Charges Violated Agreement Cause of Monday Supply Stoppage," *Silver City Enterprise*, February, 15, 1968.

80. See "Chino Mines Division, Kennecott Copper Corp., Issues Newsletter about Contract Negotiations," September 28, 1967; "Copper Strike Continues in Its 41st Day," August 24, 1967, *Silver City Enterprise*; *Silver City Enterprise*, April 18, 1968; also see "Special Strike Issue," *Chinorama* 14, no. 1 (January 31, 1968): 12.

81. "Kennecott Agreement Still Not Finalized," March 18, 1968; "Kennecott Reaches Pacts with Three Local Unions," March 25, 1968; "Last of Local Unions Ratify New Agreement," March 28, 1968; "Notice!" March 30, 1968; "Copper Workers Return to Work," April 1, 1968, all *Silver City Daily Press*.

82. Interview of Herbert Yee Toy, June 1, 2007, Silver City, New Mexico, by Chris Huggard. Herb Toy's experience at Chino is one example of the company's scholarship program. Toy, who began working for Kennecott in 1959, was given tuition scholarships during his early tenure at the company to attend Western New Mexico University. As a result, he earned a degree in chemistry in 1971.

Epilogue

MINING AND THE ENVIRONMENT

Chino's future is uncertain early in the twenty-first century. Escalating operational costs, dwindling ore grades, decline in water supplies, and environmental remediation have complicated successive owners' ability to earn a profit at the nation's fourth-largest open-pit copper mine. Still, Kennecott's successors at Chino, Phelps Dodge from 1986 until 2007 and Freeport-McMoRan since, have employed as many as 1,400 workers at the century-old open-pit

mining, milling, and smelting complex. The economic benefit to Grant County and New Mexico generally has been phenomenal despite the impending demise of the mining enterprise. Since 1999 the companies have paid out more than $300 million in wages and about that same amount in auxiliary trade and tax revenues.

The 100-year fiscal sustainability to southwestern New Mexico's economy, however, may last less than a decade longer, depending on the length of the current shutdown caused by the economic crisis of 2008–2011. As a 2001 report of the Interhemispheric Resource Center in Silver City claimed, there are three major concerns at Chino: (1) management of the economic fallout from the looming mine closure, (2) improvement of current operations to limit "damage [to] community health and the

environment," and (3) adequate planning and implementation of "mine clean-up."[1] All three features portend an ominous economic future for the county, even though the latter two alarming worries could be parlayed into a decade or more of employment (similar to Superfund cleanups). Such a strategy would require current owner Freeport-McMoRan to broker a deal with federal and state authorities and environmental groups to fund such a remediation program.

Clearly the company and its employees realize the imminent end to mining. Just as Kennecott planned for the demise of Santa Rita from the 1930s to 1960s, more recent corporate officials and Chino engineers have been preparing for the permanent shutdown of the operations. Even the most advanced processing technologies, highlighted by the solution-extraction electro-winning (SX-EW) process, cannot extend the mine's life indefinitely. Unlike the logging industry, which can regrow its natural product, mining extracts mineral resources that can never be replaced. This reality has been an industry-wide concern for at least three decades. It is reflected in the rapid decline of the US position in world copper production from first in 1990, with 32 percent of total global output, to fourth or fifth, with less than 10 percent in 2011. Latin American, Chinese, African, and Canadian production now leads the way until those sources of ore run out too. The American corporations, which have been mining decades longer than their global competitors, successfully extracted the vast deposits of North America with efficiency and an eye to profitability. That era is coming to an end.

On the environmental front, the industry, while reaping the economic rewards of its investments, was loath to address air and water pollution and the consequences of transforming the landscape. At the same time, Kennecott and other copper corporations began planning in the 1970s and 1980s for endgame strategies with new technologies and cost-cutting practices. So executives attempted to block environmental regulations. They claimed that the expense to comply with new federal and state regulations would lead to closings and they disregarded the broader society's growing concerns for the endangered environment.[2] Kennecott executive Robert Swan, in fact, testified before Congress on at least two occasions in the 1970s in an attempt to thwart federal and state restrictions formulated as part of the provisions of the Clean Air Acts of 1970 and 1977 on sulfur dioxide and particulate matter emissions. Despite obvious threats to smelter workers' and their families' health, USWA labor leaders and locals themselves vociferously complained during the 1970s "air battle" about the environmentalists. They "don't have anything better to do than to bother us."[3] While Local 890 officials like Juan Chacón and Chano Merino negotiated for safer working conditions at the Hurley smelter, they also resisted the efforts of the clean-air advocates. So John and Nancy Bartlit of the New Mexico Citizens for Clean Air & Water (NMCCA&W), Brandt Calkin of the Sierra Club, and other grassroots environmental groups countered the corporate-labor alliance by insisting on enforcement of the new regulations. They were particularly effective making public presentations in favor of compliance

at Environmental Protection Agency (EPA) and New Mexico Environmental Improvement Board hearings in Santa Fe, Silver City, and Washington, DC.

Workers did have mixed feelings about the emissions problems. Smelter worker Felix Martinez, for example, complained that Kennecott and the environmentalists failed to realize the crux of the emissions problem. "We're not concerned about some rabbit sneezing on the mountain 40 miles away," he claimed in a 1975 interview, "but we are concerned about the environmental conditions inside the plant. We're concerned about the conditions where we work." Only a few months earlier, Local 890 president Merino agreed with Martinez that "we don't need a lot of outsiders coming in and telling us what to do." Yet the labor leader sided with the company in fearing that meeting the clean-air standards threatened jobs. "We are not in love with Kennecott, but we negotiate [on things like smelter safety] in good faith. When somebody says Kennecott is not in good faith I resent it. It isn't the [smelter] smoke that bothers us. It's when we don't see smoke that it bothers us."[4] Jobs took precedence over clean air for company and labor officials.

One former smelter worker conceded that conditions could become unbearable because of the smelter smoke. The late Herb McGrath, who worked for Kennecott for nearly thirty years, did recall numerous occasions in the 1960s, 1970s, and 1980s when the "smoke invaded" the more populous parts of Grant County. "Everybody coughed, hacked, and gagged and complained." In the end, though, McGrath claimed that those pervasive layers of yellow smoke eventually disappeared except on rare occasions, especially once Kennecott came into compliance in the 1980s with emissions standards. In 1993 the former smelter worker claimed, "I haven't smelled sulfur in the morning [at my home in Silver City] in a long, long, long time. There's no question it has improved."[5]

Ultimately, the corporation and the union could not block the inevitable restrictions on emissions. Even strong support from US senator Pete Domenici (Rep., NM) and other politicos of the Copper Caucus, which led to hushed meetings with President Richard M. Nixon, did not ensure relaxed air standards. By the end of the 1970s, in fact, environmentalists like the Bartlits celebrated the implementation of new technologies. Kennecott spent hundreds of millions of dollars to put in flash furnaces, acid plants, and air-monitoring devices at Chino as well as at its Nevada, Arizona, and Utah properties. These air-quality upgrades produced fewer emissions, removed much of what was emitted before it exited the stacks, and then recorded the results in case adjustments were needed to increase the odds of meeting environmental standards. All of the Southwest's copper smelters had come into compliance except Phelps Dodge's antiquated open-hearth system in Douglas, Arizona, which finally closed in 1986. Pragmatic environmentalist John Bartlit, president of the NMCCA&W, proudly proclaimed that compliance translated into safer working conditions, recovery for plants and animals in the Desert Southwest, and a clearing of the skies above the Grand Canyon and other natural wonders in the region. "I am for industry," he proclaimed. "And I am for clean air." Even Juan Chacón changed his tune and served in the early 1980s on the American Lung Association of New Mexico's Air Quality Committee with Nancy Bartlit.[6]

The end of the air pollution problem eventually came as well. On June 5, 2007, more than 5,000 interested observers witnessed the demolition of Hurley's two smelter stacks. Local residents began to lament the loss of the smelter town's iconic smokestacks soon after Phelps Dodge announced in August 2006 plans for the destruction of the two towers. The older stack had not been used since the mid-1990s and the newer one since 2002. So the "end of an era" was materializing. Now the townspeople could anticipate the bringing down of two of the most distinctive symbols of their cherished way of life. In that moment, nearly a year later on June 5, when the stacks finally came down, only the Kneeling Nun and the pit itself remained of the four most pervasive icons of copper mining in southwestern New Mexico. The smelter stacks, like the town of Santa Rita, were now gone. John Portillo, who won the right to push the plunger that detonated the fateful blast, went through a flood of feelings as did many of those in the crowd. "My hands are sweaty. I'm feeling sentimental, excited and sad. I grew up four houses away from the stacks." His wife, Priscilla, and four of their children watched with similar sentiments as well. Thousands of others shared a rush of emotion as the tall monuments crashed to the ground (see figure 6.1).[7]

The end to smelting at Chino Mines, however, has not eliminated environmentalists' anxieties about another threatened resource, water. Throughout the history of open-pit mining, milling, and smelting at Chino, company officials realized the significance of water to

the operations. Consequently, they devised elaborate water conservation strategies as early as the 1910s, accumulating surface and underground sources and then constructing elaborate piping, storing, and recovery systems to secure enough of the precious liquid for processing the ores. Eventually, though, the great thirst of the operations—from feeding steam boilers, transporting slurry, and floating ores to leaching the dumps and pumping tailings—required increasing supplies of water in the desert climate. Since about 2000, the vast majority of Chino's copper production has come from SX-EW processing. This method requires hundreds of millions of gallons of water to leach the dumps, transport the "impregnated" copper water, and fill the massive electricity-charged cathode tanks. The results of this solution process are hundreds of thousands of tons of cathode copper ready for shipment to refineries. In 2000 alone, for example, this technology allowed Phelps Dodge to produce more than 800,000 tons of cathode copper from its various electro-winning facilities in the American West. This industrial need for the precious resource has resulted in a precipitous drop in the underground water table in southwestern New Mexico. This demand has also forced the owners to pipe water from the Mimbres and Gila–San Francisco watersheds, despite historic access to hundreds of millions of gallons from water rights on thousands of acres of land in Grant County.[8]

More disconcerting to environmental groups like the Gila Resources Information Project and Gila Watch are acid rock drainage, tailings pond spills, and groundwater contamination. Runoff from ore and waste dumps and even natural watercourses concerns them because of the release of harmful metals that normally would remain in place if not disturbed by former blasting and dumping. During the past two decades more than forty spills at Chino have threatened ecosystems and surface and groundwater sources. Among the worst of the toxic washes was the release of 185 million gallons of "process water" into Whitewater Creek in 1988, 270,000 gallons of "sulfuric acid waste" into Hanover Creek in 1990, and the discharge of 3.25 million gallons of tailings into Whitewater draw in 1999.[9]

The public's response to this water pollution, common to the mining industry generally, according to the EPA, has resulted in changes. In 1974, in fact, Kennecott established the Environment Department (ED) at Chino. The result of a 1971–1972 Environmental Task Force

FIGURE E.1. *On June 5, 2007, Phelps Dodge Corporation demolished the two smelter stacks at Hurley, New Mexico. More than 5,000 observers witnessed the spectacular explosion and then collapse of the industrial icons.*
COURTESY OF BOB PELHAM.

formed in response to the federal Clean Air and Water Acts, the ED addressed "workplace exposures and environmental hazards." Among the duties of hygienic and environmental engineers, according to longtime ED official Herb Yee Toy, was to collect data on air emissions at the crusher, concentrator, and smelter. Engineers like Toy worked to meet workplace dust standards under federal safety regulations of the Mining Enforcement and Safety Administration (MESA) and the Mine Safety and Health Administration (MSHA), and labor safety standards of the Occupational Safety and Health Administration (OSHA).

Toy and other engineers, who by the late 1970s and 1980s were receiving more sophisticated environmental science training at regional universities, focused on collecting data on air, water, and noise pollution. If they stayed vigilant with their remediation efforts, Chino would be permitted by the EPA and the New Mexico Environmental Improvement Division to continue operating. In particular, they sought air-quality, groundwater, and groundwater discharge permits. "Early on, the main concern of the company," Toy recalled, "was to get permitted. In other words, you had to determine if you were complying with regulatory standards. It required a lot of data gathering . . . [We had] to drill numerous test wells," for example, "to determine where aquifers were, to drill monitor wells." These wells drilled near Chino's water discharge sites allowed engineers to test for toxins, such as heavy metals, nitrates, sulfates, and other pollutants. As Toy described in a 2007 interview, "general mining activities discharge. Not every pound of dirt is copper ore. You're disturbing the earth. Large-scale mining like [at] Chino we were moving half million tons of ore out of the pit every day. If it rains, you have leaching . . . Because you're disturbing the earth, you have air pollution problems."[10]

The activities and actions of the ED at Chino did create a "much cleaner operation because of the changes."[11] The acid plant at the smelter eventually cleaned up 90 percent of the sulfur emissions, monitor wells resulted in water treatment, and noise monitoring protected workers' hearing. Significantly, Kennecott officials determined the fiscal viability of the implementation of environmental technology just as they would for new equipment, wage increases, and other financial concerns. This economic hardship strategy often pitted the copper companies against federal and state regulators and the environmental groups. In the end, multinational copper companies like Kennecott and Phelps Dodge had no choice but to clean up their operations. The environmental movement, like big labor in the post–World War II period, compromised corporate hegemony, in this case concerning the company's traditional treatment of nature.[12]

The result of the rise in environmental awareness and regulation was greater emphasis on water pollution by the mid-1980s. Initially, New Mexico Environmental Improvement Division litigation and then lawsuits filed by environmentalists, especially the NMCCA&W, forced Phelps Dodge beginning in the late 1980s to address its weak efforts at preventing toxic runoffs, spills, and leaching. By the mid-1990s the company invested millions of dollars to strengthen tailings dams, monitor leach dumps and other water-related processes, and employ specialized engineers to manage toxic water treatment and to report spills. As Marchelle Schuman, the New Mexico Environment Department's[13] chief regulator of Chino, stated in 2000: "They [Phelps Dodge officials] are getting better about reporting their spills. It used to be that we were threatening litigation all the time to get them to comply, but I think they are starting to learn that we mean business . . . They are also learning how to prevent spills more often, and that comes from improved company monitoring."[14]

In the 1990s and 2000s, Phelps Dodge spent millions of dollars to ameliorate the water problem. Engineers designed and then workers constructed diversion channels that, though sometimes insufficient, have reduced acid runoff. The company also drilled "interceptor wells"

to "capture groundwater and recycle it back" through existing water systems to greatly reduce pollution on non-company lands. Water engineers also continue to test new monitoring wells strategically located to measure the toxicity of groundwater and stream flows. These improvements, although heartening for some of the environmentalists and state and federal regulators, have not eliminated the dangers inherent in using such massive volumes of water infused with sulfuric acid and other contaminants.[15]

Finally, something should be said about the reordering of the landscape. After 100 years of open-pit mining, the Santa Rita Basin is nothing like it was when that first steam shovel in 1910 dug that first bucket of porphyry ore. Many of the hills below Ben Moore Mountain have literally disappeared and a giant 1.5-mile-wide and 1,500-foot-deep stairstep pit steeply descends into the multicolored bedrock (see figure 6.2). Waste and leach dumps surround the gargantuan gash appearing as well-ordered "mountains" that sit below the Kneeling Nun like a tenuous platform for the revered obelisk-like monument. This 100-hundred-year-long process of degradation, exemplified by the pit and aggradation—the re-piling of the crushed materials into human-made mountains—has reordered the landscape. Those who celebrate the mining operations and modern technology that allowed this reformulation of the landscape view the scene as an example of human achievement. Those who vilify the industry for its environmental destruction view the scene as an open wound surrounded by ugly contusions. Whatever an observer's perspective, nothing can be done to return the formerly bucolic valley to its original state. It is a fait accompli.

To come full circle, all observers of this extraordinary scene should remember the people of and the place called Santa Rita. They and their hometown were defined by copper mining. From the Mimbres peoples of millennia ago to Americans of today this place has a story. That story is one of discovery yet hardship, one of failure as well as success. In that narrative, the community of Santa Rita del Cobre weathered the difficulties of isolation and conflicts with the Apaches under the Spaniards, and then the Mexicans, only to disappear. The rich red metal, however, ensured a rebirth under the Americans who were lured to the new Santa Rita in the late nineteenth century. That town, so impermanent a place for about a century, eventually gained what seemed like

FIGURE E.2. *This June 2007 photograph of the Chino Pit shows the dramatic transformation of the landscape over the last 100 years at the southwestern New Mexico mining enterprise. In the background are pit benches that rise up the side of Ben Moore Mountain just below the Kneeling Nun. Note the two 300-ton diesel haul trucks carrying material out of the pit and another heading back for more.*
COURTESY OF CHRIS HUGGARD.

permanence when a sustained corporate presence initiated in 1909 by the Chino Copper Company emerged in the early twentieth century. Yet fate and the processes of modern open-pit technology again threatened that place's well-being. Kennecott had no choice, if mining was to continue and jobs and profits were to be had, but to dig up that place and literally remove the homes of the longtime copper camp. Santa Rita was transformed into "space." Ultimately, though, memories allow us to reflect on that place and its former inhabitants in the form of the pit, the mountain-size dumps, the ever-present Kneeling Nun, and now this volume. This narrative and photographic account is an attempt to remember Santa

Rita, its community of people, and the mining processes that both sustained it and led to its ultimate demise. For in the end, Santa Rita has been one of New Mexico's most remarkable communities.

NOTES

1. Interhemispheric Resource Center (Silver City, NM), "Copper, Phelps Dodge, and the Future of Grant County's Mining District" (Silver City: October 2001), 4–7, http://www.irc-online.org/. I calculated these amounts based on Phelps Dodge's 1999 figures of about $70 million in payroll that year and another $71 million from sales and taxes that year from both its Chino and Tyrone operations in Grant County.

2. For the full story on the smelter emissions battle, see Christopher J. Huggard, "Mining and the Environment: The Clean Air Issue in New Mexico, 1960–1980," *New Mexico Historical Review* 69, no. 4 (October 1994): 369–388; and "'Squeezing out the Profits': Mining and the Environment in the U.S. West, 1945–2000," in Richard W. Etulain and Ferenc Szasz, eds., *The American West in 2000: Essays in Honor of Gerald D. Nash* (Albuquerque: University of New Mexico Press, 2003), 105–126.

3. Tom Barry, "Kennecott and Silver City: The Rabbits Sneeze and the Workers Cough," *Seer's* (November 15–29, 1975), in the Juan Chacón Collection, Miller Library, Western New Mexico University, Silver City.

4. *Silver City Daily Press*, June 20, 1975.

5. Interview of Herb McGrath, Silver City, New Mexico, December 20, 1993, by Chris Huggard.

6. Interview of John Bartlit, Los Alamos, New Mexico, December 14, 1993, by Chris Huggard; list of the "American Lung Association of New Mexico Committee Appointments, 1981–1982," in Juan Chacón Collection, Miller Library, Western New Mexico University, Silver City, New Mexico.

7. Rene Romo, "'End of an Era': Nostalgic Residents Will Miss Hurley's Iconic Copper Smelter Smokestacks," *Albuquerque Journal*, August 6, 2006; Mary Alice Murphy, "End of an Era: Smokestacks Come Down," *Silver City Daily Press*, June 5, 2007.

8. See Christopher J. Huggard, "Reading the Landscape: Phelps Dodge's Tyrone, New Mexico, in Time and Space," *Journal of the West* 35, no. 4 (October 1996): 35–36, for a discussion of water acquisition strategies of Phelps Dodge in Grant County, New Mexico; IRC, "Copper," 15–16.

9. IRC, "Copper," 17–18.

10. Interview of Herb Yee Toy, Silver City, New Mexico, June 1, 2007, by Chris Huggard.

11. Ibid.

12. Ibid.

13. The New Mexico Environmental Improvement Board or Division was formed in the 1970s and later reorganized and renamed the New Mexico Environment Department.

14. Quoted from IRC, "Copper," 18.

15. IRC, "Copper," 19–20.

Appendices

LIST OF CHURN AND ROTARY DRILLS AT SANTA RITA, 1908–1996

Drill No.	Date in Service	Model	Notes
1	November 1908	Cyclone 4	Sanderson Cyclone Drilling Co.
2	November 1908	Cyclone 4	
3	March 1909	Cyclone 6	capable of drilling 1,000 ft.
4	July 1909	Cyclone 6	converted to gas power in 1930
5	July 1909	Cyclone 6	
6	July 1909	Cyclone 6	
7	July 1909	Cyclone 6	
8	August 1909	Cyclone 6	#1–#8 steam, 15 ft./12 hr. shift
9	May 1910	Cyclone 12	gas-powered, 8 hp engine, 4.5" bit
10	February 1911	Cyclone 12	
11	March 1911	Cyclone 12	
12	March 1911	Cyclone 12	
14	October 1912	Cyclone 12	
15	June 1913	Cyclone 12	
16	June 1913	Cyclone 12	
17	March 1914	Cyclone 12	
18	March 1916	Cyclone 14	gas-powered, 14 hp, 6" bit
19	?	Cyclone 14	
20	August 1917	Cyclone 14	
21	September 1920	?	
22	June 1923	n/a	
23	June 1923	n/a	
24	October 1924	n/a	
25	October 1924	n/a	
26	March 1925	n/a	spudder drill for blast-hole drilling
27	March 1925	n/a	spudder drill for blast-hole drilling
28	December 1925	Cyclone Class E	prospect drill
29	December 1925	Cyclone Class E	prospect drill

Drill No.	Date in Service	Model	Notes
8	August 1934	Bucyrus-Armstrong electric drill	sent to Nevada, October 1935
1	January 1937	BH Model 29T	electric-powered, 6" bit
2	January 1937	BH Model 29T	
3	February 1937	BH Model 29T	
4	April 1937	BH Model 29T	
5	April 1937	BH Model 29T	
6	May 1937	BH Model 29T	
7	May 1937	BH Model 29T	some 9" bits changed to 12" in 1945
8	May 1937	BH Model 29T	
9	May 1937	BH Model 29T	all rigs 12" bits by April 1952
10	May 1937	BH Model 29T	
11	June 1937	BH Model 29T	
12	June 1937	BH Model 29T	
14	February 1939	BH Model 29T	9" inch bit, changed to 12" in 1948
15	1947	BE Model 42T	Bucyrus Erie, 12" bit
16	1949	BE Model 42T	
17	1949	BE Model 42T	
18	1953	50T	Bucyrus Erie, 12" bit
19	1953	50T	
20	1953	50T	
22	May 1955	60 BH	Joy rotary drill, 12" bit
23	November 1956	60 BH	Joy rotary drill, 12" bit
24	November 1956	60 BH	Joy rotary drill, 12" bit

Drill No.	Date in Service	Model	Notes
25	November 1959	60 BH	Joy rotary drill, 12" bit
26	1969	60R	Joy rotary drill, 12" bit
27	1969	60R	Joy rotary drill, 12" bit
28	March 1976	n/a	Franks rotary
29	n/a	n/a	
30	July 1968	BE 60-R	out of service, October 1996
31	n/a	BE 60-R	
32	March 1974	BE 60-R III	
33	June 1975	BE 60-R III	
34	September 1979	M-4	Marion
35	February 1980	M-4	Marion
36	November 1981	M-4	Marion
37	December 1981	T-5	Ingersoll Rand
38	n/a	n/a	
39	n/a	n/a	
40	n/a	n/a	Gardner Denver
41	January 1989	T-5	Ingersoll Rand
42	n/a	T-5	Ingersoll Rand
43	July 1990	DM-H	Ingersoll Rand
44	June 1991	T-5	
45	October 1987	BE 60-R III	used from Ajo
46	August 1989	BE 60-R III	used from Tyrone
47	n/a	n/a	
48	n/a	n/a	
49	n/a	n/a	
50	April 1994	49-RII-120	Bucyrus Erie
51	April 1995	49-RII-120	Bucyrus Erie
52	July 1996	49-RIII	Bucyrus Erie
53	September 1996	49-RIII	Bucyrus Erie

LIST OF CHINO SHOVELS, 1910–2008

No. 1 Steam Shovel. Marion, Model 91. Put in service October 1, 1910. Shovel was equipped with caterpillar crawlers in September 1924 and electrified January 2, 1931. A 3.5-yard bucket (5 tons) was enlarged to 4 yards in 1937. Number was changed to No. 2 on January 1, 1937. This shovel was never modified as a one-man control shovel and was taken out of service May 1940 as the electrical Marion Model 4161 shovels began to arrive.

No. 1 Electric Shovel. Formerly the No. 14 Electric Shovel (see info for that shovel). It was renumbered in 1937 and used until 1950, when it was dismantled.

No. 2 Steam Shovel. Marion Model 91. Put in service September 23, 1910 (first shovel to dig at Chino). Had a 3.5-yard bucket. Taken out of service April 20, 1930.

No. 2 Electric Shovel. Formerly the No. 1 Steam Shovel (see info for that shovel). Parked May 1940.

No. 3 Steam Shovel. Marion Model 91. Put in service November 1910. Shovel was electrified September 17, 1928, and equipped with caterpillar tracks at the same time. The 3.5-yard bucket was enlarged to 4 yards in 1937. Changed over to a one-man control in May 1938. The shovel was parked July 1941.

No. 4 Steam Shovel. Marion Model 91. Put in service December 1910. Shovel was electrified August 6, 1931, and equipped with caterpillar tracks on January 11, 1934. The 3.5-yard bucket was enlarged to 4 yards in 1937. This shovel was parked in July 1942 but was not obsolete until 1945.

No. 5 Steam Shovel. Marion Model 40. Put in service November 1910. This was a small shovel with a 1.5-yard dipper and had traction wheels. The number was changed to No. 10 in July 1913 and called the "dimey."

No. 5 Steam Shovel. Marion Model 91. This shovel was purchased in July 1913. Electrified April 3, 1927, and equipped with caterpillar tracks at the same time. Changed over to a one-man control in March 1934. The 3.5-yard bucket was enlarged to 4 yards in 1937. The shovel was obsolete and parked July 1942.

No. 6 Steam Shovel. Marion Model 91. Put in service February 1, 1912. Had a 3.5-yard bucket. Equipped with caterpillar crawlers in April 1924 but never electrified. Parked in the boneyard on March 21, 1932.

No. 6 Electric Shovel. Formerly No. 15 Steam Shovel, which was renumbered in 1937 (see info for that shovel). Taken out of service in August 1942 and obsolete in 1945.

No. 7 Steam Shovel. Marion Model 100. Put in service December 20, 1912, with a 6-yard dipper and crawler tracks. Shovel had a 50-foot boom. The bucket was changed to a 3.5-yard in May 1917 to conform to the other shovels and then enlarged to 5 yards in January 1934. The shovel was electrified April 25, 1930, and equipped as a one-man control in July 1934. It was parked in October 1941 and obsolete in 1945.

No. 8 Steam Shovel. Marion Model 92. Began working May 27, 1913, with a 3.5-yard bucket. Equipped with crawlers July 1925. Parked in boneyard and used for parts on October 5, 1931.

No. 9 Steam Shovel. Marion Model 92. Put in service June 1913 with a 3.5-yard bucket that was enlarged to 4 yards in 1937. Electrified and equipped with crawlers June 2, 1929, and modified for one-man control in November 1933. Obsolete and parked May 1942.

No. 10 Steam Shovel. Marion Model 40. Formerly the No. 5 shovel, the number was changed to No. 10 when the No. 5 Model 91 was put in service. This small shovel with a 1.5-yard bucket was called the "dimey" and used where a smaller shovel would be more beneficial, including the excavation for the flume in 1927. Taken out of service and parked December 1932.

No. 10 Electric Shovel. Marion Model 4161. Put in service August 25, 1937, with a 5-yard bucket. This was the first of the 4161 fleet that would eventually number nine shovels by 1949. The bucket was increased to a 6-yard size in October 1957. Taken out of service February 25, 1979, after a huge boulder rolled off bank on B shift at 8:30 PM and demolished the cab. The operator lost both of his legs in this accident. The shovel was used for parts to repair the rest of the Marion fleet.

No. 11 Steam Shovel. Marion Model 92. This shovel was evidently assembled from parts and put to work August 16, 1917. It was taken out of service April 20, 1930.

No. 11 Electric Shovel. Marion Model 4161. Put in service December 18, 1937, with a 5-yard bucket. Dipper size increased to 6 yards in October 1957. This shovel was placed on display at the Santa Rita Museum in November 1981. When Phelps Dodge bought Chino in 1987, the shovel was cut up in pieces for scrap.

No. 12 Steam Shovel. Marion Model 92. Purchased December 1923 with crawlers already installed and a 3.5-yard bucket. It was taken out of service July 22, 1931.

No. 12 Electric Shovel. Marion Model 4161. Put in service June 6, 1938, with a 5-yard bucket.

No. 14 Electric Shovel. Marion Model 350 Electric with an 80-foot boom. Purchased June 1924 with an 8-yard bucket, it was Chino's first electric shovel. It was mounted on two sets of wheel trucks and took a double railroad track to work on. It was so large that it loaded cars on the level above the working area. The number was changed to No. 1 on January 1, 1937. ES 1 was used until 1950.

No. 14 Electric Shovel. Marion Model 4161. Put in service March 20, 1940, with a 5-yard bucket.

No. 15 Steam Shovel. Marion Model 92. Bought with caterpillar trucks and put into service May 14, 1925, with a 3.5-yard bucket. It was electrified in 1929 and the

number changed to ES 6 on January 1, 1937. This shovel was parked August 1942.

No. 15 Electric Shovel. Marion Model 4161. Put in service March 15, 1941, with a 5-yard bucket.

No. 16 Diesel Shovel. Bucyrus Model 30B. Put in service June 10, 1926, with a 1-yard bucket. Used for smaller jobs and cleanup in both Santa Rita and Hurley. The number was changed to DS 30 in February 1942 when a Marion 4161—ES 16—was ordered.

No. 16 Electric Shovel. Marion Model 4161. Put in service February 28, 1942, with a 5-yard bucket.

No. 17 Electric Shovel. Marion Model 92 Electric. Put in service January 1927 with a 3.5-yard bucket. It was converted to a one-man control on July 12, 1933. This was only the second shovel Chino purchased that was already electric powered. Its number was changed to ES 8 on January 1, 1937, and taken out of service in July 1941.

No. 17 Gas/Electric. Bought secondhand in 1939 with a 1.25-yard bucket. The number was changed to No. 31 in January 1942. This gas-electric shovel was converted to a straight electric shovel in July 1943.

No. 17 Electric Shovel. Marion Model 4161. Put in service October 13, 1942, with a 5-yard bucket.

No. 18 Electric Shovel. Marion Model 4161. Put in service August 1948 with a 5-yard bucket.

No. 19 Electric Shovel. Marion Model 4161. Put in service October 14, 1949, with a 5-yard bucket.

No. 33 Diesel Shovel. Northwest 2.5-yard bucket. Put in service September 6, 1954.

Note: 4.5- and 5-yard buckets were replaced by 6-yard buckets in 1957 through 1960.

ES20	P&H 1800	In service November 15, 1957	8-yard bucket	Retired
ES21	BE 280B	In service February 1969	15-yard bucket	Retired
ES22	BE ??	In service March 1970	15-yard bucket	Retired January 1993
ES121	BE 280B	From Nevada May 1972	15-yard bucket	?
ES122	BE 280B	From Nevada May 1972	15-yard bucket	?
ES123	P&H 2100	From Nevada May 1972	15-yard bucket	?
ES23	P&H 2100	In service February 1975	17-yard bucket	Retired January 1993
ES24	BE 295B	In service February 1977	20-yard with 17-yard bucket	?
ES25	BE 295B	In service February 1978	17-yard bucket	Retired September 1992
ES26	P&H 2100BL	In service November 1979	17-yard bucket	Retired September 1996
ES27	P&H 2100BL	In service December 1979	17-yard bucket	Retired 1998
ES28	P&H 2100BL	In service April 1981	17-yard bucket	Retired May 1998
ES29	P&H 2100BL	In service November 1981	17-yard bucket	Retired May 1998
ES30	P&H 2100BL	In service September 1983	17-yard bucket	Retired 2006
ES31	P&H 2100BL	In service October 1983	17-yard bucket	To Cobre January 1999
ES32	P&H 2100BL	In service January 1988	17-yard bucket	Retired February 1993
ES33	P&H 2800	In service December 1989	36-yard bucket	To Miami, AZ, 2009
ES34	P&H 2800	In service January 1993	40-yard bucket	To Miami, AZ, 2009
ES41	P&H 4100A	In service May 1996	56-yard bucket	To Morenci December 1998
ES42	BE 495 B1	In service October 1996	56-yard bucket	Parked December 2008

LIST OF CHINO LOCOMOTIVES, 1910–1970

No.	In Service	Description	Notes	Out
1	Aug. 1910	Porter 0-4-0* 15 × 24† steam	weighed 87,000 pounds, could pull 80 tons	Sep. 1920
2	Aug. 1910	Porter 0-4-0, 15 × 24 steam	weighed 87,000 pounds, could pull 80 tons	Apr. 1932
3	Sep. 1910	Porter 0-4-0, 15 × 24 steam	weighed 87,000 pounds, could pull 80 tons	Apr. 1932
4	Sep. 1910	Porter 0-4-0, 15 × 24 steam	weighed 87,000 pounds, could pull 80 tons	1921
5	Dec. 1910	Porter 0-4-0, 15 × 24 steam	weighed 87,000 pounds, could pull 80 tons	Apr. 1932
6	Oct. 1911	Porter 0-4-0, 15 × 24 steam	weighed 87,000 pounds, could pull 80 tons	1921
7	Oct. 1911	Porter 0-4-0, 15 × 24 steam	weighed 87,000 pounds, could pull 80 tons	1926
8	Oct. 1911	Porter 0-4-0, 15 × 24 steam	weighed 87,000 pounds, could pull 80 tons	Nov. 1923
9	Dec. 1911	Porter 0-4-0, 15 × 24 steam	weighed 87,000 pounds, could pull 80 tons	Mar. 1923
10	Mar. 1912	Porter 0-4-0, 15 × 24 steam	weighed 87,000 pounds, could pull 80 tons	Apr. 1923
11	Mar. 1912	Porter 0-4-0, 15 × 24 steam	weighed 87,000 pounds, could pull 80 tons	Apr. 1923
12	Sep. 1912	Porter 0-4-0, 15 × 24 steam	weighed 87,000 pounds, could pull 80 tons	1925
13	Sep. 1912	Porter 0-4-0, 15 × 24 steam	weighed 87,000 pounds, could pull 80 tons	Mar. 1923
14	Nov. 1912	Porter 0-4-0, 15 × 24 steam	weighed 87,000 pounds, could pull 80 tons	1925
15	Nov. 1912	Porter 0-4-0, 16 × 24 steam	—	1934
16	May 1913	Porter 0-4-0, 16 × 24 steam	—	1934
17	May 1913	Porter 0-4-0, 16 × 24 steam	—	1934
18	Jun. 1913	Porter 0-4-0, 16 × 24 steam	—	1934
19	Jun. 1913	Porter 0-4-0, 16 × 24 steam	—	1934
20	Sep. 1914	Porter 0-4-0, 16 × 24 steam	—	1945
21	Sep. 1914	Porter 0-4-0, 16 × 24 steam	—	1945
22	Dec. 1919	American 0-6-0, 21 × 26 steam	75-ton locomotive	May 1940
23	Dec. 1919	American 0-6-0, 21 × 26 steam	—	Sep. 1942
24	Dec. 1919	American 0-6-0, 21 × 26 steam	—	May 1940
25	Oct. 1920	American 0-6-0, 21 × 26 steam	80-ton locomotive	1942
26	Oct. 1920	American 0-6-0, 21 × 26 steam	80-ton locomotive	Oct. 1942
27	Oct. 1920	American 0-6-0, 21 × 26 steam	80-ton locomotive	Dec. 1942
28	Nov. 1923	American 0-6-0, 21 × 26 steam	80-ton locomotive	
29	Nov. 1923	American 0-6-0, 21 × 26 steam	80-ton locomotive	Dec. 1942
30	Nov. 1923	American 0-6-0, 21 × 26 steam	80-ton locomotive	
71	Aug. 1924	American 0-6-0, 21 × 26 steam	from Mesabi, MN	
72	Aug. 1924	American 0-6-0, 21 × 26 steam	—	
73	Aug. 1924	American 0-6-0, 21 × 26 steam	—	
31	?	American 0-6-0, 24 × 26 steam	90-ton locomotive	
32	?	American 0-6-0, 24 × 26 steam	90-ton locomotive	
33	Jan. 1929	Baldwin 0-6-0, 21 × 24 steam	75-ton locomotive	Nov. 1935
34	Jan. 1929	Baldwin 0-6-0, 21 × 24 steam	75-ton locomotive	Nov. 1935
35	Mar. 1929	Baldwin 0-6-0, 21 × 24 steam	75-ton locomotive	Nov. 1935
36	Mar. 1929	Baldwin 0-6-0, 21 × 24 steam	75-ton locomotive	Aug. 1940
37	1929	Baldwin 0-6-0, 21 × 24 steam	75-ton locomotive	Aug. 1940
38	1929	Baldwin 0-6-0, 21 × 24 steam	75-ton locomotive	Aug. 1940
39	1929	Baldwin 0-6-0, 21 × 24 steam	75-ton locomotive	1942
40	1929	Baldwin 0-6-0, 21 × 24 steam	75-ton locomotive	1942

No.	In Service	Description	Notes	Out
50	Apr. 1940	General Electric	GE 85-ton electric locomotive	
51	Apr. 1940	General Electric	GE 85-ton electric locomotive	
52	Apr. 1940	General Electric	GE 85-ton electric locomotive	
53	Apr. 1940	General Electric	GE 85-ton electric locomotive	
54	Apr. 1940	General Electric	GE 85-ton electric locomotive	1969
55	Apr. 1940	General Electric	GE 85-ton electric locomotive	1969
56	Nov. 1940	General Electric	GE 85-ton electric locomotive	1969
57	Nov. 1940	General Electric	GE 85-ton electric locomotive	
58	May 1941	General Electric	GE 85-ton electric locomotive	1969
59	May 1941	General Electric	GE 85-ton electric locomotive	1969
60	Mar. 1942	General Electric	GE 85-ton electric locomotive	
61	Sep. 1942	General Electric	GE 85-ton electric locomotive	
62	Sep. 1942	General Electric	GE 85-ton electric locomotive	1969
80	Jan. 1950	Baldwin diesel electric locomotive	1000 HP 125-ton switch engine	1963
63	Jan. 1952	General Electric	GE 85-ton electric locomotive	1969
64	Jan. 1952	General Electric	GE 85-ton electric locomotive	1969
100	Oct. 1954	General Electric	GE 125-ton electric locomotive	
101	Oct. 1954	General Electric	GE 125-ton electric locomotive	
102	Jun. 1958	General Electric	GE 125-ton electric locomotive	
103	Jun. 1958	General Electric	GE 125-ton electric locomotive	

* Wheels on front truck (if any), drive wheels on locomotive (middle no.), and wheels on rear truck (if any).
† Designates bore and stroke (in inches) of steam drive cylinder.

SANTA RITA WORKFORCE:
NUMBERS OF EMPLOYEES, 1910–2001

1910	764	1933	305	1956	1,192	1979	n/a
1911	1,331	1934	325	1957	1,125	1980	n/a
1912	799	1935	45	1958	992	1981	n/a
1913	1,388	1936	300	1959	n/a	1982	n/a
1914	n/a	1937	415	1960	n/a	1983	n/a
1915	863	1938	639	1961	1,977*	1984	n/a
1916	876	1939	877	1962	1,704*	1985	n/a
1917	1,007	1940	1,010	1963	n/a	1986	n/a
1918	1,068	1941	1,018	1964	n/a	1987	n/a
1919	1,089	1942	907	1965	n/a	1988	n/a
1920	983	1943	816	1966	n/a	1989	n/a
1921	660	1944	842	1967	n/a	1990	n/a
1922	522	1945	770	1968	n/a	1991	n/a
1923	780	1946	810	1969	n/a	1992	1,200*
1924	1,040	1947	842	1970	n/a	1993	n/a
1925	1,171	1948	884	1971	n/a	1994	n/a
1926	1,390	1949	889	1972	n/a	1995	n/a
1927	1,535	1950	999	1973	n/a	1996	1,200*
1928	1,053	1951	1,087	1974	n/a	1997	n/a
1929	1,285	1952	1,141	1975	n/a	1998	1,200*
1930	919	1953	1,119	1976	n/a	1999	898*
1931	660	1954	1,182	1977	n/a	2000	n/a
1932	515	1955	1,205	1978	n/a	2001	990*

* Includes Hurley workers.

FATALITIES AT CHINO MINE PROPERTIES, 1881–2005

Name	Age	Date	Description of Death	Company
John MacDonald	—	11/12/81	Cave-in at the Romero shaft	SRC&ICo.
Thomas Wall	—	04/20/83	Cave-in at Santa Rita	SRC&ICo.
Manuel Lucero	—	01/30/99	Rockfall at the Ivanhoe Mine	Leasing
Ramon Martinez	—	07/05/00	Bad air at the Herbert Dawson lease	Leasing
Aurelio Delgado	—	07/12/00	Cave-in at the Mother Lode Mine	Leasing
Francisco Alvillar	—	01/27/01	Rockfall at the Pedro Mine	Leasing
G. H. "Paddy" Welch	—	05/21/01	Explosion at the Hearst Mine	SRMCo.
Viterbo Cordoba	—	07/29/02	Rockfall	SRMCo.
Amado Robledo	—	11/30/02	Accidently killed by car falling down Chino shaft	Leasing
Jose Jimenez	—	12/??/02	Rockfall at the Cortez shaft	Leasing
Calisto Jimenez	—	12/??/02	Rockfall at the Cortez shaft	Leasing
Jay Hiler	19	01/10/04	Fell and struck head at the SR concentrator	SRMCo.
Narcisso Lascaro	—	04/03/04	Rockfall in a mine at Santa Rita	Leasing
Dionicio Portillo	—	12/24/04	Fell down shaft at Shaft No. 7	SRMCo.
Felix G. Burns	47	08/10/05	Fell down the Wildcat shaft	Lessee
Juan Duran	—	08/22/07	Rockfall at the Ivanhoe Mine	Leasing
Comodore Perry Crawford	62	08/23/07	Fell down the Rita shaft	Lessee
Jose Ladesma	22	07/29/08	Hit with shaft bucket in Whim Hill Mine	Leasing
Guadalupe Espinosa	—	05/27/09	Died from gas in the Messenie Mine	Leasing
John Shannon	49	03/26/10	Crushed in Santa Rita concentrator	CCC
Eugenio Molina	—	06/24/11	Blown up by black powder	CCC
John Stapleton	47	08/21/11	Scalded to death after locomotive collision	CCC
Felis Maris	—	10/07/11	Shovel hit an unexploded blast hole	CCC
Tomas Alvarado	—	02/16/12	Run over by waste train	CCC
Pablo Mendez	—	03/25/12	Smothered in Hurley ore bins	CCC
Alfonso Diaz	64	04/25/12	Run over by ore train in front of Santa Rita Saloon	Not employed by mine
Juan Monarez	—	04/26/12	Suffocated in Hurley fine ore bins	CCC
J. H. Johnson	23	05/12/12	Run over by train in Santa Rita only five hours after being hired	CCC
Joe Hull	60	05/??/12	Killed in cave-in at Copper Rose Mine	Not employed by mine
L. Vanlandinham	—	10/05/12	Electrocuted in Hurley	CCC
Emilio Escarsiga	30	04/15/13	Push car he was riding collided with waste train	CCC
Juan Ventura	21	05/19/13	Run over by ore train in lower yard	CCC
Evarado Huerta	42	06/19/13	Killed by blast of rock	CCC
Porfirio Rodriquez	32	06/21/13	Crushed in cave-in of old mine workings	CCC
Anastacio Vega	22	06/26/13	Died from explosion on June 19	CCC
T. J. McManus	45	07/22/13	Brakeman run over by ore train	CCC
Martin Whitman	19	08/28/13	Crushed when blast pushed shovel back	CCC
M. H. Longnickle	34	08/29/13	Crushed when blast pushed shovel back	CCC
William Byrd	22	08/30/13	Crushed when blast pushed shovel back	CCC
Casper J. Thomas	62	10/31/13	Run over by ore train	CCC
Teodoro Sedillos	28	11/08/13	Run over and killed by Chino Engine No. 2	Not employed by CCC
Lee M. Talbot	22	12/21/13	Brakeman run over by waste train	CCC

Name	Age	Date	Description of Death	Company
Walter E. Bell	30	07/14/15	Loco fireman, run over by Chino train	CCC
H. M. Derr	40	07/26/15	Santa Rita barber, fell into shaft on his property	Not employed by CCC
Dionicio Navarete	—	09/28/15	Fell in Hurley crude ore bin and suffocated	CCC
Francisco Acosta	—	02/01/16	Killed by a stick of dynamite	CCC
Robert S. Herrington	—	02/28/16	Crushed between two railcars	CCC
Demetrio Ortiz	—	04/04/16	Killed while loading a blast hole	CCC
Guillermo Rodriquez	—	05/01/16	Run over in Hurley–Santa Fe yards	CCC
Pedro Rios	—	05/03/16	Powderman killed while loading hole	CCC
T. H. Porter	45	08/26/16	Shovel engineer, dug into a missed hole	CCC
Allen E. Butcher	30	08/27/16	Shovel craneman, dug into a missed hole	CCC
Alvino Alvarado	—	11/22/16	Injured November 18 (no details)	CCC
T. A. Whitley	—	12/27/16	Caught between Engine No. 20 and 20-yard Clark car	CCC
Cruz Serceres	—	02/24/17	Electrocuted at Santa Rita crusher	Contractor
Jose Olorino	—	05/04/17	Suffocated from carload of waste dumped on him	CCC
Ausencio Herrera	—	08/21/17	Struck by rock that rolled down bank	CCC
Ed C. McDowell	—	12/23/17	Steam-shovel foreman killed	CCC
W. J. Gibson	55	01/11/18	Run over by outbound passenger train	CCC
Joseph Worrell	27	02/12/18	Scalded to death in locomotive cab	CCC
Juan Rodriquez	26	06/21/18	Struck by lightning	CCC
David R. Peck	—	02/20/19	Crushed between cars of ore train	CCC
Rafael Herrera	—	04/28/19	Crushed between two locomotive engines	CCC
Jesus H. Martinez	—	05/26/19	Fell from one level to the next, crushing head	CCC
Jose Trinidad Valera	—	07/08/19	Fell under wheels of train and killed	CCC
Harry O. Bayless	38	12/16/20	Powder foreman, accidental powder explosion	CCC
Joseph Barta	40	06/23/22	Brakeman, crushed under string of ore cars	CCC
Robert Logan	—	03/24/23	Electrician, head crushed by steam shovel No. 8	CCC
Ramon Reynosa	—	10/03/23	Track walker, run over by ore train	CCC
Jesus Guiterrez	23	06/01/24	Crushed by winch when truck slid into arroyo	RCCC
Ramon Perez	20	06/03/24	Struck on head by waste car dumping	RCCC
Roberto Castenada	21	07/17/24	Pitman, fell off boom of steam shovel No. 5	RCCC
Jose Chavez	—	12/04/24	Trammer killed by cave-in at Hearst tunnel	RCCC
Ernest Cooke	32	06/11/25	Brakeman, fell from railcar and run over	RCCC
James S. Brady	—	07/17/25	Locomotive engineer, crushed when loco turned over	RCCC
Clyde Lucas	—	08/19/25	Struck by broken swing chain on steam shovel	RCCC
Felipe Gonzales	21	10/09/25	Dumpman, run over by loco engine and killed	RCCC
Nicolas Delgado	—	10/31/25	Machineman, killed ahead of No. 5 shovel	RCCC
Jay F. Bennett	30	01/01/26	Fireman, locomotive No. 15 boiler blew up	RCCC
Francisco Herrera	30	02/06/26	Run over by train near Vanadium	RCCC
Macario Nevarez	—	07/21/26	Dumpman, killed on No. 13 dump	NCCC
Nicanor Losoya	30	10/08/26	Pitman, crushed by rock slide at shovel No. 7	NCCC
Aubrey Bibb	51	10/18/26	General foreman, killed by rock slide at shovel No. 15	NCCC
Manuel Reyes	—	03/14/27	Struck by falling boulder while drilling	NCCC
Ysario Padilla	—	06/02/27	Oiler, killed by rock slide at steam shovel No. 1	NCCC
Macario Lopez	—	01/06/28	Shovel turned over and crushed him	NCCC

Name	Age	Date	Description of Death	Company
Guadalupe Marquez	—	05/26/28	Pitman, killed by locomotive crane No. 2	NCCC
Alfredo Lujan	24	07/19/28	Struck by lightning on the way to work	NCCC
Alejandro Telles	—	10/01/28	Knocked out of bucket in ore shaft	NCCC
Juan Gonzales	36	02/09/29	Struck on head by falling block	NCCC
Refugio Ortega	—	02/11/29	Killed by blasting shot	NCCC
Jose Trejo	—	03/03/29	Cupboard fell on him in machine shop	NCCC
Lafayaette Heinrich	—	10/04/29	Fell on high-tension wire and electrocuted	Contractor
Miguel Parra	—	11/26/29	Fell down man shaft	NCCC
Francisco P. Lucero	29	02/22/37	Struck by rail thrown by locomotive	KCC
William Mosley	49	11/23/37	Driller, died from injuries received November 17	KCC
Roy Brown	21	09/05/38	Electrocuted when No. 5 drill hit high-voltage line	KCC
Luz Rodriquez	39	07/31/39	Dumpman, killed by lightning on No. 11 dump	KCC
Rudolph Montoya	22	11/10/39	Track foreman, killed by falling poles	KCC
Jose Mendoza	57	02/21/40	Car dropper, run over by loco in lower yards	KCC
Thomas Forehand	39	04/25/40	Shovel foreman, caught between two cars of ore train	KCC
Dionicio Apodaca	65	01/31/44	Trackman, run over by waste train	KCC
Thomas P. Lozano	—	07/17/44	Miner, broke back at Ivanhoe	KCC
Eugene B. Thwaits	50	11/15/44	Brakeman, run over by waste train	KCC
Oscar L. Saxon	46	12/31/44	Crane operator, crushed when crane tipped over	KCC
Frutoso Andazola	49	11/03/47	Run over by loaded Euclid truck	Isbell
John Adkins	21	06/06/50	Brakeman, run over by train on New York dump	KCC
Paul Maxwell	25	04/29/51	Crushed when seven-car ore train derailed	KCC
Trinidad Udero	29	10/24/51	Struck by trolley pole being moved	KCC
Manuel Quevedo	25	12/13/51	Killed at Oswaldo No. 2 while riding cage	KCC
William Richardson	52	11/17/53	Driller, hit on head by sheave wheel	KCC
Enrique Pedraza	27	03/31/54	Powder truck blew up	KCC
Andres Gonzales	48	03/31/54	Powder truck blew up	KCC
Jose Portillo	48	03/31/54	Powder truck blew up	KCC
Chesley Chappell	35	03/31/54	Decapitated by powder truck wheel	KCC
Magdaleno Chavez	43	04/04/54	Powder truck blew up, lived four days	KCC
Manuel R. Rodriquez	35	05/11/55	Electrocuted at precipitation plant	KCC
Alfredo C. Correa	34	02/15/59	Track foreman, struck in head by rail	KCC
Saturnino M. Godoy	46	06/06/61	Became entangled in conveyor belt at Hurley	KCC
Natividad Perez	48	12/23/63	Haul truck backed over berm	KCC
Phillip Lustig	51	05/01/65	Contract driller killed on drill rig	Joy
Thurman L. Hook	42	02/08/66	Contract driller killed on drill rig	Lyner-TX
Robert Beck	—	04/11/77	Fell at Hurley acid plant on April 7	KCC
Raul O. Torres	26	01/08/78	Driver of van that was run over by 150-ton truck	KCC
Thomas S. Cisneros	30	01/08/78	Passenger of van run over by 150-ton truck	KCC
Ester Nunez	47	08/09/87	Heavy metal stairway fell on her at Hurley	PD
Felix Nunez	45	02/23/88	Acid pipe broke and covered him at Hurley	PD
Rob McSherry	28	06/08/97	Electrocuted at the SX-EW plant in Santa Rita	PD
Reynaldo M. Delgado	33	07/22/98	Died from slag burns at Hurley smelter	PD
Wm. Scott Johnson	42	11/29/05	Pressurized pontoon exploded at SX-EW plant	PD

PRODUCTION AND PROFITS AT CHINO, 1801–2005

Year	Production in Net Pounds of Copper	Profit in Dollars	Year	Production in Net Pounds of Copper	Profit in Dollars
1801–1904	80,000,000*	n/a	1950	126,815,089	n/a
1904–1909	21,000,000*	n/a	1951	141,602,301	n/a
1910	n/a	n/a	1952	146,345,056	n/a
1911	986,375	n/a	1953	139,055,996	n/a
1912	27,776,080	2,300,000	1954	116,393,958	n/a
1913	50,511,661	2,800,000	1955	126,979,775	n/a
1914	53,999,928	3,100,000	1956	139,258,000	n/a
1915	64,887,788	6,900,000	1957	126,908,000	n/a
1916	72,319,508	12,700,000	1958	108,684,000	n/a
1917	79,636,235	9,500,000	1959	75,070,000	n/a
1918	75,655,641	4,700,000	1960	125,450,000	n/a
1919	40,488,706	1,300,000	1961	148,930,000	n/a
1920	44,051,849	1,300,000	1962	147,366,000	n/a
1921	9,137,282	1,300,000	1963	162,258,000	n/a
1922	28,406,314	800,000	1964	162,744,000	n/a
1923	54,261,228	1,300,000	1965	181,886,000	n/a
1924	66,574,486	1,600,000	1966	210,546,000	n/a
1925	67,385,620	included with RCCC	1967	nine-month strike	n/a
1926	73,569,456	2,600,000	1968	189,468,000	n/a
1927	63,739,523	1,600,000	1969	191,032,000	n/a
1928	79,416,071	3,600,000	1970	160,106,000	n/a
1929	85,127,931	6,100,000	1971	142,938,000	n/a
1930	56,702,719	1,400,000	1972	216,806,000	n/a
1931	52,988,067	(500,000)	1973	135,672,000	n/a
1932	23,235,445	(1,000,000)	1974	121,114,000	n/a
1933	22,165,620	(323,000)	1975	196,386,000	n/a
1934	19,456,905	(52,000)	1976	114,404,000	n/a
1935	none	(23,000)	1977	114,526,000	n/a
1936	none	n/a	1978	119,104,000	n/a
1937	55,605,074	2,010,000	1979	123,996,000	n/a
1938	32,509,324	849,000	1980	101,410,000	n/a
1939	73,088,394	profits after this always	1981	73,408,000	n/a
1940	129,762,270	include all KCC properties	1982	65,744,000	(189,000,000)
1941	128,124,673	n/a	1983	142,000,000	(91,000,000)
1942	146,002,473	n/a	1984	129,200,000	(160,000,000)
1943	144,425,552	n/a	1985	n/a	n/a
1944	129,851,602	n/a	1986	n/a	61,400,000
1945	107,965,697	n/a	1987	217,600,000	205,700,000
1946	95,480,353	n/a	1988	215,466,000	420,200,000
1947	113,078,253	n/a	1989	233,866,000	267,000,000
1948	143,196,111	n/a	1990	293,000,000	454,900,000
1949	105,332,274	n/a	1991	314,800,000	272,900,000

Year	Production in Net Pounds of Copper	Profit in Dollars	Year	Production in Net Pounds of Copper	Profit in Dollars
1992	n/a	221,700,000	2000	271,000,000	73,200,000
1993	319,000,000	187,900,000	2001	n/a	(275,000,000)
1994	319,000,000	271,000,000[†]	2002	107,600,000	(341,600,000)
1995	337,400,000	n/a	2003	79,800,000	n/a
1996	337,400,000	n/a	2004	n/a	n/a
1997	332,000,000	444,100,000	2005	210,000,000	n/a
1998	315,800,000	190,900,000	2006	n/a	n/a
1999	n/a	(257,800,000)	2007	n/a	n/a

* Estimated.
[†] Phelps Dodge net profit.

DAILY WAGE RATES AT CHINO, 1912–1996

Date	Trackman, Pitman, Laborer	Fireman	Brakeman	Crafts*	Churn Driller	Loco Engineer	Shovel Engineer
Apr. 1912	2.00	3.00	3.00	5.00	5.25	4.00	5.85
Feb. 1916	2.25	3.30	3.30	5.25	5.50	4.95	6.20
Jul. 1918	2.75	3.80	3.80	5.75	6.00	5.45	6.70
Feb. 1919	2.00	3.00	3.00	5.00	5.25	4.75	5.85
Apr. 1920†	2.65–3.70	4.00–5.00	4.15–5.15	5.50–6.50	5.50–6.50	5.25–6.25	7.00–8.00
Jan. 1923	2.65	4.00	4.00	5.50	5.25	5.25	7.00
Jan. 1924	2.50	4.00	4.00	5.50	5.25	5.25	7.00
Feb. 1937	—	—	—	—	5.75	—	—
Apr. 1937	—	—	—	—	6.25	—	—
Jan. 1941	3.85	5.30	5.30	6.75	6.75	6.75	9.40
Jan. 1943	5.10	6.56	6.81	8.00	8.00	8.00	10.15
Jan. 1945	5.36	6.81	7.56	8.26	8.26	8.26	10.41
Jan. 1948	8.32	9.47	9.47	10.70	10.70	10.70	13.85

Date	Trackman, Pitman, Laborer	Fireman	Brakeman	Truck Mechanic	Churn Driller	Truck Driver	Shovel Engineer
Oct. 1950	10.48	—	12.10	12.86	11.66	12.10	15.01
Jul. 1951	11.12	—	13.92	14.72	14.72	13.52	16.72
Jan. 1953	12.06	—	14.86	15.66	15.66	14.86	17.66
Aug. 1955	13.98	—	17.14	18.08	16.92	18.55	20.43

Date	Trackman, Pitman, Laborer	Fireman	Brakeman	Crafts*	Rotary Driller	Loco Engineer	Shovel Engineer
Jul. 1957	15.26	—	18.70	19.72	20.74	—	23.80
Dec. 1959	—	—	20.04	—	22.16	—	—
Mar. 1968	21.84	—	23.23	28.86	—	28.32	—
Mar. 1970	23.92	—	27.64	31.98	36.13	31.36	32.60
Jul. 1973	29.92	—	34.58	31.36	38.46	39.23	40.78
Jul. 1977	50.80	—	—	64.39	60.50	60.50	66.33
Jul. 1979	54.16	—	—	68.71	64.55	66.63	70.79
Jul. 1982	77.32	—	—	95.91	89.00	92.19	100.86
Jul. 1986	62.80	—	—	96.00	79.52	84.64	102.00
Jul. 1991	71.60	—	—	114.00	98.80	102.80	118.80
Jul. 1996	78.00	—	—	122.40	106.40	110.40	127.20

Note: All wages are in dollars.
* Crafts included boilermakers, machinists, carpenters, welders, and electricians.
† These wages were based on a sliding-scale depending on copper prices (from 16 cents to 27 cents per pound).

Bibliography

MANUSCRIPT COLLECTIONS

Stanford University, Stanford, California

Daniel C. Jackling Papers. Special Collections. Stanford University Library.

University of Colorado, Boulder

Papers of the International Union of Mine, Mill, and Smelter Workers, Local 890. Western History Collection Archives.

University of Texas at El Paso

Archivos del Ayuntamiento de Chihuahua.
Archivos del Historicos de Janos.

University of Texas, Austin

Archivos Historicos de Janos.

Western New Mexico University, Silver City

Dale Giese Collection.
Fort Webster Post Returns.
Jack Cargill Empire Zinc Collection.
Juan Chacón Collection.

PHOTOGRAPH COLLECTIONS

New Mexico Bureau of Mines and Mineral Resources Photograph and Slide Collection. Socorro, New Mexico.

Palace of the Governors Photograph Archives, Museum of New Mexico. Santa Fe, New Mexico.

Private Collection of Christopher J. Huggard. Fayetteville, Arkansas.

Private Collection of Don Turner. Silver City, New Mexico.

Private Collection of Terrence M. Humble. Bayard, New Mexico.

Rio Grande Historical Collection, New Mexico State University Library, Archives and Special Collections. Las Cruces, New Mexico.

Silver City Museum Photograph Collection. Silver City, New Mexico.

NEWSPAPERS AND JOURNALS

Albuquerque Tribune.
Chinorama (Kennecott Copper Corporation, Chino Mines Division publication, 1955–1971).

The Daily Southwest (Silver City, NM).

The Grant County Herald (Silver City, NM).

Mesilla Times (Mesilla, NM).

Mining Life (Silver City, NM).

New Mexico Miner and Prospector (Santa Fe, NM).

New Southwest (Silver City, NM).

Pay Dirt (Bisbee, AZ).

Phelps Dodge Today (New York, NY).

Santa Fe Weekly Gazette.

The Sentinel (Silver City, NM).

The Silver City Daily Press and Independent.

The Silver City Enterprise.

The Southwest (Silver City, NM).

Southwestern Mines (city unknown).

Southwest Sentinel (Silver City, NM).

This Is Chino (Kennecott Copper Corporation, Chino Mines Division publication, 1960s–1980s).

Transactions of the American Institute of Mining Engineers (New York, NY).

The Tribune (Silver City, NM).

Zinc (New Jersey Zinc newsletter, 1916–1929, 1939, 1948–1949, 1955–1956).

ATLASES AND MAPS

Beck, Warren A., and Ynez D. Haase. *Historical Atlas of New Mexico*. Norman: University of Oklahoma Press, 1969.

Chino Copper Company Claim Map Showing Ore Bodies, Stripped Areas, Ore and Waste Dumps and Railroad Tracks, January 1, 1916. Chino Copper Company, Annual Report. 1917.

"Map of Buildings and Water Lines on Company Property." Kennecott Copper Corporation, Chino Mines Division. Santa Rita, New Mexico, 1957.

"Map Showing the Property of the Santa Rita Copper & Iron Co. and the Carrasco Copper Co., Grant County, New Mexico, 1881." Grant County Clerk's Office, Silver City, New Mexico.

Sketch of Survey of Santa Rita del Cobre Mineral Claim, 1869. Historic American Buildings Survey, National Park Service, Santa Fe, New Mexico.

US Department of Agriculture. *Gila National Forest Map*. Albuquerque: US Forest Service, Southwest Region, 1990.

Walker, Juan Pedro. "Mapa Geografico, Que comprehende los Terrenos de las P de I, Coahuala, Nueva Biscaya y NM, que de orden del Sor Comte Gral de todas las ynternas de N España, Don N. Salcedo y Salcedo, ha reconocido el Alferez de la Co. Presidial de Janos, 1805." Map Collection, University of Texas at El Paso.

Williams, Jerry L., ed. *New Mexico in Maps*. Albuquerque: University of New Mexico Press, 1986.

PUBLISHED SOURCES

Ackerly, Neal W. *A History of the Kneeling Nun: Evidence from Documentary Sources*. Silver City, NM: Dos Rios Consultants, 1999.

Ailman, Harry B. *Pioneering in Territorial Silver City, 1871–1892*. Albuquerque: University of New Mexico Press, 1983.

Alexander, Bob. "And We Say, Well Done Dan!" *National Association for Outlaw and Lawman History Quarterly* 27, no. 1 (January–March 2003): 7–12.

Allen, James B. *The Company Town in the American West*. Norman: University of Oklahoma Press, 1966.

Almada, Francisco R. *Diccionario de Historia, Geografia y Biografia Chihuahuenses*. 2nd ed. Cuidad Juárez: Impresora de Juárez, S.A., 1968.

———. *Gobernadores del Estado de Chihuahua*. Mexico City, DF: Imprinta de la H. Camara de Deputados, 1950.

Arny, W.F.M. *Interesting Items Regarding New Mexico: Its Agriculture, Pastoral and Mineral Resources, People, Climate, Soil, Scenery, Etc*. Santa Fe, NM: Manderfield and Tucker Printers, 1873.

Arrington, Leonard J., and Gary B. Hansen. *"The Richest Hole on Earth": A History of the Bingham Copper Mine*. Logan: Utah State University Press, 1963.

Arrowsmith, Rex. *Mines of the Old Southwest: Early Reports on the Mines of New Mexico and Arizona by Explorers Abert and Others*. Santa Fe, NM: Stagecoach Press, 1963.

Bahre, Conrad J., and Charles F. Hutchison. "The Impact of Historic Fuelwood Cutting on the Semidesert Woodlands of Southeastern Arizona." *Journal of Forest History* 29 (October 1985): 175–186.

Baker, Ellen. *On Strike and on Film: Mexican American Families and Blacklisted Filmmakers in Cold War America*. Chapel Hill: University of North Carolina Press, 2007.

Bakewell, Peter J. *Silver Mining and Society in Colonial Mexico: Zacatecas, 1546–1700*. Cambridge: Cambridge University Press, 1971.

Ball, Eve. *In the Days of Victorio: Recollections of a Warm Springs Apache*. Tucson: University of Arizona Press, 1970.

———. *Indeh: An Apache Odyssey*. Provo, UT: Brigham Young University Press, 1980.

Bancroft, Hubert Howe. *History of Arizona and New Mexico, 1530–1888*. San Francisco: The History Company, 1888.

Barrett, Elinore. *The Mexican Colonial Copper Industry*. Albuquerque: University of New Mexico Press, 1987.

Bartlett, John Russell. *Personal Narrative of Explorations and Incidents in Texas, New Mexico, California, Sonora, and Chihuahua, Connected with the United States and Mexican Boundary Commission, during the Years 1850, '51, '52 and '53*. New York: D. Appleton and Co., 1854.

Basso, Keith, ed. *Western Apache Raiding and Warfare: From the Notes of Grenville Goodwin*. Tucson: University of Arizona Press, 1987.

Bateman, Richard. *James Pattie's West: The Dream and the Reality*. Norman: University of Oklahoma Press, 1986.

Bender, A. B. "Frontier Defense in the Territory of New Mexico, 1846–1853." *New Mexico Historical Review* 9 (1934): 249–272.

Bennett, James A. "Diary of Corporal James Bennett." *New Mexico Historical Review* 22 (1947): 84–176.

———, Clinton E. Brooks, and Frank D. Reeve, eds. *Forts and Forays: A Dragoon in New Mexico*. Albuquerque: University of New Mexico Press, 1948.

Berry, Susan. *Built to Last: An Architectural History of Silver City, New Mexico*. Santa Fe: New Mexico Historic Preservation Divisions, 1986.

Bradfield, Wesley. "The Cameron Creek Village." *El Palacio* 24 (1928): 150–160.

Briggs, Walter. *Without Noise of Arms: The 1776 Dominguez-Escalante Search for a Route from Santa Fe to Monterey*. Flagstaff, AZ: Northland Press, 1976.

Brinsmade, Robert B. "Mining and Milling near Silver City, New Mexico." *Mining World* (December 26, 1908): 942–950.

Brody, Jacob J. *Mimbres Painted Pottery*. Santa Fe, NM: School of American Research, 1977.

Brough, P. V. "Concentrating Lead-Zinc Ore at the Bayard Mill." *Mining and Metallurgy* 29 (1948): 562–566.

Brown, Ronald C. *Hard-Rock Miners: The Intermountain West, 1860–1920*. College Station: Texas A&M University Press, 1979.

Bryan, Bruce. "The Galaz Ruin in the Mimbres Valley." *El Palacio* 23 (1927): 323–337.

Byrkit, James W. *Forging the Copper Collar: Arizona's Labor-Management War of 1901–1921*. Tucson: University of Arizona Press, 1982.

Calvin, Ross, ed. *Lieutenant Emory Reports: A Reprint of Lieutenant W. H. Emory's Notes of a Military Reconnaissance*. Albuquerque: University of New Mexico Press, 1951; reprint of 1848 edition.

Campbell, Robert F. *The History of Basic Metals: Price Control in World War II*. New York: Columbia University Press, 1948.

Cargill, Jack. "Empire and Opposition: 'Salt of the Earth' Strike." In *Labor in New Mexico: Unions, Strikes, and Social History since 1881*, edited by Robert Kern, 183–267. Albuquerque: University of New Mexico Press, 1983.

Chamberlain, Kathleen P. *Victorio: Apache Warrior and Chief*. Norman: University of Oklahoma Press, 2007.

Clark, Ira G. *Water in New Mexico: A History of Its Management and Use*. Albuquerque: University of New Mexico Press, 1987.

Cleland, Robert Glass. *A History of Phelps Dodge, 1834–1950*. New York: Alfred Knopf, 1952.

Clifford, James O. "Interesting Review of Chino's Mines and Methods." *Mines and Methods* (August 1912): 547–552.

Cremony, John C. *Life among the Apaches*. Lincoln: University of Nebraska Press, 1983.

Cutts, James M. *The Conquest of California and New Mexico*. Albuquerque: Horn & Wallace, 1965.

Davis, Carolyn O'Bagy. *Treasured Earth: Hattie Cosgrove's Mimbres Archaeology in the American Southwest*. Tucson, AZ: Sanpete Publications, 1995.

Dempsey, Stanley, and James E. Fell. *Mining the Summit: Colorado's Ten Mile District*. Norman: University of Oklahoma Press, 1986.

Deutsche, Sarah. *No Separate Refuge: Culture, Class, and Gender on an Anglo-Hispanic Frontier in the American Southwest, 1880 1940*. New York: Oxford University Press, 1987.

Dinsmore, Charles A. "Mining in the Santa Rita District, New Mexico." *Mining World* (November 20, 1909): 1028–1029.

Douglas, James. "The Copper Resources of the United States." *Transactions of the American Institute of Mining Engineers* 19 (May–February 1890–1891): 700–701.

Duriez, Leo H., and James V. Neuman. "Geology and Mining Practice at the Bayard, New Mexico, Property." *Mining and Metallurgy* (October 1948): 559–561.

Elliott, Russell. *Nevada's Twentieth-Century Mining Boom: Tonopah-Goldfield-Ely.* Reno: University of Nevada Press, 1965.

Emmons, David M. *The Butte Irish: Class and Ethnicity in an American Mining Town.* Urbana: University of Illinois Press, 1989.

Faulk, Odie B. "The Presidio: Fortress or Farce." *Journal of the West* 8, no. 1 (1969): 20–32.

———. *Too Far North . . . Too Far South.* Los Angeles: Westernlore Press, 1967.

Fell, James E. *Ores to Metals: The Rocky Mountain Smelting Industry.* Lincoln: University of Nebraska Press, 1979.

Francaviglia, Richard V. *Hard Places: Reading the Landscape of America's Historic Mining Districts.* Iowa City: University of Iowa Press, 1991.

Gardner, E. D., and C. H. Johnson. *Copper Resources of the World.* Vol. 2. Wenasha, WI: George Bantor Publishing Co., 1935.

Giese, Dale F. *Forts of New Mexico, Echoes of the Bugle.* Silver City, NM: Dale Giese, 1991.

Goetzmann, William H. *Army Exploration in the American West, 1803–1863.* New Haven, CT: Yale University Press, 1959.

———. *Exploration and Empire: The Explorer and the Scientist in the Winning of the American West.* New York: Knopf, 1966.

Golley, Frank B. "James Baird: Early Santa Fe Trader." *Missouri Historical Society Bulletin* 5, no. 3 (1959): 181–196.

Graf, William L. "Mining and Channel Response." *Annals of the Association of American Geographers* 69 (June 1979): 262–275.

Grauman, Melody Webb. "Kennecott: Alaska Origins of a Copper Empire, 1900–1938." *Western Historical Quarterly* 9 (April 1978): 197–211.

Greever, William S. *The Bonanza West: The Story of the Western Mining Rushes, 1848–1900.* Norman: University of Oklahoma Press, 1963.

Griffin, William B. "The Compás: A Chiricahua Apache Family of the Late 18th and Early 19th Centuries." *The American Indian Quarterly* 7, no. 2 (1983): 22–45.

Gutiérrez, David G. "Significance to Whom? Mexican Americans and the History of the American West." *Western Historical Quarterly* 24 (November 1993): 519–539.

Halleck, Henry W., ed. *A Collection of Mining Laws of Spain and Mexico.* San Francisco: University of California, 1859.

Harper, David C. *Coins and Prices: A Guide to U.S., Canadian, and Mexican Coins.* Iola, WI: Krause Publications, 1993.

Hays, Samuel P. *Beauty, Health, and Permanence: Environmental Politics in the United States, 1955–1985.* Cambridge: Cambridge University Press, 1987.

———. *Conservation and the Gospel of Efficiency: The Progressive Conservation Movement, 1890–1920.* Boston: Harvard University Press, 1959.

Herfindall, Ovis Clemens. *Copper Costs and Prices, 1870–1957.* Baltimore: Johns Hopkins University Press, 1954.

Hildebrand, George H., and Garth L. Mangum. *Capital and Labor in American Copper, 1845–1990: Linkages between Product and Labor Markets.* Cambridge, MA: Harvard University Press, 1992.

Hollon, W. Eugene. *The Lost Pathfinder: Zebulon Montgomery Pike.* Norman: University of Oklahoma Press, 1949.

Holmes, Jack E. *Politics in New Mexico.* Albuquerque: University of New Mexico Press, 1967.

Huggard, Christopher J. "Copper Mining in New Mexico, 1900–1945." In *Essays in Twentieth-Century New Mexico History,* edited by Judy Boyce DeMark, 43–62. Albuquerque: University of New Mexico Press, 1994.

———. "The Impact of Mining on the Environment of Grant County, New Mexico, to 1910." *Annual of the Mining History Association* (1994): 2–8.

———. "Mining and the Environment: The Clean Air Issue in New Mexico, 1960–1980." *New Mexico Historical Review* 69, no. 4 (October 1994): 369–388.

———. "Reading the Landscape: Phelps Dodge's Tyrone, New Mexico, in Time and Space." *Journal of the West* 35, no. 4 (October 1996): 29–39.

———. "'Squeezing Out the Profits': Mining and the Environment in the U.S. West, 1945–2000." In *The American West in 2000: Essays in Honor of Gerald D. Nash*, edited by Richard W. Etulain and Ferenc Szasz, 105–126. Albuquerque: University of New Mexico Press, 2003.

Huginnie, A. Yvette. "A New Hero Comes to Town: The Anglo Mining Engineer and 'Mexican Labor' as Contested Terrain in Southeastern Arizona, 1880–1920." *New Mexico Historical Review* 69, no. 4 (October 1994): 323–344.

Humble, Terrence M. "The Chino Bandits." *Quarterly of the National Association for Outlaw and Lawman History* 29, no. 1 (January–June 2005): 8–14.

———. "The Pinder-Slip Mining Claim Dispute of Santa Rita, New Mexico, 1881–1912." *The Mining History Journal* 3 (1996): 90–100.

Huttl, John B. "Chino Today." *Engineering and Mining Journal* 140, no. 9 (September 1939): 29–33.

Hyde, Charles K. *Copper for America: The United States Copper Industry from Colonial Times to the 1990s.* Tucson: University of Arizona Press, 1998.

Interhemispheric Resource Center. "Copper, Phelps Dodge, and the Future of Grant County's Mining District." Silver City, NM: IRC, October 2001. Available at http://www.irc-online.org/.

Jackson, Donald, ed. *The Journals of Zebulon Montgomery Pike; With Letters and Related Documents.* Norman: University of Oklahoma Press, 1966.

Jackson, W. Turrentine. *Wagon Roads West: A Study of Federal Road Surveys and Construction in the Trans-Mississippi West, 1846–1869.* New Haven, CT: Yale University Press, 1964.

Jensen, Vernon. *Heritage of Conflict: Labor Relations in the Non-ferrous Metals Industry up to 1930.* Ithaca, NY: Cornell University Press, 1934.

———. *Nonferrous Metals Industry Unionism: 1932–1954.* Ithaca, NY: Cornell University Press, 1954.

Jones, Fayette A. *New Mexico: Mines and Minerals.* Santa Fe: New Mexico Printing Co., 1904.

———. *Old Mining Camps of New Mexico, 1854–1904.* Santa Fe, NM: Stagecoach Press, 1964.

Jones, Paul M. *Memories of Santa Rita.* Silver City, NM: Silver City Enterprise, 1985.

Joralemon, Ira B. *Copper: The Encompassing Story of Mankind's First Metal.* Berkeley, CA: Howell North Books, 1973.

Kern, Robert W., ed. *Labor in New Mexico: Strikes, Union, and Social History, 1881–1981.* Albuquerque: University of New Mexico Press, 1983.

Kiner, Ralph, with Joe Gergen. *Kiner's Korner, at Bat and on the Air: My 40 Years in Baseball.* New York: Arbor House, 1987.

Kiner, Ralph, Danny Peary, and Tom Seaver. *Baseball Forever: Reflections on Sixty Years in the Game.* New York: Triumph Books, 2004.

Kingsolver, Barbara. *Holding the Line: Women in the Great Arizona Mine Strike of 1983.* Ithaca, NY: ILR Press, 1989.

Kluger, James R. *The Clifton-Morenci Strike: Labor Difficulty in Arizona, 1915–1916.* Tucson: University of Arizona Press, 1970.

Lamar, Howard R. *The Far Southwest, 1846–1912.* New York: W. W. Norton & Co., 1966.

Lawson, Thomas W. *Report on the Property of the Santa Rita Mining Company.* Santa Rita, NM: Santa Rita Mining Co., 1906.

Lorence, James J. *The Suppression of Salt of the Earth: How Hollywood, Big Labor, and Politicians Blacklisted a Movie in Cold War America.* Albuquerque: University of New Mexico Press, 1999.

Lundwall, Helen J. *Pioneering in Territorial Silver City.* Albuquerque: University of New Mexico Press, 1983.

MacDonald, Bryce, and Moshe Weiss. "Impact of Environmental Control Expenditures on Copper, Lead, and Zinc Producers." *Mining Congress Journal* (January 1978): 45–50.

Malone, Michael P. "The Close of the Copper Century." *Montana: The Magazine of Western History* 35 (Spring 1985): 69–72.

Marcosson, Isaac F. *Anaconda.* New York: Dodd & Mead, 1957.

McBride, James. "La Liga Protectora Latina: A Mexican-American Benevolent Society in Arizona." *Journal of the West* 14, no. 4 (October 1975): 82–90.

McClure, R. C. "Condition and Administration of the Gila River Forest Reserve, New Mexico." *Twelfth National Irrigation Congress* (November 1904): 277–283.

McGraw, William Cochran. *Savage Scene: The Life and Times of Mountain Man Jim Kirker.* New York: Hastings House, 1972.

Meier, Matt S., and Feliciano Rivera. *The Chicanos: A History of Mexican Americans*. New York: Hill & Wang, 1972.

Meinig, Donald W. *The Shaping of America: A Geographical Perspective on 500 Years of History,* Vol. 2: *Continental America, 1800–1867*. New Haven, CT: Yale University Press, 1993.

———. *Southwest: Three Peoples in Geographic Change, 1600–1970*. New York: Oxford University Press, 1971.

Mellinger, Philip J. *Race and Labor in Western Copper: The Fight for Equality, 1896–1918*. Tucson: University of Arizona Press, 1995.

Melzer, Richard. "Phelps Dodge Knows Best: Welfare Capitalism in a New Mexico Camp, Dawson, 1920–1929." *Southwest Economy and Society* 6 (Fall 1982): 12–34.

Mercier, Laurie. *Anaconda: Labor, Community, and Culture in Montana's Smelter City*. Urbana: University of Illinois Press, 2001.

Miller, Darlis A. *The California Column in New Mexico*. Albuquerque: University of New Mexico Press, 1982.

Miller, Richard A. *Fortune Built by Gun: The Joel Parker Whitney Story*. Walnut Grove, CA: Mansion Publishing Co., 1969.

Montejano, David. *Anglos and Mexicans in the Making of Texas, 1836–1986*. Austin: University of Texas Press, 1987.

Moorhead, Max. "Spanish Transportation in the Southwest, 1540–1846." *New Mexico Historical Review* 32, no. 2 (1957): 107–122.

Mowry, Sylvester. *Arizona and Sonora: The Geography, History, and Resources of the Silver Region of North America*. New York: Harper & Bros., 1866.

Muñoz, Ricardo. *The Kneeling Nun: My Eternal Mother*. Las Cruces, NM: Del Valle Printing & Graphics, 1983.

Myers, Lee. "Fort Webster on the Mimbres River." *New Mexico Historical Review* 41, no. 1 (1966): 47–57.

———. "Military Establishments in Southwest New Mexico: Stepping Stones to Settlement." *New Mexico Historical Review* 43 (1968): 5–49.

Naegle, Conrad Keller. "The Rebellion of Grant County, New Mexico in 1876." *Arizona and the West* 10 (Autumn 1968): 225–240.

Nash, Gerald D. *World War II and the West: Reshaping the Economy*. Lincoln: University of Nebraska Press, 1990.

Navin, Thomas R. *Copper Mining and Management*. Tucson: University of Arizona Press, 1978.

Nymeyer, Donna. *John Vincent Mitchell*. Tucson, AZ: Nymeyer, 2001.

Ogle, Ralph. *Federal Control of the Western Apaches, 1848–1886*. Albuquerque: University of New Mexico Press, 1940.

Opler, Morris E. "Chiricuahua Apaches." In *Handbook of North American Indians*, edited by Alfonso Ortiz, 10: 399–421. Washington, DC: Smithsonian Institution, 1983.

Parsons, Arthur B. *The Porphyry Coppers*. New York: A. B. Parsons, 1933.

———. *The Porphyry Coppers to 1956*. New York: American Institute of Mining, Metallurgical, and Petroleum Engineers, 1957.

Pattie, James O. *The Personal Narrative of James O. Pattie of Kentucky*. Cincinnati: E. N. Flint, 1833.

Petrik, Paula. *No Step Backward: Women and Family on the Rocky Mountain Mining Frontier*. Helena: Montana Historical Society Press, 1987.

Quaife, Milo Milton, ed. *Kit Carson's Autobiography*. Lincoln: University of Nebraska Press, 1966.

Reichwein, George S. *John M. Reichwein: A Piece of New Mexico History*. St. Louis: Reichwein, 1990.

Rich, John L. "Recent Stream Trenching in the Semi-arid Portion of Southwestern New Mexico: A Result of Removal of Vegetation Cover." *American Journal of Science* 32 (1911): 237–245.

Richardson, J. K. "Kennecott Copper Corporation Chino Mines." *Professional Engineer* (December 1957): 8–11.

Richter, F. Ernest. "The Amalgamated Copper Company: A Closed Chapter in Corporate Finance." *The Quarterly Journal of Economics* 30, no. 2 (February 1916): 387–407.

Rickard, Thomas A. "The Chino Enterprise—Part I." *Engineering and Mining Journal-Press* 116, no. 18 (November 3, 1923): 753–758.

———. "The Chino Enterprise—Part II." *Engineering and Mining Journal-Press* 116, no. 19 (November 10, 1923): 803–810.

———. "The Chino Enterprise—Part III." *Engineering and Mining Journal-Press* 116, no. 23 (December 8, 1923): 981–985.

———. "The Chino Enterprise—Part IV." *Engineering and Mining Journal-Press* 116, no. 26 (December 20, 1923): 1113–1121.

———. "The Chino Enterprise—Part V." *Engineering and Mining Journal-Press* 117, no. 1 (January 5, 1924): 13–20.

Robertson, David. *Hard as the Rock Itself: Place and Identity in the American Mining Town.* Boulder: University Press of Colorado, 2006.

Robinson, Sherry. "Mining Slowdown Doesn't Tarnish Silver City's Prospects." *New Mexico Magazine* 72 (January 1994): 30–36.

Rose, Arthur W. *The Porphyry Copper Deposit at Santa Rita, New Mexico.* Tucson: University of Arizona Press, 1966.

Roth, Leland M. "Company Towns in the Western United States." In *The Company Town: Architecture and Society in the Early Industrial Age,* edited by John S. Garner, 173–206. New York: Oxford University Press, 1992.

Sandwich, Brian. *The Great Western: Legendary Lady of the Southwest.* Southwest Studies 94. El Paso: Texas Western Press, 1990.

Scamehorn, H. Lee. *Mill & Mine: The CF&I in the Twentieth Century.* Lincoln: University of Nebraska Press, 1992.

Shimmin, John T. "The Hurley Mill, Water Supply and Power Plant." American Mining Congress, New Mexico Chapter, Santa Rita Meeting, July 25–26, 1927.

Simmons, Marc. *New Mexico: An Interpretive History.* Albuquerque: University of New Mexico Press, 1977.

Smith, Duane A. "How It Worked: The Stamp Mill." *Montana: The Magazine of Western History* 58, no. 1 (Spring 2008): 63–65.

———. *Mining America: The Industry and the Environment, 1800–1980.* Lawrence: University Press of Kansas, 1987.

———. *Rocky Mountain Mining Camps: The Urban Frontier.* Lincoln: University of Nebraska Press, 1967.

Smith, Ralph. "Apache Plunder Trails Southward, 1831–1840." *New Mexico Historical Review* 37 (June 1962): 20–42.

Smith, Ralph Adam. *Borderlander: The Life of James Kirker, 1793–1852.* Norman: University of Oklahoma Press, 1999.

Snodgrass, O. T. *Realistic Art and Times of the Mimbres Indians.* El Paso, TX: O. T. Snodgrass, 1975.

Spence, Clark C. *Mining Engineers and the American West: The Lace-Boot Brigade.* New Haven, CT: Yale University Press, 1970.

Spude, Robert L. "The Santa Rita del Cobre: The Early American Period, 1846–1886." *The Mining History Journal* 6 (1999): 8–38.

Steinberg, Sheila L. "Santa Rita, New Mexico, Community Report." Community Studies Series Report No. 4. Department of Sociology, Humboldt State University, March 2003.

Strickland, Rex W. "The Birth and Death of a Legend: The Johnson 'Massacre' of 1837." *Arizona and the West* 18, no. 3 (1976): 257–286.

———. "Robert McKnight." In *The Mountain Men and the Fur Trade of the Far West,* vol. 9, edited by LeRoy Hafen, 259–268. Glendale, CA: The Arthur H. Clark Co., 1972.

Sully, John M. *Chino Copper Company: The Story of the Santa Rita del Cobre Grant and Its Development from Discovery to Present Date.* San Diego: Resources and Industries of the Sunshine State, Panama-California Exposition, 1915.

———. "Milling the Ore of the Chino Mine." *Mining and Scientific Press* (March 30, 1912): 464–466.

———. "The Story of the Santa Rita Copper Mine." *Old Santa Fe Magazine* 3 (1916): 133–149.

Sully, John M., II. "Copper King." *Desert Exposure* (October 2007): 18–19.

Sweeney, Edwin R. *Mangas Coloradas.* Norman: University of Oklahoma Press, 1998.

———. "Mangas Coloradas and Mid-Nineteenth-Century Conflicts." In *New Mexican Lives: Profiles and Historical Stories,* edited by Richard W. Etulain, 131–162. Albuquerque: University of New Mexico Press, 2002.

Thompson, Jerry D., ed. "With the Third Infantry in New Mexico, 1851–1853: The Lost Diary of Private Sylvester W. Matson." *The Journal of Arizona History* 31, no. 4 (Winter 1990): 349–404.

Thorne, Harry A. "Mining Methods at the Chino Copper Mines." American Mining Congress, New Mexico Chapter, Santa Rita Meeting, July 25–26, 1927: 21–38.

Thrapp, Dan. *The Conquest of Apacheria.* Norman: University of Oklahoma Press, 1967.

———. *Victorio and the Mimbres Apaches.* Norman: University of Oklahoma Press, 1974.

Thurston, Herbert, and Donald Attwater, eds. *Butler's Lives of the Saints*. New York: P. J. Kennedy & Sons, 1962.

Titley, Spencer R., ed. *Advances in Geology of the Porphyry Copper Deposits, Southwestern North America*. Tucson: University of Arizona Press, 1982.

Twitchell, Ralph E. *The Leading Facts in New Mexico History*. Cedar Rapids, IA: The Torch Press, 1912.

Vargas, Zaragosa. *Labor Rights and Civil Rights: Mexican American Workers in Twentieth-Century America*. Princeton, NJ: Princeton University Press, 2005.

Walker, Billy D. "Copper Genesis: The Early Years of Santa Rita del Cobre." *New Mexico Historical Review* 54 (January 1979): 5–20.

Watson, Dorothy. *The Pinos Altos Story*. Silver City, NM: Silver City Enterprise, 1960.

Weber, David J. *The Mexican Frontier 1821–1846: The American Southwest under Mexico*. Albuquerque: University of New Mexico, 1982.

———, ed. *New Spain's Far Northern Frontier: Essays on Spain in the American West, 1540–1821*. Albuquerque: University of New Mexico Press, 1979.

———. *The Spanish Frontier in North America*. New Haven, CT: Yale University Press, 1992.

Wendt, Arthur F. "The Copper Ores of the Southwest." *Transactions of the American Institute of Mining Engineers* (New York, 1887): 25–77.

West, Robert C. *The Mining Community in Northern New Spain: The Parral Mining District*. Berkeley: University of California Press, 1949.

White, James W. *The History of Grant County Post Offices*. James White, 2003. Silver City, NM.

Whitney, J. P. *The Santa Rita Native Copper Mines in Grant County, New Mexico*. Boston: Alfred Mudge & Son, 1884.

Wood, C. W., and T. N. Nash. "Copper Smelter Effluent Effects on Sonoran Desert Vegetation." *Ecology* 57 (1976): 311–316.

Worcester, Donald E. *The Apaches: Eagles of the Southwest*. Norman: University of Oklahoma Press, 1979.

Wyman, Mark. *Hard Rock Epic: Western Miners and the Industrial Revolution, 1860–1910*. Berkeley: University of California Press, 1979.

Young, George J. *Elements of Mining*. 4th ed. New York: McGraw-Hill Book Co., 1946.

Young, Otis. *Western Mining*. Norman: University of Oklahoma Press, 1970.

Zhu, Liping. *A Chinaman's Chance: The Chinese on the Rocky Mountain Mining Frontier*. Niwot: University Press of Colorado, 2000.

UNPUBLISHED SOURCES

Company Records

Chino Copper Company. *Annual Reports*. 1910–1923.

———. Monthly Reports. 1909–1926.

———. Quarterly Reports. 1911–1926.

———. Weekly Reports. February–October, 1909–1924.

Kennecott Copper Corporation. *All about Kennecott: The Story of Kennecott Copper Corporation*. 1961.

———. *Annual Reports*. 1932–1980.

———. General Manager's Reports, Chino Mines Division. 1940–1955.

Nevada Consolidated Copper Company. *Annual Reports, Chino Mines Division*. 1926–1932.

Phelps Dodge Corporation. *Annual Reports*. 1909–1930, 1976–1988.

Ray Consolidated Copper Company. *Annual Reports, Chino Mines Division*. 1923–1926.

Santa Rita Hospital. Records. 1910–1954. Terrence M. Humble Collection.

Santa Rita Mining Company. Smelter Returns. 1900–1911. Terrence M. Humble Collection.

———. Weekly Reports. 1900–1909. Terrence M. Humble Collection.

———. Yearly Reports. 1900–1908. Terrence M. Humble Collection.

United States Smelting and Refining & Mining Company. *Annual Reports, Bullfrog Concentrator Operations*. 1944–1949, 1953.

Interviews

Santa Rita Project, 1992. New Mexico Endowment for the Humanities.

Archer, Mary Lee. March 1992.

Blair, Frank. February 1992.

Blair, Jesse. February 1992.

Chacón, Elvira. November 1992.

Chacón, Jesus "Tuti" P. November 1992.

Chavez, Olga. March 1992.

Delgado, Lilly Durán. November 1992.

Enriquez, Alexandra. September 1992.

Fletcher, Guadalupe. March 1992.

Kirker, Nicanora. October 1992.

Kirker, Rafael. October 1992.

Lopez, Soccoro Herrera. November 1992.

Madrid, Victor. November 1992.

Marrujo, Daniel. March 1992.

Muñiz, Tracy. November 1992.

Ortiz, Toni. November 1992.

Valencia, Antonio. March 1992.

Interviews by Christopher J. Huggard

Ballmer, Jonette. December 17, 1993.

Bartlit, John. December 14, 1993, and March 7, 1994.

Bartlit, Nancy. December 14, 1993, and March 31, 1994.

Culhane, Jolane. March 25, 1989, and December 15, 19, 1993.

Foy, Thomas. December 18, 1993.

Galvin, Peter. December 17, 1993.

Herrera, Mike. December 22, 1993.

Himes, Larry. December 18, 19, 1993.

Humble, Pat. July 13, 1991.

Humble, Terry. July 20, 1991, December 17, 19, 1993.

Jager, Allan. July 9, 1992.

Leveille, Richard. December 18, 1993.

Little, William W. December 20, 1993.

McGee, Maggie. December 20, 1993.

McGrath, Herb. August 3, 1991, December 20, 1993.

Morales, Joe T. December 17, 1993.

Nicholson, Bruce. July 7, 1992.

Ryan, Murray. December 17, 1993.

Toy, Herb Yee. June 1, 2007.

Manuscripts

Boise, D. W. "A Brief Summary of the History of the Santa Rita Mines." Silver City, NM, 1954.

Dannelley, Claude. "Fifty Years of Memories of Hurley, New Mexico." July 13, 1960. Special Collections, Miller Library, Western New Mexico University, Silver City.

Himes, Barney. "The Unions and the Chino Mine." Laguna Niguel, CA: Barney Himes, 1986.

Lou Blachly Oral History Collection. Coronado Room, Zimmerman Library, University of New Mexico.

Lundwall, Helen, and Terrence M. Humble. "Santa Rita del Cobre, Mining in Apachería 1801 1838." Silver City, NM, 2005.

Munroe, George B. (chairman, Phelps Dodge Corp.). "Remarks of George B. Munroe." Annual Phelps Dodge Luncheon, Albuquerque, New Mexico, January 16, 1984.

Safford, Irma F. "One Hundred Fifty Years of Chino." Special Collections, Miller Library, Western New Mexico University, Silver City, 1955.

Topmiller, Vic, Jr. "Making the Kneeling Nun a National Monument in Name of Mother Teresa." September 12, 1997. Silver City, NM.

GOVERNMENT DOCUMENTS

Federal

"An Act to Promote the Development of the Mining Resources of the United States." 42 Cong., 2nd. sess., May 10, 1872, c. 152, 17 Stat., 91.

Eighth Census of the United States, 1860. Doña Ana Co., NM. Washington, DC: Government Printing Office, 1866.

Eleventh Census of the United States, 1890. "Abstract of the Eleventh Census." Washington, DC: Government Printing Office, 1894, 1896.

——. "Report on Mineral Industries in the United States." Washington, DC: Government Printing Office, 1892.

——. "Special Census Report on the Occupations of the Population of the United States." Washington, DC: Government Printing Office, 1896.

Emory, William H. *Notes of a Military Reconnaissance, from Fort Leavenworth, in Missouri, to San Diego, in California, including Part of the Arkansas, Del Norte, and Gila Rivers.* Washington, DC: Government Printing Office, 1848.

Fifteenth Census of the United States, 1930. Grant County, New Mexico, Precinct 13, Santa Rita. Washington, DC: Government Printing Office, 1931.

Fourteenth Census of the United States, 1920. Grant County, New Mexico, Precinct 13, Santa Rita. Washington, DC: Government Printing Office, 1921.

Gardner, E. D., C. H. Johnson, and B. S. Butler. *Copper Mining in North America.* US Bureau of Mines. Bulletin 405. Washington, DC: Government Printing Office, 1938.

Haywood, J. K. *Injury to Vegetation and Animal Life by Smelting Works.* Washington, DC: Government Printing Office, 1908.

Hough, Walter. *Antiquities of the Upper Gila and Salt River Valleys of Arizona and New Mexico.* Washington, DC: Government Printing Office, 1907.

———. *Culture of the Ancient Pueblos of the Upper Gila River Region, New Mexico and Arizona.* Washington, DC: Government Printing Office, 1914.

Leong, Y. S., Emil Erdreich, J. C. Burritt, O. E. Kiessling, E. C. Neighman, and George C. Heikes. *Technology, Employment, and Output per Man in Copper Mining.* Report No. E-12. US Bureau of Mines, Philadelphia: Works Project Administration, 1940.

Lindgren, Waldemar, Louis C. Graton, and Chas. H. Gordon. *The Ore Deposits of New Mexico.* Washington, DC: Government Printing Office, 1910.

McDonald, Donald F., and Charles Enzian. *Prospecting and Mining of Copper Ore at Santa Rita, New Mexico.* US Bureau of Mines. Bulletin 107. Washington, DC: Government Printing Office, 1916.

Raymond, Rossiter W. *Statistics of Mines and Mining in the States and Territories West of the Rocky Mountains.* Washington, DC: Government Printing Office, 1874.

Tenth Census of the United States. "The United States Mining Laws and Regulations Thereunder, and State and Territorial Mining Laws." Washington, DC: Government Printing Office, 1885.

Tenth–Twentieth Censuses of the United States, 1880–1970, Grant County, NM. Washington, DC: Government Printing Office, 1881–1974.

US Department of the Interior. "Gila River Flood Control." 65th Cong., 3rd sess., 1919, Senate Doc. 436.

———. *Mineral Resources of the United States.* Washington, DC: Government Printing Office, 1882–1931.

———. *Minerals Yearbook.* Washington, DC: Government Printing Office, 1932–1993.

US Department of the Interior, US Geological Survey. "Forest Conditions in the Gila River Forest Reserve, New Mexico." 58th Cong., 3rd sess., 1905, House Doc. 411.

US Senate, Committee on Public Works, Hearings before the Subcommittee on Air and Water Pollution. *Air Pollution—1970.* 91st Cong., 2nd sess., March–May 1970.

———. *Clean Air Act Oversight.* 93rd Cong., 2nd sess., May 1974.

———. *Implementation of the Clean Air Act—1975.* 94th Cong., 1st sess., April–May 1975.

———. *Implementation of the Clean Air Act Amendments of 1970—Part 2.* 92nd Cong., 2nd sess., February–April 1972.

Wheeler, George M. *Annual Report upon the Geographical Surveys of the Territory of the United States West of the 100th Meridian, in the States and Territories of California, Colorado, Kansas, Nebraska, Nevada, Oregon, Texas, Arizona, Idaho, Montana, New Mexico, Utah, Washington, and Wyoming.* Washington, DC: Government Printing Office, 1879.

State of New Mexico

Findlay, J. R. *Report of Appraisal of Mining Properties of New Mexico.* Santa Fe: New Mexico State Tax Commission, 1922.

Howard, E. Viet. *Metalliferous Occurrences in New Mexico.* Santa Fe: State Planning Office, 1967.

Jones, Fayette A. *Epitome of the Economic Geology of New Mexico.* Santa Fe: New Mexico Bureau of Immigration, 1908.

———. *Old Mines and Ghost Camps of New Mexico.* Socorro, NM: New Mexico Mines and Minerals, 1904.

The Mines of New Mexico, Inexhaustible Deposits of Gold and Silver, Copper, Lead, Iron, and Coal. Santa Fe: New Mexico Bureau of Immigration, New Mexico Printing Co., 1896.

New Mexico Inspector of Mines. *Annual Reports.* 1912–1990. Santa Fe: New Mexico State Inspector of Mines, 1911–1992.

New Mexico Mining Laws, Statutes, Etc. Albuquerque: New Mexico State Inspector of Mines, 1955.

Records of the Environment Department of New Mexico (formerly Environmental Improvement Agency and Environment Improvement Division). State of New Mexico, Santa Fe, 1970–1990.

New Mexico Bureau of Mines and Mineral Resources

Allen, John Eliot. "The Making of a Mine." Circular 41. Socorro: New Mexico Bureau of Mines and Mineral Resources, 1956.

Anderson, Eugene Carter. *The Metal Resources of New Mexico and Their Economic Features through 1954.* Socorro: New Mexico Bureau of Mines and Mineral Resources, 1957.

Benjovsky, Theodore D. *Contributions of New Mexico's Mineral Industry to World War II.* Bulletin 27. Socorro: New Mexico Bureau of Mines and Mineral Resources, 1947.

Burks, Marian R. *Bibliography of New Mexico Geology and Mineral Technology through 1950.* Bulletin 43. Socorro: New Mexico Bureau of Mines and Mineral Resources, 1955.

Christiansen, Paige W. *The Story of Mining in New Mexico.* Socorro: New Mexico Bureau of Mines and Mineral Resources, 1974.

Elston, Wolfgang E. *Geology and Mineral Resources of Grant, Luna, and Sierra Counties.* Bulletin 38. Socorro: New Mexico Bureau of Mines and Mineral Resources, 1957.

File, Lucien, and Stuart A. Northrop. *County, Township, and Range Location of New Mexico's Mining Districts.* Circular 84. Socorro: New Mexico Bureau of Mines and Mineral Resources, 1966.

Fowler, C. H. *Mining and Mineral Laws of New Mexico.* Bulletin 6. Socorro: New Mexico Bureau of Mines and Mineral Resources, 1930.

———, and S. B. Talmage. *Mining, Oil, and Mineral Laws of New Mexico.* Bulletin 16. Socorro: New Mexico Bureau of Mines and Mineral Resources, 1941.

Gillerman, Elliott. *Mineral Deposits of Western Grant County, New Mexico.* Bulletin 83. Socorro: New Mexico Bureau of Mines and Mineral Resources, 1964.

Heljeson, D. M., and C. L. Holts. *Supplemental Bibliography of New Mexico Geology and Mineral Technology through 1975.* Bulletin 108. Socorro: New Mexico Bureau of Mines and Mineral Resources, 1981.

Johnson, P. H., and R. B. Bhappn. "Chemical Mining—A Study of Leaching Agents." Circular 99. Socorro: New Mexico Bureau of Mines and Mineral Resources, 1969.

Lasky, Samuel G. *The Metal Resources of New Mexico and Their Economic Features.* Socorro: New Mexico Bureau of Mines and Mineral Resources, 1933.

New Mexico Bureau of Mines and Mineral Resources. *Annual Reports.* Socorro: New Mexico Bureau of Mines and Mineral Resources, 1930 to the present.

New Mexico Land Grants, Surveyor General Records. Santa Fe, New Mexico.

Sorensen, Earl F. "Mineral Resources and Water Requirements for New Mexico Mineral Industries." Circular 138. Socorro: New Mexico Bureau of Mines and Mineral Resources, 1973.

Trauger, Frederick D. *Water Resources and General Geology of Grant County, New Mexico.* Hydrologic Report 2. Socorro: New Mexico Bureau of Mines and Mineral Resources, 1972.

Verity, Victor H. *Laws and Regulation Governing Mineral Rights in New Mexico.* Bulletin 104. Socorro: New Mexico Bureau of Mines and Mineral Resources, 1973.

Grant County

Books of Mining Claims, Deeds, etc. (1880–1990). County Clerk's Office.

County Commissioner's Records.

Deed Books.

Grant County, Administration Building, Silver City, New Mexico.

Mining Records.

Probate Minutes.

THESES AND DISSERTATIONS

Burrage, Melissa D. "Albert Cameron Burrage: An Allegiance to Boston's Elite, 1859–1931." MA thesis, Harvard University, Cambridge, MA, 2004.

Cargill, Jack. "Empire and Opposition: Class, Ethnicity, and Ideology in the Mine-Mill Union of Grant County, New Mexico." MA thesis, University of New Mexico, Albuquerque, 1979.

Collins, Dale. "Frontier Mining Days in Southwestern New Mexico." MA thesis, Texas Western College, El Paso, 1955.

Cunningham, William J. "The Control of Mineral Resources in New Mexico." MA thesis, University of New Mexico, Albuquerque, 1950.

Deupree, Dean C. "Pinos Altos, New Mexico, 1860–1880." MA thesis, Texas Western College, El Paso, 1962.

Huggard, Christopher J. "Environmental and Economic Change in the Twentieth-Century West: The History of the Copper Industry in New Mexico." PhD diss., University of New Mexico, Albuquerque, 1994.

Milbauer, John Albert. "The Historical Geography of the Silver Mining Region of New Mexico." PhD diss., University of California, Riverside, 1983.

Naegle, Conrad Keller. "The History of Silver City, New Mexico, 1870–1886." MA thesis, Western New Mexico University, Silver City, 1943.

Seligmann, Gustav L. "The El Paso and Northeastern Railroad System and Its Economic Influence in New Mexico." MA thesis, New Mexico State University, Las Cruces, 1958.

Valliant, Maude D. "The History of the Railroads of the Southwest." MA thesis, Columbia University, New York, 1932.

Withers, Allison C. "Copper in the Prehistoric Southwest." MA thesis, University of Arizona, Tucson, 1946.

Index

Page numbers in italics indicate illustrations.